The
Edge
of the
Universe

Celebrating **10 Years** of **Math** Horizons

To all our friends at St Olaf,

Deanna Haunsperger

Stephen Kennedy

©2006 by The Mathematical Association of America, Inc.

ISBN 10: 0-88385-555-0
ISBN 13: 978-0-88385-555-3

Library of Congress Catalog Card Number: 2005937266

Current Printing (last digit):
10 9 8 7 6 5 4 3 2 1

The
Edge
of the
Universe

Celebrating **10 Years** of **Math** Horizons

Edited by

DEANNA HAUNSPERGER
and
STEPHEN KENNEDY
Carleton College

Published and Distributed by
The Mathematical Association of America

For Sam and Maggie

Yes, we're finally done with *Math Horizons*.
Now we can go to the beach.
Thanks for your patience.

SPECTRUM SERIES

The Spectrum Series of the Mathematical Association of America was so named to reflect its purpose: to publish a broad range of books including biographies, accessible expositions of old or new mathematical ideas, reprints and revisions of excellent out-of-print books, popular works, and other monographs of high interest that will appeal to a broad range of readers, including students and teachers of mathematics, mathematical amateurs, and researchers.

777 Mathematical Conversation Starters, by John de Pillis

99 Points of Intersection: Examples—Pictures—Proofs, by Hans Walser. Translated from the original German by Peter Hilton and Jean Pedersen

aha! A two volume collection: aha! Gotcha and aha! Insight, by Martin Gardner

All the Math That's Fit to Print, by Keith Devlin

Carl Friedrich Gauss: Titan of Science, by G. Waldo Dunnington, with additional material by Jeremy Gray and Fritz-Egbert Dohse

The Changing Space of Geometry, edited by Chris Pritchard

Circles: A Mathematical View, by Dan Pedoe

Complex Numbers and Geometry, by Liang-shin Hahn

Cryptology, by Albrecht Beutelspacher

The Edge of the Universe: Celebrating 10 Years of Math Horizons, edited by Deanna Haunsperger and Stephen Kennedy

Five Hundred Mathematical Challenges, by Edward J. Barbeau, Murray S. Klamkin, and William O. J. Moser

The Golden Section, by Hans Walser. Translated from the original German by Peter Hilton, with the assistance of Jean Pedersen.

I Want to Be a Mathematician, by Paul R. Halmos

Journey into Geometries, by Marta Sved

JULIA: a life in mathematics, by Constance Reid

R. L. Moore: Mathematician and Teacher, by John Parker

The Lighter Side of Mathematics: Proceedings of the Eugène Strens Memorial Conference on Recreational Mathematics & Its History, edited by Richard K. Guy and Robert E. Woodrow

Lure of the Integers, by Joe Roberts

Magic Tricks, Card Shuffling, and Dynamic Computer Memories: The Mathematics of the Perfect Shuffle, by S. Brent Morris

Martin Gardner's Mathematical Games: The entire collection of his Scientific American columns

The Math Chat Book, by Frank Morgan

Mathematical Adventures for Students and Amateurs, edited by David Hayes and Tatiana Shubin. With the assistance of Gerald L. Alexanderson and Peter Ross

MAA Service Center

P. O. Box 91112

Washington, DC 20090-1112

1-800-331-1622   fax: 1-301-206-9789

Preface

Ten years ago a skinny green magazine with a triptych of a smiling Andrew Wiles on the cover appeared unsolicited in our mailbox. Inside, it was full of great stuff: the mathematics of bar codes, career advice, tips on choosing a grad school, the story of Wiles's proof of Fermat's Last Theorem, a review of Hardy's *Apology*, brainteasers, and more. We were completely charmed.

The premier issue of *Math Horizons* had been sent to every MAA member courtesy of the Exxon Education Foundation, the William and Flora Hewlett Foundation and the National Science Foundation. It was supposed to appeal to our students (and it did), but we loved it too. The style was lighthearted, the tone was conversational, and the math was real. Don Albers, the founding editor, had for many years been mulling over exactly how to put together a magazine that would attract undergraduates and here was the fruit of that labor. You could feel his excitement in the editorial proudly titled, "This is it!" in that inaugural issue. There he outlined his vision: a lively brew of career information, articles about contemporary mathematics, profiles of mathematicians, humor, book reviews, and contributions from students. A few months later the second issue arrived and we were hooked; we couldn't wait to acquire a subscription.

Four times a year the skinny magazine arrived and gradually Albers's vision unfolded: profiles of great and interesting mathematicians: Conway, Uhlenbeck, Diaconis, Graham; exposition by the best mathematical expositors writing in English: Gardner, Apostol, Dunham, Cipra, Devlin, Peterson. When it showed up we always dropped everything, put feet up on desk and romped through it.

When, after five years of editing *Math Horizons*, Don invited us to replace him, we could hardly credit our good fortune. This was clearly the most exciting job to be had in mathematics. We laid out our aims in the Instructions for Authors in our first issue:

> Our purpose is to introduce students to the world of mathematics outside the classroom. Thus, while we especially value and desire to publish high quality exposition of beautiful mathematics, we also wish to publish lively articles about the culture of mathematics. We interpret this quite broadly—we welcome stories of mathematical people, the history of an idea or circle of ideas, applications, fiction, folklore, tradition, institutions, humor, puzzles, games, book reviews, student math club activities, and career opportunities and advice.

****

This volume is published to celebrate the first decade of *Math Horizons*. The lifeblood of a magazine is, more than anything else, a roster of talented, generous contributors. The success of *Math Horizons* is due entirely to the long list of extraordinary mathematicians who have written for her pages. It's not possible to mention them all here, but you'll meet many of them in this volume: there's the inimitably amusing frequent contributor Woody Dudley, the indefatigable and infinitely productive Joe Gallian, Martin Gardner who has been incredibly gracious and generous for the entire decade of *Math Horizons*'s existence, and Stan Wagon on whom we could always count for fantastic snow sculpture pictures. There's the incredible Jim Tanton who wrote so many articles for *Math Horizons* that we eventually made him a regular columnist; Ira Rosenholtz, possibly the nicest man in mathematics, and possessed of a gift for whimsical mathematical exposition; and Steve Abbott, our literary go-to guy, whose extraordinarily penetrating analysis of Tom Stoppard's plays is a model for writing about mathematics in the arts. All of these folks are well-represented in these pages as are regular columnists: Mark Schilling who delivered a fascinating article on statistics every other issue; Randy Schwartz and Rheta Rubenstein, the lighthearted linguists, who produced a series of articles on mathematical etymology; and Tom Apostol and Mamikon Mnatsakanian whose research program on cyclogons unfolded in the pages of *Math Horizons*. We could have made a fantastic book by stapling together the contributions of just the folks we've mentioned, but there have been so many other great articles over the years that wouldn't have been fair to you.

We owe a debt of gratitude to the incomparable Beverly Joy Ruedi who taught us nearly everything we know about magazine and book publishing (which is only a fraction of what she knows); that she did so with grace, good humor, and generous friendship only increases our gratitude. Thanks, also, to Elaine Pedreira for her warm smile, professionalism, and patience in the production of this book. Finally, one should consider this collection a tribute to Don Albers's vision in creating *Math Horizons* and his decade-long encouragement of us in our association with it. That his mentorship has evolved into an enduring friendship is a source of great pleasure.

We considered many different organizational schemes for the articles that follow, in the end we simply laid them out chronologically. Our thought was that this would allow you to experience this book as you would an issue of the magazine; a little bit of mathematics, then maybe some history, a joke or two to cleanse the palate, followed by an interview with an artist.

Our fondest hope is that this book will charm and delight you as ten years of *Math Horizons* have done us.

Enjoy,

Steve and Deanna

Contents

John Horton Conway
—Talking a Good Game

DON ALBERS
Mathematical Association of America

Quick now—what day of the week did December 4, 1602 fall on? Sorry, time's up. You have to give the answer (Saturday) in less than two seconds to compete with Professor John Conway of Princeton University. Conway enjoys mentally calculating days of the week so much that he has programmed his computer so he cannot log on until he does ten randomly-selected dates in a row. He usually does ten dates in about 20 seconds. His best time is 15.92 seconds. Conway says "the ability to do these lightning mental calculations is very important to me. You've no idea how fast you have to think to do them. The reason I do it is because it gives me a kick. The adrenaline spills all over you, and when you're thinking that quickly, it's really nice."

Conway, at age 56, is one of the world's most original mathematicians and is a member of the prestigious Royal Society of London. He is in the middle of a second career as professor of mathematics at Princeton University with his second family. It was a great coup for Princeton mathematicians when they lured Conway away from Cambridge University in 1986. We are visiting with him today (November 29, 1993) to gain a few insights into his work and what makes him tick.

Conway has made substantial contributions to several branches of mathematics: set theory, number theory, finite groups, quadratic forms, game theory, and combinatorics. He is best known, in a popular sense, for his work on the theory of games, especially the Game of "Life"

and his invention of a theory of numbers that has its origins in games. Conway's enchantment with games is reflected in the title of one of his papers, "All Games Bright and Beautiful." In Conway's theory of numbers, every two-person game is a number! Don Knuth, the noted computer scientist, was so taken with Conway's new theory of numbers that he wrote *Surreal Numbers*, a novel that explains the theory for students.

"Life"

"Life," Conway's most famous game creation to date, burst on the scene in 1970 when Martin Gardner brought it to the attention of hundreds of thousands of readers of his "Mathematical Games" column of *Scientific American* magazine. "Life's" popularity was quick and far-reaching. Its great popularity spawned "Lifeline," a newsletter for "Life" enthusiasts, which was published for many years.

Gardner later wrote that his "column on Conway's 'Life' forms was estimated to have cost the nation millions of dollars in illicit computer time. One computer ex-

pert, whom I shall leave nameless, installed a secret switch under his desk. If one of his bosses entered the room he would press the button and switch his computer screen from its 'Life' program to one of the company's projects."

Conway says that "Life" arose out of "the aim to find a system in which you can see what happens in the future.... I always thought you ought to be able to design a system that was deterministic, but unpredictable."

Although he has co-authored with Berlekamp and Guy *Winning Ways for Your Mathematical Plays*, the two-volume classic on games, he asserts that he is not very interested in playing actual games. He claims that "I can't play chess, I know the rules, but you would be amazed at how badly I play. That's not the thing that turns me on. I'm interested in the theory of it, especially if it's simple and elegant. I really like to consider the simpler games, like checkers. I used to play checkers with my first wife, and she always used to beat me. Perhaps I would win one game in ten or twenty, and I was trying very hard.... I had a similar experience with my daughter, playing the game called Reversi."

Conway's mathematical abilities, especially his rapid calculating skills, were evident as a little boy. He says that "my mother found me reciting 2, 4, 8, 16, 32, 64,... —the powers of 2 when I was four." When he was eleven, he told the headmaster of his grammar school in Liverpool that "I want to go to Cambridge and study mathematics."

Conway illustrates sphere packing.

What was it about mathematics that attracted Conway so strongly? "I can't recall what started it," he says. "It's probably just the fact that I was good at it, and that was that. If you regard it as a competitive subject, then to stand out and beat the other kids was fun.

"When I was a teenager, I thought a lot about the different departments of knowledge, in some sense, and I know what turned me on to math was this feeling of objectivity. Consider other things you might do, like law. Then you're basing your life on essentially arbitrary decisions that have been taken by individuals, or by the way society has developed as a whole. I can't develop much interest in that.... I like the idea that with philosophical, mathematical, and scientific questions, there's a chance of communicating with beings on other planets, so to speak. There's a certain universality that definitely is central to mathematics.

"When I was young, things were quite difficult. It was quite a rough district we lived in, and some terrible things happened." Conway remembers being beaten up by older boys "because I hadn't chosen the right professional soccer team. Sometimes it didn't matter whether I chose the right one or not, they would beat me up anyway." He also remembers being taken into an ancient air raid shelter and having lighted cigarettes applied to his skin.—Ouch! He eventually got to Cambridge, but Conway says that "from ages 11–13, with the onset of adolescence, puberty, and all of that, I didn't do terribly well. I started to hang around with a bunch of lay-abouts. I had a hell of a time when I was a high school student.

"Teachers and my parents were getting concerned about me," he remembers. "I was given several good talkings to by various people, and by age 16, I started going to classes again and started being on top again."

The Real Me

Conway indeed got on top again to the point that he won a scholarship to Cambridge. He clearly remembers his train trip to Cambridge, and being rather introverted, quiet, and shy at the time. "I was on the train when I said to myself, 'You don't have to be like this anymore. Nobody at Cambridge knows you.' I had stepped out of the world I was previously in. So I decided to turn myself into an extrovert, and I did. I decided I was going to laugh with people, and make fun of myself. I got there, and that's what happened. For quite a long time, I felt like a fraud. I said to myself, 'This isn't the real me.' And then it ceased to be acting. Every now and then, I still feel shy on occasions, but not very often."

Anyone who watches Conway bounce around a classroom or organize a knot theory square dance would agree that the introvert is long gone.

Conway's Princeton office is an environment that clearly would appeal to children of all ages from two to one hundred. Pleasantly cluttered with books, bric-a-brac, and mathematical models hanging from the ceiling and walls, it bears a striking resemblance to a classroom in a progressive elementary school. He even has a home-built "quaternion machine" hanging on one wall, which he gleefully operates for visitors.

During our visit, Conway tells us about his new system for clarifying the mysteries of knot theory. He brings in two undergraduate students to join him and the interviewer in a special "square dance" using two colored ropes that do indeed serve to explicate what he calls the "theory of tangles." He recalls working out a good part of his theory of tangles while still a high school student.

At one point in our discussion, he brings out a few dozen tennis balls to illustrate a problem in sphere packing. Packing the maximum number of spheres in a given space, especially higher dimensional spaces, has been one of Conway's passions for several years. The sphere packing problem in eight-dimensional space is very important to transmitting data over telephone lines. He tells us that twenty-four dimensional space is wonderful for "there is really a lot of room up there among those packed spheres." His interest in sphere packing led to his writing, with Neil Sloane, the book entitled *Sphere Packings, Lattices, and Groups*. He is quite proud of a recent review of the book which describes it as "the best survey of the best work in the best fields of combinatorics written by the best people. It will make the best reading by the best students interested in the best mathematics that is now going on." He is so proud of the review that he has displayed the "best" parts of it in large letters on one of his walls. Of the many nice reviews he has had of his work, he says this one is the "best."

Tennis balls (sphere-packing), colored ropes (knot theory), and counters on a checker board (the game of "Life") all reflect Conway's intense need to make things simple. He claims that "lots of people are happy when they've understood something. And I'm usually not. I'm only happy when I've really made it simple. Moreover, I don't understand so many deep things as a lot of other people do. But I'm interested in getting a still deeper understanding of some simpler things."

The Splash

Conway became a mathematical star in 1969 when he discovered a new simple group, now called the Conway Group. At the time he was a junior faculty member at Cambridge University who was depressed because "I had been known as a quite-bright student at Cambridge, but I had never produced anything of any significance. I was feeling guilty and was almost suicidal. In addition I was married with four little girls, and we had very little money.

"About then John Leech discovered a beautifully symmetric structure in 24-dimensional space. It was believed that a special group corresponded to Leech's structure. I decided to take a crack at finding it since I knew a bit about groups. So I set up a schedule with my wife. We agreed that I would work on the problem on Wednesday nights from six until midnight and on Saturdays from noon until midnight. I started work on a Saturday, and made progress right away. At a half hour past midnight on that first Saturday, I came out with the problem solved! I had found the group!

"That did it. I immediately got offers to lecture on my new group all over the world. I became something of a mathematical jet-setter. My discovery really was a big splash."

So Much for Guilt

Conway went on to explain another plus that followed his discovery. "I suddenly realized that it was a good idea not to feel guilty; feeling guilty didn't do any good. Guilt just made it impossible to work.... So I decided for myself that from then on I wasn't going to work on something just because I felt guilty. If I was interested in some childish game, I would think

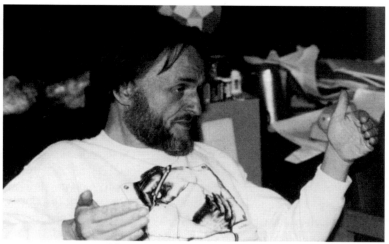
John Conway in his Princeton office.

about that childish game, whereas previously I would have sort of looked around and wondered what my colleagues were thinking."

Last June, Andrew Wiles, one of Conway's Princeton colleagues, made an even bigger splash with his announcement of a proof of Fermat's Last Theorem. (See the premiere issue of *Math Horizons* for the story on Wiles.) Conway worked in number theory for several years and has been fascinated with the work of Wiles and the ensuing excitement. He says "I have slightly different opinions about the Fermat problem from most people. What I think is that it's quite likely that Fermat proved it, not just that he believed that he'd proved it.... There was no reason for him to deceive anybody. There's not much point in writing something, writing a note to yourself, and telling a lie in it. It wouldn't work.

"And now you're faced with a real problem. Even if Fermat deceived himself, what was his proof? What was that proof? Until you can solve that problem by exhibiting something that was available to Fermat and would fool Fermat, you're not really entitled to make any judgment.... The argument that says we haven't found a proof for over 300 years is not so fascinating. We're not all that clever. Many of the other things he wrote down lasted for 200 years.... When I die, I might knock on Fermat's door and see what happened. He'd be an interesting guy to talk to. I've often toyed with the idea of talking to people from the past."

Archimedes and Euler are two other mathematicians with whom Conway would like to chat. He adds that "I wouldn't like to talk with either Newton or Gauss, because neither of them seems to be my kind of person."

Conway also has a secret to pass on about producing mathematics. He advises us to "keep several things on the board, or at least on the back burner, at all times.... One of them is something where you can probably make progress.... If you work only on the really deep interesting problems, then you're not likely to make much progress. So it's a good idea to have some less deep, less significant things, that nevertheless are not so shallow as to be insulting." ■

Photographs by Carol Baxter.

The Game of "Life"

The rules of Conway's Game of "Life" were chosen, after experimenting with many possibilities, to make the behavior of the population both interesting and unpredictable. The genetic laws are remarkably simple. The game is played on a chessboard, an infinite chessboard, where each cell has eight neighboring cells. The game begins with some arrangement of counters placed on the board (the live cells) as the initial generation. Each new generation is determined by the following rules:

1. Consider a live cell. If it has 0 or 1 live neighbors, then it dies from isolation. If it has 2 or 3 live neighbors, then it survives to the next generation. If it has 4 or more live neighbors, then it is crowded out and dies.

2. On the other hand, if a dead (unoccupied) cell has exactly 3 live neighbors, then it is a birth cell; a counter is placed on it in the next generation.

G_1

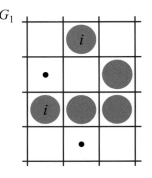

Here is a little example. The five circles in diagram G_1 indicate the live cells in the first generation. Those marked i will die in the next generation due to isolation (Rule 1). The cells with the black dots are empty cells that will become live in the next generation (Rule 2).

Diagram G_2 indicates what the second generation looks like. The cell marked c will die of crowding (Rule 1). To help you draw the pattern for the third generation we have used the marks i and • as before. You should continue to draw the pictures for several successive generations.

G_2

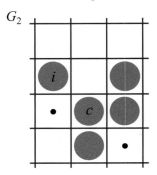

If you play the Game of "Life" with various initial generations you will find that the population undergoes unusual and unexpected changes. Patterns with no initial symmetry become symmetrical. Some initial configurations die out entirely (although this may take a long time), others become stable (still lifes), while others oscillate forever.

Conway originally conjectured that no finite initial pattern can grow without limit. But he was wrong. There is a "gun" that shoots out "gliders" and a "train" that moves along but leaves a trail of "smoke."

If you draw several more generations of the above example you will understand that it is called "glider" because it is a glide-reflection of an earlier generation (which one?).

To learn more about the Game of "Life," including the glider gun, see Martin Gardner's *Wheels, Life and Other Mathematical Amusements* (1983), which contains three chapters on "Life." You may want to write a program to play "Life" on your computer. ∎

John 'Horned' (Horton) Conway

Long Run Predictions

MARK F. SCHILLING
California State University-Northridge

It is June of 1992, and the citizens of riot-torn Los Angeles have gone to the polls to cast their ballots in the local elections. The results are announced, and a judge orders a recount for one particularly-closely contested race. Precinct by precinct, each batch of ballots is scrutinized by several officials and witnesses. It seems that one candidate's votes often occur in long runs, with several ballots in a row marked for the same individual. Could there have been sabotage?

Here is the voting record from one typical precinct:

```
CEEDAACCANAAEAADANNNA
CADDDANCBBADAAAAAECCN
ANABNAAABACNADACBENAN
ECAEBNAEDNNDNBAABAAAA
BABABAANAAAAAAAANADCA
ANANNNAAAAGEAAAAAADCA
AAAADNACABBBCNAAAACCA
DBANCAAAAADAACABNAAAN
ACACAFAABFAABCAAAACAN
FAABFBFDCBAAAAAENNAAA
AAAADDAAAANDAAAEABANA
ANAAEBNAAAAABABEECDAC
```

The seven candidates are indicated by the letters A through G; N represents a ballot with no vote marked. Notice that there are several long runs of votes for candidate A, including a series of eight in a row, seven in a row and six in a row. Do you think that the ballots were tampered with?

The fans cheered with enthusiasm as the Seattle Supersonics of the National Basketball Association trotted onto the court for the start of the 1990–91 basketball season. With talented young players such as Shawn Kemp and Gary Payton complementing an able crop of veterans, hopes were high for a stellar season. Unfortunately, inconsistency marked the team's play and the Supersonics wound up with a mediocre .500 record of 41 wins and 41 losses. Here is the game-by-game record of their performance (W = win, L = loss):

```
WWWLLLLWLLLLLLWLLWLWW
WWWWLLWWWLWLLWLWWWLWLL
WLWWLWLLLWLWWWWWLWLL
LLLWWLLLWWWLWWWWWLLWL
```

Several winning and losing streaks are evident, including one each of length six. Is it fair to say that the Seattle Supersonics of 1990–1991 were an unusually streaky team?

On August 18, 1913, at the famous Monte Carlo casino in Monaco, black came up 26 times in a row on a roulette wheel. As the run on black continued to grow, people began to bet larger and larger sums of money on red in the belief that another repeat of black was virtually impossible. The casino made several million francs that night. Was the wheel fixed? Could a run as long as 26 possibly be expected to occur on an *honest* roulette wheel?

To help answer these questions, let's compare these results to the sorts of runs that tend to occur in truly random sequences. For example, if an ordinary coin is tossed, say 250 times, how long is the longest run of consecutive heads likely

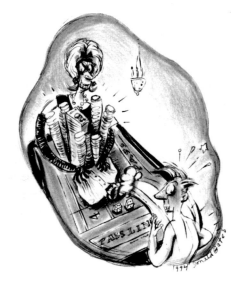

to be? A simple argument can give us a rough answer: Except on the first toss, a run of heads can only begin directly after a toss showing tails. There should be around 125 tails in 250 tosses, each (except for a tail on the last toss) providing an opportunity for a head run to start. After about half of these tails the succeeding toss will be heads, giving around 63 head runs in all. Roughly half of the time, the first head will be followed by a second one, so around 32 runs will be at least two heads long. Again, about half of these will contain at least another head. Thus we can expect around 16 head runs of length at least three, eight runs at least four heads long, four runs of length at least five, two of length six or more, and one run of seven heads or longer.

If many people each toss a coin 250 times, we can therefore expect most of them to obtain a head run of at least seven heads. In fact, precise calculations show that among all strings of 250 coin tosses, 87% will contain head runs of length at least six, 63% will have runs of seven or more, and a substantial 38% will possess runs that are at least eight heads long.

What about long runs of either heads or tails? Note that we can translate any string of 250 coin tossing outcomes into a new sequence of 249 elements in which we keep a record of whether the outcome of each toss after the first is the same as (S) or different from (D) the previous one. For instance, HTTHTTTHTHH… generates the sequence DSDDSSDDDS…. Since D's and S's occur independently with probability 1/2 for each, we can apply the same analysis as above to predict the length of the longest string of S's. The difference between 250 and 249 outcomes has a negligible effect on run lengths; since a string of k S's in a row corresponds to a string of either $k + 1$ heads or $k + 1$ tails, the results for head runs are simply shifted up by one. For instance, 63% of all sequences of 250 tosses of a fair coin will have *some* run at least eight elements long.

This is a surprising result. In fact, when asked to write down long sequences of heads and tails that would look like a typical random arrangement, most people are quite reluctant to include strings of more than four or five heads (or tails) in a row. It is therefore generally rather easy to distinguish a sequence simulated by a human from an actual random sequence by the failure of most simulated sequences to incorporate long enough runs!

A Formula for the Longest Run

We can easily generalize the informal arguments given above to predict probable run lengths for a sequence of n tosses of a coin in which the chance of heads on each toss, say p, is any value other than 0 or 1. Reasoning as before, there should be approximately $n(1-p)$ tails in the sequence, hence $n(1-p)$ possible starting points for a run of heads. Then about $n(1-p)p$ head runs of length one or more will occur, about $n(1-p)p^2$ head runs of length at least two, and so forth. In general, we will

obtain about $N_R = n(1-p)p^R$ runs of length at least R. To find a reasonable value for the typical length of the *longest* run of heads, we can solve the equation $N_R = 1$ for R to obtain

$$R = \log_{1/p}(n(1-p)).\qquad(1)$$

In the case of a fair coin ($p = 0.5$), this reduces to $R = \log_2(0.5n) = \log_2(n) - 1$, which then gives simply $R = \log_2(n)$ for the longest run of heads or tails.

We can use (1) to predict typical longest run lengths in a wide range of situations whose structure, like coin tossing, can be modeled (at least approximately) by what statisticians call Bernoulli trials. Bernoulli trials are repetitive sequences with the same two possible outcomes for each event in the sequence. Each event's outcome is independent of the others, and the probabilities remain unchanged from trial to trial.

For example in the voting data, candidate A received $p = 51.6\%$ of the $n = 252$ votes in the precinct. Formula (1) gives $R = 7.3$ votes, quite in line with the actual longest run of eight votes for A. Votes that are not for A show similar results, with two runs of seven in a row. So the observed long runs in the ballot data do not represent evidence of tampering.

In the basketball data, the chance of a Seattle win undoubtedly varied somewhat from game to game. We shall use $p = 0.5$ since the team won exactly half of its $n = 82$ games. Formula (1) predicts $R = 6.4$ for the longest run of either wins or losses. The actual longest run was six. The Supersonics' 1990–1991 performance was not unusually streaky by this measure.

Figure 1 shows the approximate chances that the longest run will be longer (area to the right of zero) or shorter (area to the left of zero) than formula (1) predicts. The shaded region represents the central 95% of the curve's area.

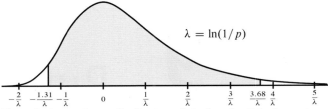

Figure 1. Approximate distribution of the length of the longest run minus its predicted value.

Thus, for example, for coin tossing, $p = 0.5$ gives $\lambda = \ln(1/p) = .69$, so there is about a 95% chance that the length of the longest run of heads will be somewhere between $\log_2(0.5n) - 1.9$ and $\log_2(0.5n) + 5.3$. A surprising feature of the curve in Figure 1 is that its scale depends on p but not on n (unless n is very small). Remarkably, therefore, one can predict the length of the longest run with the same degree of accuracy for $n = 100$ as for $n = 1,000,000$!

Long Run Theory vs. Momentum

Most people attribute streaks of successful or unsuccessful performances in sports to "momentum." For example, a team that has won several games in a row is considered "hot," and is considered more likely to win its next game than its overall record would predict. The opposite kind of momentum is believed to hold for "cold" teams in losing streaks. Similar "hot" and "cold" periods are believed to affect baseball hitters, basketball shooters, and so forth. Many people believe that momentum also applies in gambling, producing both "lucky" and "unlucky" streaks during which betting should be increased or decreased accordingly.

One way to evaluate the case for "momentum" is to compare calculations from long run theory to actual records from human experience, such as those that have occurred in sports and gambling. Perhaps the most famous of these is Joe DiMaggio's accomplishment of managing at least one hit in 56 consecutive baseball games. Was DiMaggio a "hot" hitter whose proficiency during the streak was greatly increased over his normal ability? Or could such a streak have been predicted by long run theory? Although DiMaggio's record was clearly an exceptional accomplishment even after accounting for his superb ability to hit a baseball, the real question we need to ask is whether in the absence of momentum we could have expected anyone in the history of major league baseball to achieve a hitting streak as long as 56 games.

It seems highly probable that the record would be set by a player who is a very good hitter. If we place end-to-end career records of the top 20% of all major league baseball players from 1901 to the present, we obtain a string of about $n = 500,000$ player-games. (The possibility that the longest run in this list overlaps two different players' career records is fairly remote and will be ignored.) The overall batting average (hitting percentage) of these players is roughly .300; assuming four hitting opportunities per game, the chance that a .300 hitter would get no hits in a given game is $(1 - .300)^4 = .24$, thus $1 - .24 = .76$. Applying formula (1) gives $R = 43$, while the 95% prediction interval of 38 to 56 just reaches the DiMaggio record.

Similarly adjoining all of the spins of honest roulette wheels in history (estimating a total of half a billion spins) predicts a longest run of $R = 27$ of the same color. The remarkable run at Monte Carlo is in fact quite reasonable when viewed in this context.

Table 1 presents several prominent record streaks in sports and gambling and compares them to predictions derived from the long run theory above. In most cases, the figure used for p is an average, as the value typically varies among the trials. The values of n are also approximate, in some cases "ballpark figures" obtained from rough calculations. However, moderate changes in n do not greatly alter the predictions.

The record streaks in the sports categories of Table 1 are in some cases somewhat above the length predicted from formula (1), but all lie within the 95% prediction intervals. The runs in gambling, while at first glance amazing, are completely in line with what is expected for

Event	Record	p	n	Predicted Longest Run	95% Prediction Interval
Baseball					
Consec. games with hit (top 20% of batters)	56 (DiMaggio, 1941)	.76	500,000	43	(38, 56)
Consecutive hits (top 20% of batters)	12 (Higgins, 1938; Dropo, 1952)	.30	2,000,000	12	(11, 15)
Consec. wins, team (top 20% of teams)	26 (NY Giants, 1916)	.60	55,000	20	(17, 27)
Consec. wins, pitcher (top 20% of pitchers)	24 (Hubbell, 1936-7)	.62	40,000	20	(17, 28)
Basketball					
Consec. wins, team (teams winning >70%)	33 (L.A. Lakers, 1971–2)	.73	6,000	23	(19, 35)
Consec. losses, team (teams winning < 30%)	24 (Cl. Cavaliers, 1982)	.73	6,000	23	(19, 35)
Free Throws	97 (Williams, 1993)	.90	40,000	79	(66, 114)
Roulette					
Same Color	26 (Monte Carlo, 1913)	.48	5×10^8	27	(26, 32)
Craps					
Consec. passes	28 (Las Vegas, 1950)	.49	5×10^8	27	(25, 33)

Table 1

run lengths in very long strings of independent Bernoulli trials. Although a run of 26 in a row in roulette, for example, has only a 1 in 68,411,592 chance of occurring in a specified set of 26 rolls at Monte Carlo, our analysis has shown that it is quite a reasonable thing to have occurred on some wheel, at some time.

The data therefore refute the idea of momentum in games of chance, while giving less conclusive results for baseball and basketball. Although it can be argued that player attitudes and emotions must surely cause significantly longer runs in sports than those expected by chance alone, the empirical support for this claim is weak. Detailed statistical analyses indicate, in fact, that contrary to the strong prevailing opinions of fans, sports reporters, and the players themselves, momentum may be merely an illusion of the human mind [1–4].

The almost universal tendency to regard long runs as having underlying, nonrandom causes is probably due to the selectiveness of human perception, which is geared towards pattern recognition—streaks tend to stand out and be remembered—coupled with an unawareness of the

extent to which long runs arise by chance alone in merely random progressions of data. When a randomly generated sequence is placed side by side with an actual sequence reflecting human performance, gambling outcomes, etc., the patterns in the two sequences are quite often indistinguishable. Additional information on the theory of longest runs can be found in [5] and in the references cited therein. ■

References

1. Albright, C. (1992), "Streaks and slumps," *OR/MS Today*, vol. 19, 94–95.
2. Albright, S.C. (1993), "A statistical analysis of hitting streaks in baseball (with discussion)," *Journal of the American Statistical Association*, vol. 88, no. 424, 1175–1183.
3. Gilovich, T. and Tversky, A. (1989), "The cold facts about the hot hand in basketball," *Chance: New Directions for Statistics and Computing*, vol. 2, no. 1, 16–21.
4. Gilovich, T. and Tversky, A. (1989), "The 'hot hand': statistical reality or cognitive illusion?," *Chance: New Directions for Statistics and Computing*, vol. 2, no. 4, 31–34.
5. Schilling, M.F. (1990), "The longest run of heads," *The College Mathematics Journal*, vol. 21, no. 3, 196–207.

The
Art Gallery
Problem

ALAN TUCKER
SUNY Stony Brook

The Art Gallery Problem asks, what is the least number of guards required to watch over all paintings in an art gallery with n walls? The guards are positioned at specified locations and collectively must have a direct line of sight to every point on the walls. We seek a formula for the number $g(n)$ defined so that:

(i) for any art gallery with n walls, $g(n)$ guards can provide surveillance of all walls; and

(ii) for some galleries with n walls, $g(n) - 1$ guards cannot provide complete surveillance.

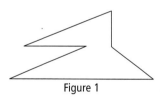

Figure 1

Figure 1 shows a simple 6-walled gallery. The reader should be able to determine the minimum number of guards needed to watch all walls—the answer is 1. The reader might want to try to solve this problem for the more complex gallery shown in Figure 2, before reading our general solution to the problem.

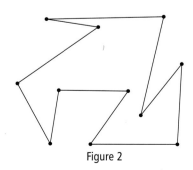

Figure 2

The walls are assumed to form some sort of polygon, such as in Figure 1 or Figure 2. A guard at a corner is assumed to be able to see the two walls that end at that corner. We note that this is a problem in a field of computational geometry which lies in the intersection of mathematics and computer science. Typical problems in this field include routing a robot through an obstacle-filled room and recognizing particular shapes in a digitized picture. A practical example of the latter problem arises in computerized colorization of old black-and-white movies, in which a computer program needs to identify, say, a sofa in a sequence of frames and give it a specified color.

We shall develop an answer to the Art Gallery Problem using graph theory. While the term "graph" is most commonly used to describe the graph, or locus of points, of a function plotted in the plane or higher dimensions, the word "graph" has another, very different usage in mathematics. This other type of graph consists of a set of elements called vertices and a set of edges joining pairs of vertices. Figures 2 and 3 display two such graphs. The graph of Figure 2 is a polygon consisting of the walls of an hypothetical art gallery. This polygon graph has 11 edges, representing the gallery's 11 walls, and 11 vertices, representing the gallery's 11 corners.

Triangulation of a Polygon

For our analysis of this problem we need to create a *triangulation of a polygon.* Such a triangulation is obtained by add-

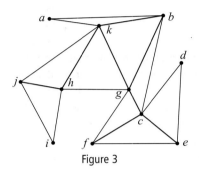

Figure 3

ing a set of (straight-line) chords between pairs of vertices of a polygon so that all interior regions of the resulting graph are bounded by a triangle; these chords cannot cross each other nor can they cross the sides of the polygon. In each interior region with 4 or more boundary edges, a chord can always be drawn between some pair of (non-consecutive) boundary vertices to split the region into two smaller regions. This process of region splitting can be continued until all interior regions have just 3 boundary edges, that is, until all interior regions are triangles. Figure 3 shows one possible triangulation of the polygon in Figure 2. The reason for triangulating the art gallery graph will be apparent shortly.

Graph Coloring

Next we introduce the concept of coloring a graph. We color a graph by assigning colors to the vertices with the requirement that adjacent vertices must have different colors. Figure 4 displays a graph that has been colored with colors R (red), W (white), and B (blue).

Reprinted from Spring 1994, pp. 24–26

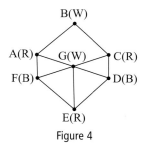

Figure 4

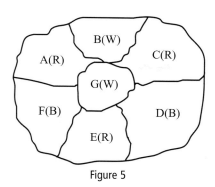

Figure 5

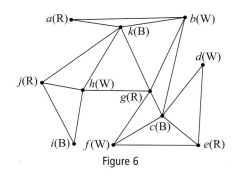

Figure 6

Graph coloring arose from the problem of coloring a map with the standard requirement that bordering countries be given different colors. The map coloring problem can be converted into a graph coloring problem by making a vertex for each country in a map and joining two vertices when they represent countries with a common border. Figure 5 shows a map with countries labeled A, B, C, D, E, F, and G which have been properly colored R (red), W (white), B (blue). The corresponding graph for this map is the graph in Figure 4.

One of the most famous problems in all of mathematics for many years was the Four Color Conjecture, which stated that any map (or its associated graph) could always be colored with four colors. This conjecture was shown to be true in 1976 by Appel and Haken using an extremely lengthy proof much of which was computer-generated.

We want to color a triangulated polygon, such as the triangulated art gallery graph in Figure 3, using just 3 colors. We shall prove the following theorem shortly.

Theorem. *The vertices in a triangulation of a polygon can be 3-colored.*

For the moment, assume the validity of this Theorem. Figure 6 shows a 3-coloring of the triangulated-polygon graph in Figure 3.

The Art Gallery Problem Solved

We can use the Theorem to solve the Art Gallery Problem as follows.

1. Make the triangulation of the polygon formed by the walls of the art gallery (as in Figure 3). Observe that a guard at any corner vertex of an interior triangle has all sides of the triangle under surveillance.

2. Obtain a 3-coloring of this triangulation (such a 3-coloring is possible by the Theorem). Note that each triangle will have one corner of each color.

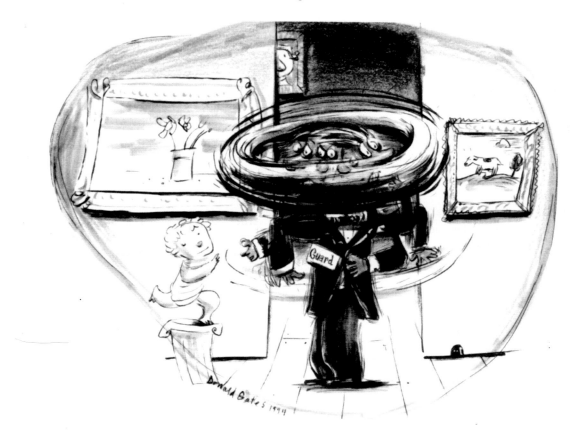

3. Take one of the colors, say 'red,' and place a guard at every corner named 'red.' This places a guard at a corner on every triangle. Hence the sides of the triangles, and, in particular all the gallery walls, will be under surveillance.

A polygon with n walls has n corners. If there are n corners and 3 colors, some color is used on $[n/3]$ or fewer corners, where $[r]$ denotes the largest integer less than or equal to r. In the 3-coloring in Figure 6, there are 11 vertices and so some color is used at most $[11/3] = 3$ times. We observe that color B (blue) is used just 3 times in Figure 6. Thus the B corners—c, i, k—are where we should station the 3 guards. No fewer guards will suffice since corners a, d, and i in this figure each require a separate guard to watch them.

The bound $[n/3]$ is the best possible, since there exist n-walled galleries that require $[n/3]$ guards, that is, $3k$-walled galleries that require k guards. Figure 7 illustrates such a gallery for $k = 4$.

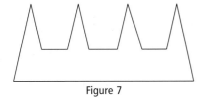

Figure 7

Our solution of the Art Gallery Problem will be complete once we prove the Theorem.

Proof of Theorem

Our proof is by induction on n, the number of edges of the polygon. For $n = 3$, give each corner a different color. Assume that any triangulated polygon with fewer

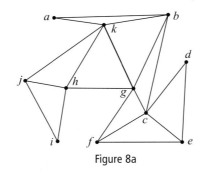

Figure 8a

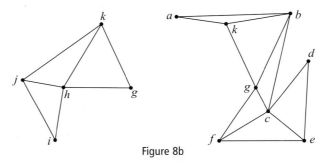

Figure 8b

than n boundary edges, $n \geq 4$, can be 3-colored and consider a triangulated polygon T with n boundary edges.

Pick a chord edge ε, as illustrated by chord (g, k) in Figure 8a. We note that T must have at least one chord edge, or else since $n \geq 4$, T would not be triangulated.

This chord ε splits T into two smaller triangulated polygons, as shown in Figure 8b, each of which can be 3-colored by the induction assumption. The 3-colorings of the two subgraphs can be combined to yield a 3-coloring of the original triangulated polygon by picking the names for the colors in the two subgraphs so that the end vertices of chord ε have the same color in each subgraph. In Figure 8b, this would mean making the color of k be the same in the two subgraphs and making the color of g be the same. ∎

Readers who want to learn more about the Art Gallery Problem and related topics in computational geometry should refer to *Art Galley Theorem and Algorithms*, by J. O'Rourke, Oxford University Press, New York, 1987.

Army Beats Harvard
in
Football and Mathematics

DAVID C. ARNEY
Army Research Office

On a Saturday afternoon in the fall of 1932 on Soldier's Field in Cambridge, Massachusetts, Army's powerful football team trounced Harvard 46-0. This lopsided victory avenged a surprising 1-point win by Harvard the year before. During a luncheon following the game at the home of Harvard's President Abbott Lawrence Lowell, Mrs. Elizabeth Lowell Putnam suggested to General Connor, the Superintendent of the United States Military Academy (USMA), a contest between the two schools in mathematics. Her late husband, William Lowell Putnam, had previously suggested the development of intercollegiate academic competitions between teams of undergraduate students. He believed that the motive of winning laurels for their college in team competition would make students more interested in their studies.

After returning to West Point, General Connor wrote to President Lowell: "There is one matter that I have had in my mind ever since my visit and that is the mathematical contest your sister contemplated. I would very much like to test our method of teaching mathematics against that of your institution. I, frankly, think our method is superior to yours, and would like to try it out." Just two days later President Lowell responded: "Your challenge is a very interesting one which we will be glad to accept."

After agreeing to the contest, the leaders of the two schools turned over the responsibility for the details to the chairmen of their mathematics departments,

On November 5, 1932, Army scored in every period to crush Harvard at Cambridge by a score of 46 to 0 before 40,000 fans. It was Harvard's worst defeat since 1884. This action shot is from the collection of H. W. Riley (USMA class of 1932).

Colonel Harris Jones and Professor William Graustein. While several important issues were resolved quickly (e.g., the competitors would be sophomores, the topics examined would be analytic geometry and calculus, and the test would be given in May 1933), some potential problems surfaced. Both teams wanted to be the home team and there was disagreement on the number of men on each team and the length of the test.

Apparently, USMA thought it had an advantage in depth of quality students, while Harvard believed its star performers could go on the road and win. Therefore, in his first letter to Graustein, Jones called for ten to twenty men on a side and stated that "it will probably not be practicable for cadets to leave West Point" for the contest.

Graustein replied that he did not have in mind teams as large as that. Jones responded that it was probable that the Army team could go to Harvard if the exam date was May 20. Graustein remained adamant that the number of men on each team should be restricted to ten, but offered to "play the role of the visiting team," since Army had been the visitors at the football game. On the issue of the number and length of the tests, Graustein offered the possibility of two tests of about three hours with the second examination "a real test of the men's capacity." It was finally decided that the team size would be ten and Harvard

"The Team of Mathematics" was caricatured in The Pointer *with the Army in dress-grey on the right and Harvard in suits on the left (honestly, they really did dress that way then). The event was famous enough to be remembered two years later in the cadet yearbook,* The Howitzer: *"On the nineteenth of May, back in '33, ten classmates, armed with slide rules, trigonometry tables, and a formula "poop sheet," sallied forth to meet the cream of Harvard's math majors. West Point out-logarithmed, out-differentiated, and out-integrated the Cambridge aggregation by a score of 98 to 112."*

would travel to West Point to take two three-hour tests, one each on May 19 and 20, 1933. Mrs. Putnam agreed to pay for the travel expenses and prizes in the form of books, certificates, and medals for the winning team. Graustein and Jones further agreed on a third party who could write and grade the examination. Graustein wrote: "The man I have in mind as the third member of the committee is the President of the Mathematical Association of America, the body which stands for collegiate mathematics in this country. A new President, Professor Arnold Dresden, was elected at the beginning of the year, and fortunately he is at Swarthmore. Dresden seems almost 'ordained' for the job." Jones wholeheartedly endorsed Professor Dresden as "obviously the man to make up the examinations and grade them." Dresden agreed and set the scoring rules similar to cross-country track. The exams would be ordered and ranked from best to worst, and the ranks of the ten team members would be summed. The team with the lower sum would win.

The mathematics curricula for the first two years at both schools were very similar. The order of topics, pace of coverage, and method of teaching differed slightly. All cadets at USMA were required to take four semesters of mathematics: algebra, solid geometry and trigonometry in their first semester; analytic geometry in the second semester; differential calculus in the third semester; and integral calculus in the fourth semester. USMA used the Thayer method of teaching, which required drill work at the blackboard, and recitation on the board work in every class. Since the entire curriculum of the Academy was required of all cadets, there were no mathematics majors. The 10-man team was selected from the 300 sophomore cadets doing the best in their core mathematics courses. While all 500 freshmen at Harvard took a required common mathematics course, Harvard students began concentrating in special subjects at the beginning of their sophomore year, working under the guidance of tutors. The Harvard 10-man team was comprised of

the best from the pool of 150 students taking sophomore-level mathematics, mostly from the 40 sophomores beginning their concentration in mathematics.

News of the contest reached the media. During the month of May, the Associated Press released several articles describing the contest and its inception from the football game the previous fall. On May 18, John Kieran, sports columnist for the *New York Times,* reported the impending clash with football analogies and great metaphor in his "Sports of the Times" column entitled "The Coordinate Clash, or Block that Abscissa." Kieran reported that the Army coaches had cut their team on the basis that "some men were decidedly too light for heavy line duty in Analytics and others were not fast enough for back-field work in Differential and Integral Calculus." Kieran related some mathematical history to cite the similarities of this contest with a previous mathematics battle waged by Cardan and Tartaglia. But his most revealing coverage came in the form of an introductory poem entitled "A Logarithmic Lilt."

The Harvard horde is plotting,
    under the cover of dark,
A fight to make the Crimson Chord
    subtend the Army arc.
The coefficient Corps has drilled
    with sharpened pencil tips
And plans to drive the enemy
    from the ellipse.
The Harvard cry is "Break the square
    and take the cube away!"
While at the Point "Abscissa!"
    is the watchword of the day.
And high upon the turret top
    the sentry turns his head
And hears the Cambridge legions
    come with logarithmic tread.
"Advance and give the cosine!"
    rings the challenge through the air.
The Crimson host advances —
    and we hope the fight is fair.

The Pointer, the West Point cadets' own magazine, presented their view of the contest, including the line-up, coaches, theme song, cheer, as well as the illustration on page 12. Anticipating an action packed, football-like contest, *The Pointer* parodied:

"The carrier of the ellipsoid is swinging wide on the arc of a cubical parabola, whose equation is: $4y = x^3$. Oh, that is too bad, someone has committed a numerical error. That now makes it last series with still two terms to go. The teams are lined up on the line $x - 2y = 3$, over near the y-axis. The defense finds itself completely unprepared, and their line has been neatly bisected for a gain of approximately 16.24 centimeters. Wait, wait a minute, please—there will be a penalty. Someone has omitted a constant of integration."

Each exam had eleven questions from which ten were to be answered. What were the exam questions like? Four typical questions from the two exams are given for your own enjoyment:

- (a) A function $u(x)$ being given, it is required to determine a formula giving all the functions $v(x)$ for which the derivative of the product u and v is equal to the product of their derivatives.
  (b) Work out the special cases of (a), obtained when
  (1) u = constant, different from zero
  (2) $u = 0$
  (3) $u = e^{ax}$, a is a constant different from zero
  (4) $u = \sin x$

- Determine the total area enclosed by the curve described by the point (x,y) as t varies from 0 to 2, where
$$x = \sin t + \cos t$$
$$y = \sin t - \cos t$$

- Determine the moment of inertia with respect to its axis of a right circular cylinder of height h and radius a, in which the density along a line parallel to the axis and at a distance x from the axis is equal to $b + kx$, b and k being constants.

- A point moves on a line subject to an attractive force inversely proportional to the square of its distance from a fixed point O on the line. It is held at a point A at a distance a from O and then released. How much time does it require to reach O? To reach a point midway between O and A?

After the chalk dust had cleared, Dresden graded the examinations and reported the results. Army once again had defeated Harvard. The score was Army 98 – Harvard 112. Cadet George R. Smith wrote the top paper with B. Feldman of Harvard placing second. Besides the prizes awarded by Mrs. Putnam, the cadet team received special congratulations from the Army Chief of Staff, General Douglas

This cover of the West Point cadet magazine anticipates that Army, whose mascot is a mule, will lick Harvard, mockingly represented here by the pilgrim in the stocks.

MacArthur. The yearbook noted the efforts of all "mathletes" on the Army "math squad," with the write-up for winner George Smith one of the most humorous:

"Well, while my roommate is over seeing the Rhodes Scholarship committee about getting three more years of education, I'll give the world the lowdown on him… It was, I believe, the inducement of a Snicker Bar that persuaded George to go into the Mathematical Contest with Harvard and take first place—so you see to what heights delicacies inspire him."

While it is unlikely Smith received any endorsement opportunities for his winning performance, he could have been a spokesman for an advertising campaign touting the Snicker Bar as the "snack food" of math champions.

The Harvard–Army challenge match was not repeated. However, this match was the precursor of and the inspiration for the annual William Lowell Putnam Mathematical Competition that began in 1938 and is still going strong today. The Putnam Contest is open to individuals and three-person teams from American and Canadian colleges with almost 400 schools and 2500 competitors annually. The exam is given in two parts with 12 challenging problems presented to the competitors. Harvard has won this competition 18 times. While Army has never won the Putnam Contest, its teams have done very well in the Mathematical Contest in Modeling (MCM), which began in 1985. The format for the MCM is different than the Putnam Contest. In the MCM, teams of three people solve an application problem over a weekend and submit a report on their solution. In 1988, both USMA and Harvard were winners, receiving the "outstanding" rating. ■

All the illustrations for this article are courtesy of the Special Collections Division, USMA Library, West Point, NY.

Mathematical Contest in Modeling

The Mathematical Contest in Modeling (MCM) was the brainchild of Ben Fusaro of Salisbury State University. It began in 1985 with 90 teams and had over 400 last year from around the world. It is a true team competition; teams of three work over a four-day weekend and can use any and all inanimate sources. They receive two questions and choose one to work on. The problems are open-ended; there are no correct answers. These are real problems submitted by experts in industry, government and academe.

The "outstanding" papers are published in *The UMAP Journal* each year and the teams are invited to make presentations at the annual meetings of ORSA (The Operations Research Society of America), SIAM (The Society for Industrial and Applied Mathematics), and the Joint AMS/MAA (American Mathematical Society/Mathematical Association of America) meetings.

For more information check out the Student pages on the MAA website — www.maa.org.

The Putnam Competition

Since 1938, the MAA has annually conducted a mathematical competition for undergraduate students supported by the William Lowell Putnam Prize Fund for the Promotion of Scholarship.

Students in the United States and Canada are eligible to participate and win honors for themselves and for their institutions. Each student works independently on the examination. A college or university with at least three entrants designates three of them as a team.

Prizes are awarded to the ten highest ranking individuals, to the mathematics departments of the top five teams, and to members of those teams. A graduate fellowship at Harvard University is awarded to one of the top five students.

Contest questions and solutions are published in the *American Mathematical Monthly* together with a listing of high-ranking individuals and teams. Three volumes of Putnam Problems and Solutions have been published by the MAA covering the years 1938 through 2000.

For more information check out the Student pages on the MAA website — www.maa.org.

Fermat Faces Reality

A Diophantine Drama in One Act

KENNETH M. HOFFMAN
**Massachusetts Institute
of Technology**

Student with new calculator says to Teacher out of the blue: "Is it really true that $27^2 + 18^3 = 9^4$?

Teacher to Student: "Why, ahem! Yes. [Ducking under the desk to check.] How'd you notice that? Has a nice symmetry to it, doesn't it? Relies on $1 + 2^3 = 3^2$, probably a coincidence of sorts."

Teacher, having burned the midnight oil, goes right on the Next Day: "It was. But, you know, this power equation you mentioned works with other numbers. For instance: $648^2 + 108^3 = 36^4$. Try factoring out a few 6's and using a little algebra to check this one, instead of just multiplying things out and adding. In fact, the only addition you should need to perform is $1 + 3 = 4$."

Student, having burned the midnight oil, continues on the Third Day: "Hey! A little factoring and exponents and you were right about 648^2 plus 108^3 equaling 36^4. It just says 1 plus 3 equals 4, all multiplied by 419,904—which is 9 times 6 to the 6th power. This is cool. Tell me more."

Teacher, getting wiser: "Well, if you really want to be convinced algebra is useful, try checking a larger one like 110,592 squared plus 4608 cubed equals 576 to the fourth. See what use your calculator is to you on that. But, you can fiddle with those big numbers at home tonight. Right now, be amazed. Take two minutes to check that $28^2 + 8^3 = 6^4$ too, but for a different reason."

Student "Fantasmic! But whaddaya mean, different reason?"

Teacher mumbles off-handedly: "Oh, not like $1+3 = 4$ or $1 + 2^3 = 3^2$ more like $7^2 + 2^5 = 3^4$."

Student says: "Huh?"

Illustration by Gregory Nemec

Teacher continues unabated: "After you've checked the triple $(28, 8, 6)$, use a little algebra and your calculator (you shouldn't really need it, though) to figure out what $1,176^2 + 49^3$ is the fourth power of—excuse me, of what this sum is the fourth power. You'll discover that you were wrong in thinking that the Pythagorean triple $(7, 24, 25)$ had no practical applications."

Student says: "Huh? I thought what?"

Teacher fires the supposed zinger, even further over the student's head: "But, if you want a really interesting one for tonight, check that $433^2 + 143^3 = 42^4$. You probably won't get much help from exponents on this one, since 433 is prime. For you, it may be a calculator job all the way. And it works for mysterious reasons known only to the Great Number Wielder in the Sky."

Student "Hey, Teach! How do you know all of this stuff? I mean, does it just come to you?"

Teacher "Vast experience and years of mathematical study, O inquisitive one—experience and study. Not to mention a large dose of the mathematician's secret weapon, C & P."

Student "What's C & P?"

Teacher "Calculation and Perseverance. It's usually applied with HuTSPA, which is the maxim that keeps us looking smart: **H**ide **u**re **T**ons of **S**cratch **P**aper **A**fterward."

Student "Yeah, yeah. A real knee-slapper. Listen I gotta go, but could I ask just one more question? This guy Fermat: After he and his buddies noticed that these Pythagorean triples you told us about were so easy and then showed that $x^3 + y^3 = z^3$ was a loser, why didn't he next take up $x^2 + y^3 = z^4$, like we're doing, instead of going to all fourth powers — another loser?"

Teacher "He lacked imagination, I guess."

Student, missing the joke: "Yeah. His equation was sorta dumb. Boring. No solutions at all. Ours is good for something — for us. We've already found five solutions (x, y, z) with y less than 150."

Teacher "Yup, 'we' have. There are six more with $y \leq 150$ that you should try to find. One that has $y = 128$ is really not new—it's $(28, 8, 6)$ in disguise. Of the five additional "primitive" solutions, one has y prime, and in another z is prime—types not easy to come by. But 11 is all there are.*" ∎

---

*Pssst! Ecoutez moi! The Teach has found a truly unremarkable proof of this. Fortunately, the space down here is much too small to contain it. Vive le Professeur Wiles! PdF.

Why History?

UNDERWOOD DUDLEY
Depauw University

Actually, the title should be "Why should a mathematician or a student of mathematics care about the history of mathematics since history doesn't prove theorems and anyway Henry Ford said 'History is bunk,' so the history of mathematics doesn't matter, right?" But thirty-seven word titles, however descriptive, are not customary. Nevertheless, the question is there: why bother with the history of mathematics?

The answer is *not* the terrific usefulness of history. That saying about those who do not know history being doomed to repeat it is correct enough, but it does not have much content since it is also correct that those who *do* know history are doomed to repeat it. Whether we know history or not, we make the same mistakes over and over again, as individuals, groups, and nations, from $\int \sin x\, dx = \cos x + C$ all the way on up. Do not look to history to increase your paycheck or happiness or decrease the number of your errors. It won't.

However, there are other answers. One is that some mathematicians have had interesting lives, and it is good to know interesting things. Think of Galois! A genius, a revolutionary embroiled in politics, his paper lost by Fourier, his work not understood—Poisson said that one paper duplicated in parts some results of Abel and that the rest was incomprehensible—and dead at the age of twenty years and seven months from wounds received in a duel. The only reason that there has been no blockbuster movie about Galois is that solvable polynomial equations lack visual appeal. Besides Galois, there is Gerbert, the only mathematician ever to become Pope (Sylvester II). And Sophie Germain, who, when circulating her mathematical results, did not let on that she was a woman, one reason for that being how Gauss's attitude toward her changed when he found out. And Gerhard Gentzen, starved to death by the Russians in Prague after the end of World War II. And Kurt Gödel, so shy that he delivered his lectures to the blackboard. And those are only among the *G*s! Going back to the *A*s, there is Abel's pathetic search for a job, and Archimedes running down the street naked, yelling "Eureka!" (The point of that anecdote, by the way, is not that Archimedes was *naked*—male nudity was no big thing for the ancient Greeks— but that he, the dignified and renowned sage, was *running* and *yelling*.) There is the witch of Maria Agnesi, which is not a witch at all. And so on, all the way to the *Z*s—Zolotarev fell under a train, with fatal results. People are fascinating, and mathematicians are people.

You may dismiss all that as mere gossip, unworthy of serious attention. You would be wrong to do that, but even if you do there are other reasons for finding out about the history of mathematics. One is that it reminds us of something that many people ought to be reminded of, namely that mathematics is a human activity, done by people, and not something that was engraved on stone tablets by some deity and handed down once and for all. God did not decree that the ratio of the circumference of a circle to its diameter be denoted by π. It was not until 1706 that William Jones thought of using that symbol. Similarly the square root of a number n has not always been designated by $\sqrt{n}$. Irrational and imaginary numbers have not always been as natural as they seem to us—their names show how disagreeable they were once found. Mathematics looks very smooth in the textbooks but that smoothness came only with a struggle, and with a lot of human effort put into ironing out the rough spots. Calculus has not always been universally accepted as incontrovertible truth. Bishop Berkeley said that derivatives were not to be trusted, being made up of "the ghosts of departed quantities." Mathematics has changed and evolved. Quaternions were once hailed as a revolutionary advance, but it is now possible to get a degree in mathematics and know nothing at all about them. As we struggle with mathematics, and change as a result, we are in a sense recapitulating the history of the subject. We are at the end of the long chain of humans who have made mathematics what it is, and mathematics will not be finished with us. It is nice to realize that we are part of history.

Another advantage of knowing that history exists is the chance it gives to read the masters of the past. To see Euclid explain the Euclidean algorithm all in terms of line segments, with not an x or a y in sight (Book VII, Proposition 2) will, if nothing else, add to your admiration of Euclid. For another example, in his *Elements of Algebra,* Euler writes (part II, chapter 5, section 64)

> The formula cz^2 can become a square only when c is a square; for the quotient arising from the division of a square by another square being likewise a square, the quantity cz^2 cannot be a square, unless cz^2/z^2, that is to say, $c,$ be one. So that when c is not a square, the formula cz^2 can by no means become a square; and, on the contrary, if c be itself a square, cz^2 will also be a square, whatever number be assumed for z.

Reprinted from November 1994, pp. 10–11

Illustration by Greg Nemec

"Yes, yes, Leonhard," you think, "that's all quite clear, you don't have to beat it to death." But then, only four pages further on, without having encountered anything less clear, you find that you have shown that $5t^2 + (5n + 2) u^2$ is never a square for any n. Euler has carried you along, so effortlessly that you are tempted to think that this must be really easy stuff, but the temptation should be resisted. What is actually the case is that Euler was a master at exposition. It can be instructive to see a master at work. Further examples could be given—those were taken just from the *E*s.

The reason for bothering with history that includes all the others is that you will live in a larger world. It is acceptable to think, as I am afraid many people unconsciously do, that mathematics has always been the way it is today, and that everyone has always studied groups and rings and ideals. But it is better to know that ideals originally were ideal numbers, devised by Kummer so that unique factorization would hold and progress could be made on Fermat's Last Theorem. It is good to know about mathematical objects,

but it is better to know where they came from and why they are studied.

It is better, by far, but it is not *necessary*. Squirrels don't know history and squirrels, if asked, would probably say that, on the whole, they had fine lives. But—observe squirrels closely. Twitch! Jump! Scurry! Squirrels live in a world of random potential disasters. Squirrels are constantly in a state of panic. *Everything* comes as a surprise to a squirrel. The reason is that squirrels do not know history. They live in the present, they live in a small world. When they dash for the nearest tree when there is a clap of thunder they do not know that their hysteria is unnecessary because in the entire history of squirrels, not one squirrel has ever been injured by thunder. I'm glad that I'm not a squirrel. I'm glad that I know a little about the panorama and people of mathematics. I like living in a large world, one that extends into the past.

But, as I said, knowing about the history of mathematics is not necessary. You can ignore it if you want to and no one will mind, or even notice. Your career will be unaffected. The number of theorems you prove will probably be the same. You

won't even know that you're missing anything. If you want to be a squirrel all your life, go right ahead. ∎

FURTHER READING. When a book reaches its sixth edition, it must have something going for it. *An Introduction to the History of Mathematics,* by Howard Eves (6th ed., 1990, Saunders College Publishing) is a fine introduction, filled with good things. For all its flaws, *Men of Mathematics* by E. T. Bell (1986, Simon and Schuster), remains intensely readable. It was first published in 1937 (when its title was more natural) and a book that stays in print for more than fifty years also must have something good about it. William Dunham's *Journey through Genius* (1990, Wiley; 1991, Viking Penguin), presenting episodes in the history of mathematics, has been a critical and popular success. *Whom the Gods Love,* by Leopold Infeld (1948; reprinted in 1978 by the National Council of Teachers of Mathematics), the story of Galois, is one of the few historical novels written about a mathematician. The only other one I know of is Charles Kingsley's *Hypatia*, but it is no longer in print and no longer readable. Euler's *Elements of Algebra,* translated into English by John Hewlett and published in 1840 was reprinted by Springer-Verlag in 1984. Finally, browsing through a library in the vicinity of the QA21s can do no harm and may do some good.

Carving Mathematics

DON ALBERS
Mathematical Association
of America

Over the past 30 years, Dr. Helaman Ferguson has moved slowly and steadily toward his current place in the sun—the preeminent artist-mathematician in the world. His career path underscores the difficulties in simultaneously working in two fields. Serious multidisciplinary work is exceedingly hard to accomplish, but Ferguson has done it. His magnificent sculptures are visual feasts for all who love art. Their beauty is further enhanced for mathematicians when they learn that each of his creations is inspired by mathematics. Non-mathematicians who react warmly to his sculptures are pleasantly surprised when they learn it has been inspired by mathematics. His work is now found in many institutions around the world, and his influence is growing.

But Ferguson's path has not been an easy one. His beginning was so difficult that it is a wonder that he even finished first grade. "Why us? Why me?" were the plaintive cries of Helaman Ferguson's father after seeing his newborn son in a Salt Lake City hospital. Helaman Ferguson had entered the world seriously disfigured with a cleft palate. He recalls, "It was severe enough that I had to be fed through a tube in my nose, and I must have looked awful. Imagine the scene. My artist father and my mother were in the category of beautiful people. My appearance had to have been a big blow." Many operations over the next few years corrected the problem so that today one would be surprised to learn that the handsome sculptor-mathematician once was disfigured. Ferguson says that "the fact that I

Ed Bernik/Meridian Creative Group

Sculptor Ferguson likes stone ... a lot. "If I start thinking about a nice piece of stone, my hands sometimes start sweating."

can talk at all is of interest to many doctors."

At age three, disaster visited young Helaman again. He and his sister were watching their mother remove laundry from the clothesline in the backyard when suddenly she was struck and killed by lightning. Two weeks later, his father was drafted for service in World War II. Ferguson's grandmother cared for him

Reprinted from November 1994, pp. 14–17

Eine Kleine Rock Music III
Bronze #1/8

Claire Ferguson

and his sister for the next two years. As might be expected, Helaman's emotional state was not very good during those years. He remembers having nightmares, grinding his teeth at night, and being rather withdrawn. "An aunt remembers me coming home from kindergarten, smashing my fist into the door jamb until it was bloody, and asking "What's a hare-lip?" After one year in kindergarten, "my teacher concluded that I would definitely have to be institutionalized, that I had no future and that I was unteachable."

At age six, Helaman's life began to improve. A distant cousin and her husband, a stonemason, in Palmyra, New York, were childless and they adopted Helaman and his sister. "My stepparents gave me a stable, orthodox Mormon family life. I did not remain withdrawn. I did well in high school, and had lots of friends."

The unique blending of artistic and mathematical talents possessed by Ferguson first manifested itself on the artistic side. In elementary school, his painting ability was noted and praised by Mr. Black, a roving art teacher. "Mr. Black reported to my adopted parents that I was a joy to have in class." He continued to be involved with art in grade school, and

in ninth grade his world opened up further when he encountered a great mathematics teacher, Florence Deci. According to Ferguson, "she was one of the first people who understood my dual [artistic-mathematical] components. She really liked the art I was doing at the time, clay heads. And she talked about the beauty of mathematics in terms of career possibilities for someone who did art and math-

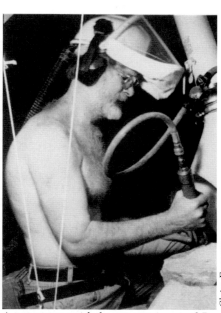

A computer-guided system is part of Dr. Ferguson's stone carving tools.

Claire Ferguson

ematics. She let me teach the class and do lots of interesting math projects, even some combined with art. Once, a friend and I put together a wire-frame model of a hyperbolic paraboloid. She accepted my creative thrusts, whether they were art or math and was positive about it. She encouraged me. I was very fortunate to have a teacher like Florence Deci."

Ferguson won several scholarships, and chose Hamilton College because it offered a combination of the sciences and the arts. He continued to do art and mathematics at Hamilton. At one point, he and a friend painted all the walls in the dorms. James Penny was an artist-in-residence while Ferguson was an undergraduate, and he recalls "learning a lot of professional artist things from him. He also gave me some sculpture training. I was interested in stone and did one plaster sculpture there. Since Hamilton didn't have any stone-carving tools, it really was moot."

Ferguson's work in mathematics at Hamilton was so good that he was awarded an NSF Fellowship to do graduate work in mathematics at the University of Wisconsin. He remembers his first reaction to getting the fellowship— "Great, now I can paint! The fellowship, of course, was for mathematics. It was a time of considerable indecision and turmoil for me.

"After a semester at Wisconsin, I gave back my NSF Fellowship, because I thought I'm not going to do math, I'm going to do art."

During his brief stay at Wisconsin, he met Claire, a young art student. He proposed marriage, she accepted, and off they went to Salt Lake City to be married and lead artistic lives. In no time flat, the newlyweds came to understand the term "poor, starving artists."

His aunt, who was aware of their modest circumstances, helped them out from time to time by buying some of his paintings. "One day," he remembers, "my aunt came to visit, took note of the bare cupboards, and asked, 'Do you have anything to eat?' I responded, 'Oh, yes.' She then walked over to the refrigerator and swung open the door. It was totally empty. She then said, 'You are going to get a job." He soon got a job as a computer programmer.

While working as a programmer, Ferguson completed a master's degree at Brigham Young University in mathematics. "One of the reasons I earned a master's degree at BYU was to tidy up my Wisconsin scholastic record, which I left in disarray. I also took graduate courses in sculpture while I was at BYU and did a number of wood carvings." His thesis advisor at BYU urged Ferguson to continue his education.

In 1966, Ferguson and his growing family moved to the University of Washington, where he would finish his doctorate five years later. "At Washington, I low-profiled my artistic thing, at least within the mathematical community. I did some commissioned paintings and that supported our graduate school existence to some degree."

After receiving his doctorate in 1971, specializing in harmonic analysis, Ferguson joined the mathematics department of BYU. In 1972, his stone carving began in earnest. "I picked up some limestone while camping, and carved some pieces using a carpenter's hammer and railroad spike." What's so special about stone? According to Ferguson, "I like carving stone. I prefer a subtraction process and the accompanying challenge. Once you remove a piece of stone, you can't put it back. I like the fact that stone is hundreds of millions of years old. Stone really interests me. If I start thinking about a nice piece of stone, my hands sometimes start sweating.

"The mathematical ideas I work with often are old and timeless, too. So there's a natural blending. Stone is permanent. Stone is worthless; it doesn't have military value. Consequently it's more likely to survive the vicissitudes of our civilization.

"I'm drawing from mathematical resources that go back to the very beginnings of our culture. With my carving, I'm also sending these mathematical messages into the future.... I like the idea of

*Costa II Minimal Surface,
Bronze and Aluminum*

celebrating mathematics with art. It's something to live with. We tend to get the idea in our early education that mathematics is not something to live with. Mathematics is something that someone creates. Maybe you have to reproduce parts of it at different times in your life. It's not usually construed as something that's part of your visual, artistic, and cultural life. I think it could be. Our family lives with it and other people do, too."

At BYU, Ferguson settled in and rose through the ranks to become a full professor. Throughout his seventeen years at BYU, his sculpting continued, and in 1988 he took a leave of absence in order to work in the East more intensively on his sculpture. In 1991, he gave up the security of his tenured professorship and established his home in the East because he says "it gave me freedom to be in a context where my work could be appreciated more fully. There are more resources for a sculptor here in the East than there are in Utah. My sculpture had developed in Utah to the point where I had a few important collectors there, but there is a lot more here in terms of resources. And the culture here is 200 years older than in Utah. People have been doing art here for a long time, and it's acceptable. I don't have to defend myself because I'm an artist. But if you're in a community of people who did not grow up with artistic traditions, they wonder why you aren't being a responsible member of society doing a real job."

Ferguson now works as a research mathematician for the Supercomputing Research Center (SRC) of the Institute for Defense Analysis. At SRC, he designs computer-based numerical algorithms for operating machinery and for visualizing data. He's especially interested in "sculpting data," i.e., developing methods for shap-

*Helaman and Claire with Michael Paul and Alexander, two of their seven children.
They are sitting on a lovely piece of marble which Ferguson plans to carve.*

ing data into forms that are useful to researchers who often are flooded with data from super-computers. His other source of income comes from sales of his sculpture, but most of that income goes back into supporting his sculpture.

One of the pleasures in visiting with Ferguson is his willingness to let you touch his creations. Although museum personnel may disagree, Ferguson says, "I want people to touch my art." For students, it's a special experience to feel *Umbilic Torus NC, Eine Kleine Rock Musik III,* or *Double Torus Stonehenge: Continuous Linking and Unlinking.* In the recent book, **Helaman Ferguson: Mathematics in Stone and Bronze**, by his wife Claire Ferguson, her description of *Figure Eight Knot Complement I* begins, "Like a small bird, the red alabaster form is friendly, easily cradled in both hands while curious fingers explore its intricate curves and hollows." Clearly, the artist's wife is moved by his art and loves to touch it, too. ■

The images are reprinted, courtesy of **Helaman Ferguson: Mathematics in Stone and Bronze**, written by Claire Ferguson, copyright 1994, Meridian Creative Group. For further information about the book, please contact:

Meridian Creative Group
5178 Station Road
Erie, Pennsylvania 16510
(814) 898-2612

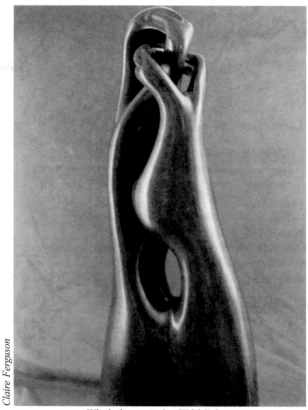

Claire Ferguson

Whaledream ii & i Wild Sphere

Word Ladders
Lewis Carroll's Doublets

MARTIN GARDNER

Doublet tasks consist of changing one word to another by altering single letters at each step to make a different word. The two words at the beginning and end of such a chain must, of course, be the same length, and they should be related to each other in some obvious way. They must not have identical letters in the same positions. All words in the chain should be common English words, proper names excluded. A "perfect" solution has a number of steps equal to the number of places the given words differ. For example: Cold, Cord, Card, Ward, Warm. If a perfect change is not possible, the best solution is the shortest chain. For playing doublets as a game with two or more players, Lewis Carroll invented a set of scoring rules to determine who wins.

The first mention of the game in Carroll's Diary is on March 12, 1878, when Carroll reports teaching "Word Links" (his original name for the game) to guests at a dinner party. He had invented the game, he tells us in a pamphlet, on Christmas Day in 1877 for two little girls who "found nothing to do."

Carroll's hand lettered Word-Links: A Game for Two Players, or a Round Game was written in April 1878. Later that year he printed a revised version as a 4-page pamphlet. Starting with the March 29, 1879, issue of *Vanity Fair,* Carroll contributed a series of articles on doublets. The first was followed by an article announcing a doublet competition and a third article giving a new method of scoring. In

1879 Macmillan gathered the *Vanity Fair* articles into a 39-page book, with red cloth covers, titled *Doublets: A Word Puzzle.* An 1880 second edition was enlarged to 73 pages. *The Lewis Carroll Picture Book* reprints part of this edition. Later that same year Macmillan published a third edition, revised and enlarged to 85 pages.

Carroll took the name doublets from a line of the witches incantation in Shakespeare's *Macbeth*: Double, double, toil and trouble—a line Carroll placed on the title page of his book.

On May 11, 1885, Carroll mentions in his *Diary* that he has extended his list of

seven-letter word pairs that can be linked together to more than 500.

Doublets became a parlor craze in London, and has been a much practiced form of word play ever since. They have been called by other names, such as "word ladders," and (in Vladimir Nabokov's novel *Pale Fire*) "word golf." Enormous energies have been expended on finding shortest ladders for a given pair of words. Computer software containing all English words is now obtainable, and programs have been written for finding minimum chains in just a few seconds. The task is equivalent to finding the shortest routes connecting two points on a graph.

Donald Knuth, Stanford University's noted computer scientist, has constructed a graph on which 5,757 of the most common five-letter English words (proper nouns excluded) are represented by points, each joined by a line to every word to which it can be changed by altering just one letter. The graph has 14,135 lines. Once in a computer's memory, programs can be written that will determine in a split second the shortest word ladder joining any two words on the graph. Knuth found three-letter words too simple, and six letter words less interesting because not too many can be connected.

Most pairs of five-letter words on Knuth's list can be joined by ladders. Some—Knuth calls them "aloof" words because one of them is the word aloof— have no neighbors. The graph has 671 aloof words, such as earth, ocean, below, sugar, laugh, first, third, ninth. Two words, bares and cores, are connected to 25 other words; none to a higher number. There are 103 word pairs with no neighbors except each other, such as odium-opium, and monad-gonad. Knuth's 1992 Christmas card featured the smallest ladder (eleven steps) that changes sword to peace by using only words found in the Bible's Revised Standard Version.

I have taken the above information from the eight pages devoted to doublets in the first chapter of Knuth's book *The Stanford GraphBase* (Addison-Wesley,

1993). Knuth will cover the topic more fully in his forthcoming three-volume work on combinatorics in his classic *Art of Computer Programming* series. For hints on how to solve doublets without a computer see his article "Lewis Carroll's Word, Ward, Ware, Dare, Dame, Game," in *Games* magazine (July-August, 1978).

It has been pointed out that doublets resemble the way in which evolution creates a new species by making small random changes in the "genes" that are intervals along the helical DNA molecule. Carroll himself, although a skeptic of Darwin's theory, evolved Man from Ape in six steps:

APE ARE ERE ERR EAR MAR MAN.

When I gave this solution in a *Scientific American* column on Mathematical Games (the column is reprinted as Chapter 4 in my *New Mathematical Diversions*), two readers produced a shorter solution:

APE APT OPT OAT MAT MAN.

In a letter of March 12, 1892 (See Morton Cohen's *The Letters of Lewis Carroll*, Volume 2, page 896) Carroll added a rule that allows one to rearrange the letters of any word, counting this as a step. With such increased freedom, he pointed out, many impossible doublets, such as changing Iron to Lead, can be achieved: IRON, ICON, COIN, CORN, CORD, LORD, LOAD, LEAD.

It is difficult but not impossible for a word chain to form a sentence. In *Vanity Fair* (July 26, 1879), one of Carroll's doublets asked "WHY is it better NOT to marry?" to change WHY to NOT he added this proviso: "the chain made [WHY to NOT] ... should embody the following observation: that lovers, during the temporary insanity of courtship, too often fail to recognize the grave prudential reasons which should deter them from taking this fatal step." Here is Carroll's clever solution: WHY, WHO, WOO, WOT, NOT.

The mathematician and writer of science fiction, Rudy Rucker, has likened doublets to a formal system. The first word is the given "axiom." The steps obey "transformation rules" and the final word is the "theorem." One seeks to "prove" the theorem by the shortest set of transformations.

Many papers on doublets have appeared in the journal *Word Ways,* a quarterly devoted to linguistic amusements. An article in the February 1979 issue explored chains that reverse a word, such as TRAM to MART, FLOG to GOLF, LOOPS to SPOOL, and so on. The author asks if an example can be found using a six-letter word.

Is there a closed chain, I wonder, that changes SPRING to SUMMER to AUTUMN to WINTER, then back to SPRING? If so, what is the shortest solution?

A.K. Dewdeny, in a Computer Recreations column in *Scientific American* (August 1987), calls a graph connecting all words of *n*-letters, a "word web." He shows that all 2-letter words are easily joined by such a web, and asks if anyone can construct a complete word web for 3-letter words.

Lewis Carroll: a drawing by Harry Furniss, illustrator of Carroll's Sylvie and Bruno.

If you want to improve your doublets skills, in the box below are fifteen from Lewis Carroll. Donald Knuth's improvements on Carroll's best results can be found on p. 29. ■

ROGUE	QUELL	KETTLE	COSTS	SHOES
vogue	quill	settle	posts	sloes
vague	quilt	settee	pests	floes
value	guilt	setter	tests	floss
valve	guile	better	tents	gloss
halve	guide	betted	tenth	glass
helve	glide	belted	tench	class
heave	glade	bolted	teach	crass
leave	grade	bolter	peach	cress
lease	grave	bolder	peace	crest
least	brave	HOLDER	PENCE	CRUST
BEAST	BRAVO			
BLACK	BEANS	GRASS	STEAL	WHEAT
clack	beams	crass	steel	cheat
crack	seams	cress	steer	cheap
track	shams	tress	sheer	cheep
trick	shame	trees	shier	creep
trice	shale	frees	shies	creed
trite	shall	freed	shins	breed
write	shell	greed	chins	BREAD
WHITE	SHELF	GREEN	COINS	
FURIES	TEARS	PITCH	FLOUR	RAVEN
buries	sears	pinch	floor	riven
buried	stars	winch	flood	river
burked	stare	wench	blood	riser
barked	stale	tench	brood	MISER
barred	stile	tenth	broad	
BARREL	SMILE	TENTS	BREAD	

Professor

of ~~Magic~~

Mathematics

DON ALBERS
Mathematical Association of America

Persi Diaconis, professor of mathematics at Harvard, has just turned 50, but the energy and intensity of the 14-year old Persi who left high school to do magic full time for the next ten years of his life still burns brightly.

His work in mathematical statistics was so good that he was awarded a $200,000 MacArthur Foundation Fellowship, tax free and no strings attached. The purpose of the awards, for which applications are neither solicited nor accepted, is to free creative people from economic pressures so they can do work that interests them. In spite of his mathematical achievements, Diaconis insists that he is better at magic, his first career, than he is at statistics. After ten years of doing magic on the road, he decided to try college. At twenty-four, he enrolled as a freshman. Five years later he had earned his PhD from Harvard.

Diaconis applies mathematics to a wide range of real-world problems, claiming that "I can't relate to mathematics abstractly. I need to have a real problem in order to think about it."

Not long ago he established a major result about card shuffling that is of importance to anyone who plays cards and who would like assurance that the cards in a deck are in random order. Diaconis proved that a deck of cards needs to be shuffled seven times in order for the cards to be in random order. He says "You might think as you shuffle a deck more and more times it just gets more and more random. That is not

the way it works at all. It is a theorem that this phenomenon of the order of cards being intact as you go from one, two, three shuffles… to being essentially random happens right at seven shuffles."

Diaconis is ranked among the top three "close up" magicians in the world. Close up magic is done tableside as opposed to on a stage. How much does Professor Diaconis love magic? His response is crystal clear: "If I could have had a professorship in magic, and if the world recognized magic the way it does mathematics, I probably would be doing magic full-time and never would have done mathematics or statistics."

His background in magic and statistics has also proved useful in exposing psychics, including Uri Geller. He is cur-

rently working on books about coincidences and mathematical magic.

A Magical Beginning

ALBERS: At the age of 14, you left your New York City home and spent the next ten years on the road practicing magic. What made you do that?

DIACONIS: That's simple. The greatest magician in the United States was a man named Dai Vernon. He called me up one day and said, "How would you like to go on the road with me?" I said, "Great," and he said, "Meet me at the West Side Highway two days from now at two o'clock." So with what money I could pick up and one suitcase, I went on the road. It was simply a question of a mag-

Strange looking dice! What is this, Persi, mathematics or magic?

Photograph by Cathy Saloff-Coste

Reprinted from February 1995, pp. 11–15

netic, brilliant expert in the field calling on me, just as a guru calls on a disciple. I was quite honored and excited to do it.

A: What did your parents say to your leaving home to practice magic?

DIACONIS: I didn't ask them. I just left home. My parents were upset at my leaving, but somehow they found out that I was okay. For a long time I was the black sheep of the family. Only when I started graduate school at Harvard did my family begin to think that I wasn't terrible.

A: So they felt very bad about your going off to practice magic.

DIACONIS: Sure they did, I was being groomed to be a virtuoso musician. I went to Julliard from the ages of 5 to 14. After school and on weekends I played the violin. All of my family members (mother, father, sister, and brother) are professional musicians. They thought I was going to become a violinist and having me desert music for magic was not very appealing to them. I think they have come to accept it all now. They never came to accept the magic, even though I was good at it. I was better at magic than I am at what I do now.

A: How did you get into magic?

DIACONIS: When I was five years old, I found the book *400 Tricks You Can Do* by Howard Thurston. I picked it up and figured out that I could do a few tricks. I soon did a little magic show at my mother's day camp. I clearly remember that show. I was the center of attention. I wasn't horrible apparently, and magic became a hobby. I sent in my dimes for mail-order catalogs on magic, and for my birthday I would ask for tricks as presents. When I got to public school I met other kids who were magicians and I joined the Magic Club. I threw myself into it with a real fury. All the energy that I didn't put into doing homework or anything else connected with school I put into magic. On many days I would cut school and hang out at the magic store until closing time.

A: Who would assemble at the magic store?

DIACONIS: Older magicians and other kids who were interested in magic. In New York City there was a big, lively magic community. When I was 12, I met Martin Gardner at the cafeteria where magicians used to hang out. He was the kindest, nicest man, and he took time out to show me some lovely, little tricks that I could do. (Gardner, in addition to being a great writer, also is an accomplished magician.) He saw that I was a troubled kid and took a liking to me. He told me to call him if I had any questions. So I used to call him and talk about magic, and he got me interested in working on mathematical tricks because he would warm to that.

Professor Diaconis posed in front of one of his favorite paintings in his Harvard office.

Photograph courtesy of Harvard University

A: Did you know that Martin Gardner was a big name?

DIACONIS: Sure. I knew who the other magicians respected, who was famous and who was not so famous. He was obviously a very special guy, the kind of guy who could go on and on about things and remain interesting and never be pompous, just kind and instructive. He also was genuinely delighted if I showed him a new twist on a trick that he might know. He didn't try to put someone down because it was a trivial twist on some-

thing. When I showed him a new little idea, he would make a note of it. Every once in awhile he would put something of mine into his "Mathematical Games" column in *Scientific American* magazine and that was a great thing for me.

On the Road

A: You went on the road at age 14. What were those years like?

DIACONIS: During the first few years I was in very good company. I was being shepherded around by Dai Vernon, a brilliant man, the magician's magician and the best inventor of subtle sleight-of-hand magic of the century. He taught me magic: we talked magic morning, noon, and night. Since he was sort of old, and since I could do the sleight-of-hand very well, when he would give magic lessons, he would have me demonstrate tricks, and then he would explain them. So my experience was vaguely structured and very colorful—a lot more colorful than I choose to put into any interview. I met all kinds of interesting street people, was often broke, hitchhiked, and so forth. I left Vernon when I was about 16 and was on my own. He went on to Hollywood to found what is now known as the Magic Castle, which is a fabulous magic club, a private, wonderful magic place where movie stars hang out. I decided I didn't want to do that and would stay on my own. So I stayed in Chicago, lived in a theatrical hotel, and played club dates, usually for $50 a night. I did pretty well that way. I eventually drifted back to New York, doing magic and pursuing it as an academic discipline, inventing tricks, giving lessons, and collecting old books on magic, which I still do. It was just my life. I did it with all my energy.

A: Magic very often has card tricks associated with it and perhaps card playing. Were you playing cards at the same time?

DIACONIS: No, not at the beginning. Much later somehow I got a copy of Feller's famous book on probability, and I got interested in probability that way.

Photograph courtesy of Harvard University

Diaconis illustrates a point. He claims that "inventing a magic trick and inventing a theorem are very similar activities."

A: How did that happen?

DIACONIS: It was due to another friend of mine, Charles Radin, who is a mathematical physicist at the University of Texas. He was in college on the straight and narrow while I was still doing magic. We had been kids together in school. One day he went to Barnes and Noble Bookstore to buy a book and I went along for the ride. He said Feller was the best, most interesting book on probability, and I started to look at it. It looked as if it was filled with real-world problems and interesting insights, and so I said, "I'm going to buy it." He said, "You won't be able to read it," I said, "Oh, I can do anything like that." Well, in fact, I couldn't; I tried pretty hard to read Volume 1 of Feller, and it's one of the big reasons I went to college, for I realized that I needed some tools in order to read it.

A: What college?

DIACONIS: I started at City College at night. They wouldn't take me during the day because I was something of a strange person, so I went for a couple of years at night taking one or two courses.

I discovered that I liked college, and I decided to try for a degree. I finished up in two and a half years. It was a short time after I started college that I dropped magic as a vocation.

Martin Gardner and Graduate School

A: How did you end up at Harvard?

DIACONIS: I graduated from City College in January, and decided to start graduate school in mid-year. It turned out that some places, including Harvard, did accept mid-year applications. Harvard's mathematics department hadn't taken anyone from City College in 20 years. All of my teachers said Harvard didn't accept any students from City College, even the really good ones. So, I decided not to apply in mathematics. Instead I applied in statistics; it was the only statistics department I applied to. At the time, I didn't very much care about statistics, but I thought it would be fun to go to Harvard. I thought I would try it for six months and see if I liked it. I did like it, they liked

me, and I stayed on to finish a PhD. Because of my strange background I probably wouldn't have gotten into Harvard had it not been for the intervention of Martin Gardner. I was talking to Martin a lot during that time, asking his advice as to where to go, and he was, of course, professing to know nothing about mathematics. I said I was thinking of applying to the Harvard statistics department, and he said that he had a friend there named Fred Mosteller. Now Fred Mosteller is a great statistician, who in his youth had invented some very good magic tricks. There is, for example, a trick called the Mosteller Spelling Trick, which is still being used today. Martin wrote a letter in which he said something like, "Dear Fred. I am not a mathematician, but of the ten best card tricks that have been invented in the last five years, this guy Diaconis invented two of them, and he is interested in doing statistics. He really could change the world. Why don't you give him a try?" Fred later told me that I would not have been admitted if it had not been for that letter.

Statistics is the Physics of Numbers

A: You have spent most of your professional life working in statistics. What is statistics to you?

DIACONIS: Statistics, somehow, is the physics of numbers. Numbers seem to arise in the world in an orderly fashion. When we examine the world, the same regularities seem to appear again and again. In more formal terms, statistics is making inferences from data. It is the mathematics associated with the application of probability theory to real-world problems, and deciding which probability measure is actually governing.

A: Do you think of statistics as part of mathematics?

DIACONIS: Yes. It is part of applied mathematics. There is something about making inferences that goes beyond mathematics. In mathematics you must have something that is correct and beautiful, and that is enough to qualify as mathematics. In statistics, however, there is the question of trying to decide what is true in the world, and that is somehow going beyond any formal system.

Nothing pleases me more than being able to take some mathematical idea and apply it to solve a problem. But the bottom line for me has to be that I actually get an answer to the problem. In the case of the card shuffling, how many times do I have to shuffle a deck of cards? The answer is seven for real shuffles of a deck with fifty-two cards. Without the number seven at the end, all of the underlying mathematical ideas wouldn't mean as much to me.

A: The group theory is more beautiful for you as a result.

DIACONIS: Absolutely! I can't relate to mathematics abstractly. I need a real problem in order to think about it, but given a real problem I'll learn anything it

Professor Laurent Saloff-Coste, right, of the University of Toulouse has been Diaconis's main collaborator for the past several years. Here we see them discussing a problem of finite Markov chains, and perhaps where to have dinner.

takes to get a solution. I have taken at least thirty formal courses in very fancy theoretical math, and I got A's and wrote good final papers, and it just never meant anything. It didn't stick at all; that's something about me.

My PhD thesis involved a very concrete problem, namely the crazy first digit phenomenon. If you look on the front page of *The New York Times,* and observe all of the numbers which appear there, how many of them do you think will begin with one? Some people think about a ninth. It turns out empirically that more numbers begin with one, and in fact it is a very exact proportion of numbers that seem to begin with one; it is .301. Now that's an empirical fact, and it's sort of surprising. It comes up in all kinds of real data. If you open a book of tables, and look at all of the numbers on the page, about 30% of them begin with one. Why should that be? It's always been that way for me. There is some question and some set of mathematical tools, and often the question has been asked several times, and eventually the question drives you on to understand the set of tools, and then for me the game isn't finished until the set of tools yields the answer. This can take years. There are questions I have worked on for 30 years. Until I get the right answer, I don't stop.

The Art of Finding Real Problems

A: How do you find real problems?

DIACONIS: That's probably what I'm best at. What makes somebody a good applied mathematician is a balance between finding an interesting real-world problem and finding an interesting real-world problem which relates to beautiful mathematics. In my case, I browse an awful lot, sit in on courses, and read a lot of mathematics. As a result, I have a rather superficial knowledge of very wide areas of mathematics. Also, I am reasonably good at talking to people and finding out what ails them problemwise.

Psychics and ESP

A: How did you become involved with psychics and ESP research?

DIACONIS: ESP is a nice example of an area where my background in magic and my interest in statistics come together. It's a marvelous, clear example of a nice applied math problem. Any respectable proof of parapsychology by the standards of today is statistical in nature, and therefore in order to be a good investigator you have to know about statistics. One of the big problems for parapsychology investigators is that sometimes they work with people who cheat, deliberately or subconsciously, or both.

My involvement began when *Scientific American* reviewed a book that contained a report of a psychic in Denver who purported to make psychic photographs with his mind. Investigators would bring their Polaroid cameras and snap a picture of this guy's head, and usually they would get a picture of his head; but once in a while the photographs would look something like a fork, or a biplane, or Cro-Magnon Man or something like that. Martin Gardner arranged for me to go to Denver to investigate him; and while I was there I caught him cheating unquestionably. Over the years I have investigated several so-called psychics, as a kind of hobby and also as a source of interesting problems. I guess it's also a service to the scientific community. It's hard for ordinary scientists to do a good job at debunking psychics. We may all feel that it is baloney, but it's very hard to determine why.

Debunking

A: Why is it hard for scientists to debunk psychics?

DIACONIS: It's because most people (a) don't know the tricks, and (b) don't have the statistical background. It is very easy for the tricks to be concealed in poor statistics. A combination of (a) and (b) can be devastating. You can be a terrific physicist or mathematician, but if you don't have experience in running experiments with human subjects and with cueing, etc., you may have a very tough time. Having the experience often makes it very obvious what's wrong, and when you point out the trick or statistical fallacy to somebody else, they say "aha." It's hard for people to spot it on their own.

A: The public's interest, in ESP, astrology, and numerology is very high. How do you explain their fondness for it?

DIACONIS: It is a basic human reaction to wonder at something surprising such as an unusual coincidence. That seems to be a hard-wired reaction in people. Perhaps it is wired in there for protection. I think it is unquestionable that we have a pattern-detecting mechanism that works and is alerted and delighted by surprising coincidences.

When I was a performer, I learned that it is much easier to entertain people by pretending that your tricks are real magic, than to do wonderful tricks and just present them as tricks. People, if you let them, are quite willing to believe the most outlandish things, and the fact that you can do a little sleight-of-hand and actu-

ally make something happen in addition to creating a spell of wonder makes it all the more believable. Large proportions of our undergraduates believe that parapsychology is a demonstrated fact.

I read very thoroughly for ten years all of the refereed, serious parapsychology literature. There is not a single, repeatable experiment in that literature. Most people don't seem to know that.

The business card of the professional magician, Persi Warren (Diaconis), who left home at age fourteen and performed professionally for the next ten years.

A: Do you still do music?

DIACONIS: I don't do music anymore, but I still do magic. The way I do magic is very similar to mathematics. I do it seriously as an academic discipline. I study its history. I invent tricks, and I write material for other magicians. I meet with them, do tricks occasionally and practice. That's an activity that is not very different from mathematics for me. I subscribe to 20 magic journals. You might say I do magic as a hobby, but for me it's quite close to math.

Inventing a magic trick and inventing a theorem are very similar activities in the following sense. In both subjects you have a problem that you're trying to solve with constraints. In mathematics, it's the limitations of a reasoned argument with the tools you have available, and with magic it's to use your tools and sleight-of-hand to bring about a certain effect without the audience knowing what you're doing. The intellectual process of solving problems in the two areas is almost the same. When you're inventing a trick, it's always possible to have an elephant walk on stage, and while the elephant is in front of you, sneak something under your coat, but that's not a good trick. Similarly with mathematical proof, it is always possible to bring out the big guns, but then you lose elegance, or your conclusions aren't very different from your hypotheses, and it's not a very interesting theorem. One difference between magic and mathematics is the competition. The competition in mathematics is a lot stiffer than in magic.

A Professorship in Magic

A: Why did you leave magic as an occupation?

DIACONIS: I left the performing part of it. Show business is very different from being a creative magician. In fact, the reason I left it is because you can't be too creative. There is tremendous pressure to do the same 17-minute act: it works and it gets laughs. I can remember very clearly changing the closing trick of my act, a trick with butterflies. I took the butterfly trick out to do something else. After my performance, my agent rushed up to me backstage, and said I couldn't take the butterfly trick out of my act. He said, "That's what I book you on." At that point, I wondered if I was going to end up doing the same seventeen minutes for the next twenty years.

Magic can be done as a very academic and creative discipline; it's very similar to doing mathematics, except for the fact that the world treats you more seriously if you're a mathematician. If you say that you're a professor at Harvard, people treat you respectfully. If you say that you invent magic tricks, they don't want to introduce you to their dog.

A: When you were doing magic, you said that you were following the wind. Are you still following the wind?

DIACONIS: When I was young and doing magic, if I heard that an Eskimo had a new way of dealing a second card using snowshoes, I'd be off to Alaska. I spent ten years doing that, traveling around the world, chasing down the exclusive, interesting secrets of magic. Since then I've worked in number theory, classical mathematical statistics, philosophy of statistics, psychology of vision and pure group theory. What happens now is that if I hear about a beautiful problem, and if that means learning some beautiful math machine, then, boy, I'm off in a second to learn the secrets of the new machine. I'm just following the mathematical wind. ∎

Don Knuth's Solutions to the Word Ladder Problem
(See p. 23 for Gardner's article on Word Ladders)

ROGUE	**SHOES**	**SHOES**	**BLACK**	**BLACK**	**BLACK**
vogue	shots	shops	slack	slack	brack
vague	slots	chops	slick	shack	brace
value	sloth	crops	slice	shark	trace
valve	slosh	cross	spice	share	trice
calve	slush	cress	spine	shale	trite
carve	blush	crest	shine	whale	write
carte	brush	**CRUST**	whine	while	**WHITE**
carts	crush		**WHITE**	**WHITE**	
parts	**CRUST**				
parks					
perks	**BEANS**	**BEANS**	**COSTS**	**COSTS**	**COSTS**
peaks	beats	beams	coots	coats	coats
leaks	seats	seams	clots	chats	boats
leaps	spats	shams	plots	chaps	blats
leapt	spars	shame	plats	claps	plats
least	spare	shale	plate	clans	plate
BEAST	share	shall	place	plans	place
	shale	shell	peace	plane	peace
ROGUE	shall	**SHELF**	**PENCE**	place	**PENCE**
vogue	shell			peace	
vague	**SHELF**			**PENCE**	
value					
valve	**GRASS**	**GRASS**	**GRASS**		
calve	grabs	gross	crass		
carve	crabs	grows	cress		
carte	cribs	grown	tress		
carts	cries	groan	trees		
cants	cried	groat	treed		
cents	creed	great	greed		
bents	greed	greet	**GREEN**		
beats	**GREEN**	**GREEN**			
beaus					
beaut					
BEAST					

Weird Dice

JOSEPH GALLIAN
University of Minnesota-Duluth

Imagine you are in the middle of a game of Monopoly® and someone substitutes a pair of dice labeled with 1, 2, 2, 3, 3, 4 and 1, 3, 4, 5, 6, 8 for the standard dice. Does this change the game in any way?

The first thing that comes to mind is that the probabilities for the various sums might be different. Surprisingly, this is not the case. Figure 1 reveals that the frequencies of the possible sums are exactly the same for either pair of dice (1 way to roll a sum of 2; 2 ways to roll a sum of 3, etc.)

Because of this one might conclude that the change in dice does not effect the game. But recall that one of the rules of Monopoly® is that if a player is in jail (not visiting) and he/she rolls doubles then the player advances the number of spaces shown on the dice. With the standard dice each of the sums 2, 4, 6, 8,

+	⚀	⚁	⚂	⚃	⚄	⚅
⚀	2	3	4	5	6	7
⚁	3	4	5	6	7	8
⚂	4	5	6	7	8	9
⚃	5	6	7	8	9	10
⚄	6	7	8	9	10	11
⚅	7	8	9	10	11	12

+	⚀	⚁	⚂	⚃	⚄	⚅
⚀	2	3	3	4	4	5
⚂	4	5	5	6	6	7
⚃	5	6	6	7	7	8
⚄	6	7	7	8	8	9
⚅	7	8	8	9	9	10
⚅	9	10	10	11	11	12

10 and 12 has a 1/36 chance of occurring as the result of a double. With the weird dice the sums 2 and 8 have a 1/36 chance of occurring; the sums 4, 10 and 12 have a 0 chance; and the sum 6 has a 1/18 chance.

Consequently, a person with hotels on the orange property St. James Place (6 spaces past jail) is better off with the weird dice! Moreover, a person that owns the maroon property Virginia (4 spaces past jail) is worse off. In fact, this small change in the probabilities for certain doubles using weird dice causes St. James Place to move up from the 10th most frequently landed-on space to the 6th most landed-on space whereas Virginia drops from 24th place to 27th place in the rankings.

These weird dice raise two interesting mathematical questions: How were the labels derived and are there other weird labels consisting of positive integers? It is possible to answer these questions with a simple analysis. To do so we begin by finding a way to model summing the faces of a pair of dice.

First, let us ask ourselves how we may obtain a sum of 6, say, with an ordinary pair of dice. Well, there are five possibilities for the two faces: (5, 1), (4, 2), (3, 3), (2, 4) and (1, 5). Next, we consider the product of the two polynomials created by using the ordinary dice labels as exponents:

$$(x^6 + x^5 + x^4 + x^3 + x^2 + x) \times (x^6 + x^5 + x^4 + x^3 + x^2 + x).$$

Observe that using the distributive property we pick up the term x^6 in this produce in precisely the following ways: $x^5 \cdot x^1$, $x^4 \cdot x^2$, $x^3 \cdot x^3$, $x^2 \cdot x^4$, $x^1 \cdot x^5$. Notice that the correspondence between pairs of labels whose sums are 6 and pairs of terms whose products are x^6. This correspondence is one-to-one, and it is valid for all sums and all dice— including our weird dice and any other dice that yield the desired probabilities. So, let a_1, a_2, a_3, a_4, a_5, a_6 and b_1, b_2, b_3, b_4, b_5, b_6 be any two lists of positive integer labels for a pair of cubes with the property that the probability of rolling any particular sum with these dice is the same as the probability of rolling that sum with ordinary dice labeled 1 through 6. Using our observation about products of polynomials, this means that

$$(x^6 + x^5 + x^4 + x^3 + x^2 + x) \times (x^6 + x^5 + x^4 + x^3 + x^2 + x)$$
$$= (x^{a_1} + x^{a_2} + x^{a_3} + x^{a_4} + x^{a_5} + x^{a_6}) \times (x^{b_1} + x^{b_2} + x^{b_3} + x^{b_4} + x^{b_5} + x^{b_6})$$

Now all we have to do is solve this equation for the a's and b's. How can we solve one equation in 12 unknowns? You didn't learn this in high school but you did spend much of your efforts factoring. So, let's factor the left-hand side of the equation. The polynomial $x^6 + x^5 + x^4 + x^3 + x^2 + x$ factors uniquely into irreducibles as

$$x(x + 1)(x^2 + x + 1)(x^2 - x + 1)$$

so that the left-hand side of the equation has the irreducible factorization

$$x^2(x + 1)^2(x^2 + x + 1)^2(x^2 - x + 1)^2.$$

Reprinted from February 1995, pp. 30–31

Illustration by Greg Nemec

This means that these factors are the only possible irreducible factors of $P(x) = x^{a_1} + x^{a_2} + x^{a_3} + x^{a_4} + x^{a_5} + x^{a_6}$. Thus, $P(x)$ has the form

$$x^q(x + 1)^r(x^2 + x + 1)^s(x^2 - x + 1)^t, \text{ where } 0 \leq q, r, s, t \leq 2.$$

To further restrict the possibilities for q, r, s, and t we evaluate $P(1)$ in both forms for $P(x)$. That is,

$$P(1) = 1^{a_1} + 1^{a_2} + \cdots + 1^{a_6} = 6$$

and

$$P(1) = 1^q 2^r 3^s 1^t.$$

Clearly, this means that $r = 1$ and $s = 1$. What about q? Evaluating $P(0)$ in two ways just as we did for $P(1)$ shows that $q \neq 0$. On the other hand, if $q = 2$, the smallest possible sum one could roll with the corresponding labels for dice would be 3 since q is the smallest of the a's and the smallest permissible b is 1. Because the sum of 2 is possible with ordinary dice this violates our assumption about the dice having equal probabilities. Thus, $q = 2$ is not permissible. We have now reduced our list of possibilities for q, r, s, and t to $q = 1$, $r = 1$, $s = 1$, and $t = 0, 1, 2$. Let's consider each of these possibilities for t in turn.

When $t = 0$,

$$P(x) = x^4 + x^3 + x^3 + x^2 + x^2 + x,$$

so the die labels are 4, 3, 3, 2, 2, 1— one of our weird dice.

When $t = 1$,

$$P(x) = x^6 + x^5 + x^4 + x^3 + x^2 + x,$$

so the die labels are 6, 5, 4, 3, 2, 1— an ordinary die.

When $t = 2$,

$$P(x) = x^8 + x^6 + x^5 + x^4 + x^3 + x$$

so the die labels are 8, 6, 5, 4, 3, 1— the other weird die.

Thus we have *derived* the weird dice labels and *proved* that they are the *only* other pair of dice that have this property.

We invite the reader to investigate the analogous questions for the tetrahedron, octahedron, dodecahedron, and icosahedron dice. For the answers see [1] and [2].

References

1. Broline, Duane, (1979), "Renumbering the faces of dice," *Mathematics Magazine,* vol. 52, 312–315.

2. Gallian, Joseph A. and David Rusin, (1979), "Cyclotomic polynomials and nonstandard dice," *Discrete Mathematics,* vol. 27, 245–259.

The
Chinese Domino
Challenge

DON KNUTH
Stanford University

Dominoes look different in China than they do in the West. In fact, we might well think of them as "trominoes," because their ratio of width to height is approximately 3 to 1 instead of 2 to 1. Each domino has two sets of spots, which range from 1 to 6; but some spot pairs are repeated, so that there are 32 dominoes, not 21, in a complete set:

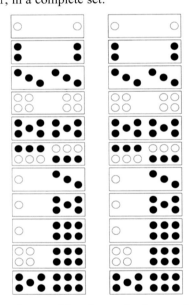
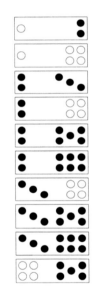
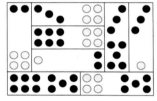

Some of the pieces also have colorful names; for example,

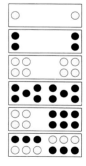

is called "earth."

is called "bench,"

is called "man,"

is called "plum flower,"

is called "redhead ten,"

is called "heaven,"

and so on. [See Stewart Culin, *Games of the Orient,* page 115. A well-made set can be seen in *Games of the World* by Frederic V. Grunfeld (Holt, Rinehart and Winston, 1975), page 107.]

The Challenge Problem is to take a complete set of Chinese dominoes and pack them into an 8 × 12 box, in such a way that four dominoes never meet at their corners, and so that the spots total exactly 11 at every point where three dominoes come together. For example, on a smaller scale, the arrangement

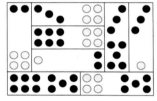

satisfies this condition: There are six interior points where three dominoes touch, and the corresponding spot sums are $2 + 3 + 6 = 4 + 4 + 3 = 4 + 6 + 1 = 4 + 2 + 5 = 4 + 1 + 6 = 5 + 2 + 4 = 5 + 1 + 5 = 11$.

Warning: Once you start on this problem, you may find it addictive.

Solution to the Chinese Domino Challenge

There are only three solutions, one of which is

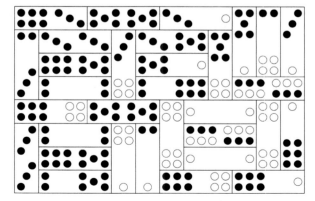

Reprinted from April 1995, pp. 8–9

The other solutions are obtained from this one by changing

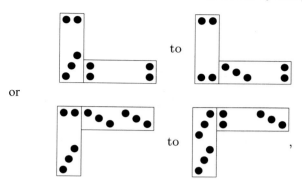

respectively.

The only other way to pack 1 × 3 rectangles into an 8 × 12 rectangle without four corners touching is obtained from the arrangement above by slightly adjusting a 3 × 4 configuration near the lower left:

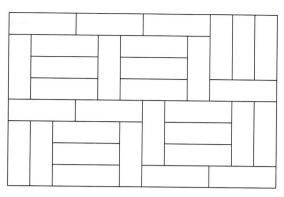

But 32 dominoes cannot be placed in this configuration without violating the sum-to-11 condition at least once.

Incidentally, the average sum of three randomly chosen spots on Chinese dominoes is exactly 3 × 227/64 = 10.640625. It turns out to be impossible to satisfy the challenge problem with spot sums all 10 instead of 11.

Making Connections

DON ALBERS
**Mathematical Association
of America**

A Profile of Fan Chung

"Don't be intimidated!" is Dr. Fan Chung's advice to young women considering careers in mathematics. "I have seen many people get discouraged because they see mathematics as full of deep incomprehensible theories. There is no reason to feel that way. In mathematics whatever you learn is yours and you build it up—one step at a time. It's not like a real-time game of winning and losing. You win if you benefit from the power, rigor and beauty of mathematics. It is a big win if you discover a new principle or solve a tough problem."

Chung regards herself as being luckier than many mathematicians. "As an undergraduate in Taiwan, I was surrounded by good friends and many women mathematicians. We enjoyed talking about mathematics and helping each other. A large part of education is learning from your peers, not just the professors. Seeing other women perform well is a great confidence builder, too!" Following that logic, Fan must have built up the confidence of many other women, for she has performed extremely well as a student and researcher.

As a young high school student growing up in Kaoshiung, Taiwan in the sixties, Fan Chung knew that she would be a mathematician. Her engineer-father told her, "in math all you need is pencil and paper." Attracted to combinatorics because it was fun, Chung recalls that "many problems from combinatorics were easily explained, you could get into them quickly, but getting out often was very hard…. Later on I discovered that there were all sorts of connections to other branches of mathematics as well as many applications."

After twenty years of work at Bell Labs and Bellcore, which contributed powerfully to her mathematical world view, Chung went back to school as a professor of mathematics at the University of Pennsylvania, where she had earned her doctorate in 1974.

She is convinced that students profit enormously from establishing connections to other branches of mathematics "It is like playing a game of GO or HEX. If your territory is all connected together, then every piece is strong and useful. On the other hand, if the parts are separated, then they are weak and not effective." She aims to make them aware of the power of connections and applications. "If you learn lots of theorems without actually using them, it is like a rich man who never spends his money. There is

Reprinted from September 1995, pp. 14–18

Good versus bad is the dividing line in mathematics according to Fan Chung.

no difference between that and having no money at all." She believes that mathematics students are short-changed if they aren't exposed to such connections during their college years. She argues that such experience will make them more employable—better researchers and better teachers.

Catching Problems

Chung's appreciation for making connections dramatically increased when she started working in information technology in 1974. With her new doctorate in hand, she joined the technical staff of Bell Labs, which was richly populated with research scientists and mathematicians, including Nobel Prize winners. She remembers being intimidated by "all of those great name tags"— Pollak, Graham, Sloane, etc.—during her first days on the job. "But I got over it," she said, "and very soon I discovered that if you just put your hands out in the hallway, you'd catch a problem."

How did Fan Chung "catch problems in the hallway"? She says that the hardest part was establishing communication. "At Bell, our office doors are always open and anyone can walk in," she recalled. "You need to have a willingness to find out what problems they are working on. Finding the right problem is often the main part of the work in establishing the connection. Frequently a good problem from someone else will give

you a push in the right direction and the next thing you know you have another good problem. You make mathematical friends and share the fun!" Over the past twenty years, she has been remarkably successful working with others. Nearly half of her 180 papers have been done in collaboration.

Math Noses vs. Physics Noses

Dr. Chung has a good nose for problems. She contrasted the differences between math noses and physics noses. "In physics there are clearly defined central problems, driven by our desire to understand the universe in which we live. Thus physicists have a clear notion of judging what the important problems are. In mathematics, by contrast, you can create your own paradigm and that can be wonderful. You can make your own rules and play your own game in small universes here and there."

She warns of dangers to be found in smaller free-wheeling mathematical universes. "Because of the large number of mathematical papers published each year (around 50,000), it is not easy to separate the wheat from the chaff. This makes it harder to determine the central problems." Chung claims that seeing a problem occur in several different settings and in different guises is key to identifying central problems.

Other keys to Chung's success in catching problems can be found in her enormous energy and determination. She says that "mathematics is on my mind all the time, sometimes even in the middle of the night." As a measure of her energy, consider her approach to raising a family. She had her first child during her last year of graduate school. "That is a wonderful time to have a child," she said. "You don't have to attend classes; you only have to write your thesis." After working at Bell Labs for only three years, she was pregnant again. Henry Pollak, her manager, wondered what she would do.

"I told him that I would work until the day I went to the hospital. Since I already had one at home, I thought what's the problem with one more? I didn't even take maternity leave; there was too much paperwork associated with that. So I just took four weeks vacation and wrote one paper in between."

Revise, Revise, Revise

In 1983, after the divestiture of Bell Telephone, came the creation of several "Baby Bells" and Bellcore, as well as a new research unit headed by Henry Pollak. He asked Chung to work with him in developing a new mathematics research group at Bellcore. She became a manager, and recruited mathematicians. "For the next seven years, in addition to my research, I had to write reports, attend meetings, and read the research papers of mathematicians I supervised," she said. When she arrived at Bell Labs and wrote her first paper under Pollak, she discovered something important. "It was absolutely clear that he was reading my paper with care," she said. "I really appreciate what he's done for my papers and for the example he set. Without Henry, I would not have done as well as I did when I became a manager."

Many years later, when Chung told Pollak that her first paper was rewritten eight times, Pollak replied that his first paper had been revised twenty-four times. It's clear that the approaches to writing and reading at Bell are different than those found in many mathematics departments. It's no surprise Chung believes that both faculty and students would profit greatly from increased attention to writing and reading.

It was also at Bell that Fan Chung met her second husband, mathematician Ron Graham. She says that they do, in fact, 'talk shop,' at home—a lot. "Mathematics is an unlimited source of adventures. It is quite wonderful to have your better half stand by you through the ups-and-downs of your journey."

In 1990, Fan went to Harvard on her leave as one of the first Bellcore Fellows. "It is not easy for some people to leave

Game or Problem?

In the world in which we live today, there are many problems that look like games but, in fact, contain new and interesting mathematics. Here is an example. Suppose we are given some number of nodes interconnected by various edges. Furthermore, at each node there sits a pebble marked with a specified destination (node) so that all destinations are distinct.

A natural problem is to find good routes for all pebbles simultaneously so that each edge is not overused. To be specific, let us consider a network with nodes labelled by strings of 0, 1's of the same length, where two nodes are joined by an edge if their strings differ in exactly one bit. Suppose we play the above pebbling game for given assignments of destinations. How do you find a good strategy for choosing routes so that each edge is in at most two routes? Suppose we introduce a new rule so that at each tick of the clock, two pebbles at the endpoints of an edge can interchange places and such interchanges can happen simultaneously at many edges (as long as they are node-disjoint).

Then, more questions can be asked:

- How long does it take for all pebbles to reach their destinations? What is the best strategy for choosing routes dynamically?

- If some edges are faulty, what is a good alternative strategy?

Many such problems come up in connection with communication networks, parallel computation and algorithms. Problems like these can be quite difficult. However, the mathematics involved can help identify the "bottleneck" and "access" of a network as well as contribute to a deeper understanding of the fundamental nature of the discrete universe.

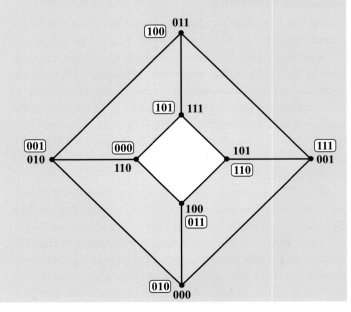

management, but it was not so hard for me," she said. "Usually with positions in management you obtain more influence and you certainly have more power to make decisions. But I do not want people to respect me because of that power. I'd rather win their admiration because of the mathematics I'm doing."

At Harvard—using her tested method of catching problems—she continued making connections with new mathematicians and new mathematics. "The interaction of combinatorics and other areas of mathematics opens up many exciting directions. It is like opening an old treasure box at the same time you find modern power tools. So you have precious crystals in one hand, a laser gun in another, and the light can go much further."

Mathematics by Chung from the hand of Chung.

The Dangers of Cloning

Her work as a visiting professor at Harvard was so stimulating that she decided to return to academe, and in 1994 she accepted an offer of an endowed professorship at the University of Pennsylvania. But Professor Chung is deeply concerned about academic mathematics, in particular what she calls the "cloning process."

"Professors train students to work in their areas and thus there is a danger of narrowing down instead of broadening and making connections with the new information technologies and other emerging areas of mathematics," she said. "In most of our universities, the mathematics curriculum has changed very little over the past twenty years. It's comfortable to teach what has been taught for twenty years, but look at the progress of technology during that same time frame. Of course, mathematical principles are unchanged. However, there has been significant growth in particular in discrete mathematics. Mathematics is more important than ever in dealing with all the hard problems arising from the advances of technology.

"There are many wonderful ideas from discrete mathematics that students need to know about. Because we live in the information age, many challenging problems arise in our binary universe. It is essential for students to be able to connect the mathematics they learn in the classroom to problems we face in this information age!"

Bad Mathematics

In spite of spending most of her professional life working in an applications environment, Chung says that a large part of her drive to solve problems comes from the beauty of mathematics. She says that "the dividing line in mathematics is not 'pure' versus 'applied'; it is 'good' versus 'bad.'" Good mathematics, according to Dr. Chung, is characterized by its impact. "Bad mathematics is cooked up artificially, perhaps in isolation, and it will vanish from view very quickly."

Photographs of Fan Chung by Martin Griff.

Math on Money

JOSEPH GALLIAN
University of Minnesota, Duluth

A regular pentagon can be rotated 0°, 72°, 144°, 216° and 288° and still coincide with itself. Likewise, a regular pentagon can be reflected across its five axes of symmetry and coincide with itself. Interestingly, a (nonzero) rotation followed by a reflection does not result in the same orientation as the reflection followed by the rotation (to keep track of the orientation we label the vertices—see Figure 1). In this article we explain how this simple fact is the basis for an idea that has recently been used by the German Bundesbank to assign serial numbers to German banknotes.

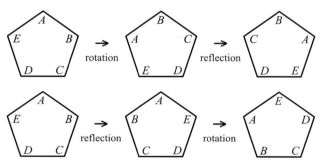

Figure 1. A rotation followed by a reflection; a reflection followed by a rotation.

Although U.S. paper currency serial numbers and social security numbers do not include an extra digit for the purpose of error detection, it is now standard practice to include such a digit on identification numbers. Typical of the genre is the Universal Product Code (UPC) found on grocery and retail items. A UPC identification number consists of twelve digits, each of which can be from 0 to 9. (For many items, the check digit is not printed, but it is always bar coded.) The check digit a_{12} for the UPC number $a_1 a_2 \ldots a_{11}$ is chosen so that $3a_1 + a_2 + 3a_3 + a_4 + 3a_5 + a_6 + 3a_7 + a_8 + 3a_9 + a_{10} + 3a_{11} + a_{12}$ is evenly divisible by 10. For instance, the number 05074311502 has the check digit 8 because $3 \cdot 0 + 5 + 3 \cdot 0 + 7 + 3 \cdot 4 + 3 + 3 \cdot 1 + 1 + 3 \cdot 5 + 0 + 3 \cdot 2 = 52$ and $52 + 8$ is evenly divisible by 10. Notice that any single error, say $a_1 a_2 \ldots a_i \ldots a_{12} \to a_1 a_2 \ldots a_i' \ldots a_{12}$, is detectable since neither $3a_1 + a_2 + \cdots + a_i' + \cdots + a_{12}$ nor $3a_1 + a_2 + \cdots + 3a_i' + \cdots + a_{12}$ is evenly divisible by 10 if $a_i \neq a_i'$.

Other widely-used error detection schemes such as used on credit cards, airline tickets, travelers checks and books utilize division by 7, 9, 10, or 11 (see [1]). However, none of the schemes that use ordinary arithmetic are capable of assigning an extra digit to all possible numbers in such a way that all single-digit errors and all transposition errors are detectable (see [1]).

In contrast, in 1969 J. Verhoeff [2] devised a method utilizing a non-commutative algebraic system with 10 elements that detects all single-digit errors and all transposition errors involving adjacent digits without the necessity of avoiding certain numbers or introducing a new character (books sometimes use an "X" as a check "digit"). To describe this method consider the function $\sigma(0) = 1$, $\sigma(1) = 5$, $\sigma(2) = 7$, $\sigma(3) = 6$, $\sigma(4) = 2$, $\sigma(5) = 8$, $\sigma(6) = 3$, $\sigma(7) = 0$, $\sigma(8) = 9$, $\sigma(9) = 4$ and the algebraic system defined by Table 1. The algebraic system defined in Table 1 is called the *dihedral group of order 10* and is denoted by D_{10}. This system is known to chemists, geologists and physicists as the dihedral point group with 10 elements. Scientists use it to describe the 5 rotational symmetries and 5 reflectional symmetries of a regular pentagon. For example, the symbol 1 represents a rotation of 72 degrees whereas the symbol 5 represents a reflection across one of the five axes of symmetry. Then, $5*1$ is a 72° rotation followed by the reflection represented by 5 whereas $1*5$ is the reflection represented by 5 followed by the 72° rotation. Note from the table that $9 = 5*1 \neq 1*5 = 6$.

Table 1. Multiplication for D_{10}.

*	0	1	2	3	4	5	6	7	8	9
0	0	1	2	3	4	5	6	7	8	9
1	1	2	3	4	0	6	7	8	9	5
2	2	3	4	0	1	7	8	9	5	6
3	3	4	0	1	2	8	9	5	6	7
4	4	0	1	2	3	9	5	6	7	8
5	5	9	8	7	6	0	4	3	2	1
6	6	5	9	8	7	1	0	4	3	2
7	7	6	5	9	8	2	1	0	4	3
8	8	7	6	5	9	3	2	1	0	4
9	9	8	7	6	5	4	3	2	1	0

Reprinted from November 1995, pp. 10–11

Table 2

	0	1	2	3	4	5	6	7	8	9
σ	1	5	7	6	2	8	3	0	9	4
σ^2	5	8	0	3	7	9	6	1	4	2
σ^3	8	9	1	6	0	4	3	5	2	7
σ^4	9	4	5	3	1	2	6	8	7	0
σ^5	4	2	8	6	5	7	3	9	0	1
σ^6	2	7	9	3	8	0	6	4	1	5
σ^7	7	0	4	6	9	1	3	2	5	8
σ^8	0	1	2	3	4	5	6	7	8	9
σ^9	1	5	7	6	2	8	3	0	9	4
σ^{10}	5	8	0	3	7	9	6	1	4	2

Table 3

A	D	G	K	L	N	S	U	Y	Z
0	1	2	3	4	5	6	7	8	9

Verhoeff's idea is to view the digits 0 to 9 as the elements of the group D_{10} and to replace ordinary addition with calculations done in D_{10}. In particular, to any string of digits $a_1 a_2 \ldots a_{n-1}$, we append the check digit a_n, so that $\sigma(a_1) * \sigma^2(a_2) * \cdots * \sigma^{n-2}(a_{n-2}) * \sigma^{n-1}(a_{n-1}) * a_n = 0$. (Here $\sigma^i(x) = \sigma(\sigma^{i-1}(x))$.) Since σ has the property that $\sigma^i(a) \neq \sigma^i(b)$ if $a \neq b$, all single-digit errors are detected. Also, because

$$a * \sigma(b) \neq b * \sigma(a) \quad \text{if } a \neq b,$$

it follows that all transposition errors involving adjacent digits are detected (since the above equation implies that $\sigma^i(a) * \sigma^{i+1}(b) \neq \sigma^i(b) * \sigma^{i+1}(a)$ if $a \neq b$).

In 1990 the German Bundesbank began using the Verhoeff check digit scheme to append a check digit to the serial numbers on German banknotes. Table 2 gives the values of the functions $\sigma, \sigma^2, \ldots, \sigma^{10}$ needed for the computations. (The functional value $\sigma^i(j)$ appears in the row labeled with σ^i and the column labeled j.) Since the serial numbers on the banknotes are alpha-numeric, it is necessary to assign numerical values to the letters to compute the check digit. This assignment is shown in Table 3.

To trace through a specific example consider the banknote (featuring the mathematician Gauss!) shown in Figure 2 with the number AU3630934N7. To verify that 7 is the appropriate check digit we observe that $\sigma(0) * \sigma^2(7) * \sigma^3(3) * \sigma^4(6) * \sigma^5(3) * \sigma^6(0) * \sigma^7(9) * \sigma^8(3) * \sigma^9(4) * \sigma^{10}(5) * 7 = 1*1*6*6*6*2*8*3*2*9*7 = 0$ as it should be. (To illustrate how to use the multiplication table for D_{10} we compute $1*1*6*6 = (1*1)*6*6 = 2*6*6 = (2*6)*6 = 8*6 = 2$.) (Here are two more serial numbers that you can try yourself: AG0614294U and DA6819403G. The answers are given below.)

A shortcoming of the German banknote scheme is that it does not distinguish between a letter and its assigned numerical value. Thus, a substitution of 7 for U (or vice versa) and the transposition of 7 and U are not detected by the check digit. This shortcoming can be avoided by using D_{36}, the dihedral group of order 36, to assign every letter and digit a distinct value together with an appropriate function σ (see [1] or [3]). Using this method to append a check character, all single position errors and all transposition errors involving adjacent characters will be detected.

Although the error detecting schemes using non-commutative systems are more effective than the schemes that use division, the German banknote application is the only one I know that uses a non-commutative system. ∎

Acknowledgments. I am grateful to R.H. Schulz for providing me with the information in Tables 2 and 3 and to Ruth Berger for alerting me about the possibility that German banknotes might use the dihedral group for calculating the check digit.

References

1. Gallian, J.A., "The mathematics of identification numbers," *The College Mathematics Journal 22* (1991) 194–202.
2. Verhoeff, J., *Error Detecting Decimal Codes,* Mathematical Centre, Amsterdam, 1969.
3. Winters, S., "Error detecting codes using dihedral groups," *UMAP Journal,* 11(1990) 299–308.

Figure 2. German banknote with serial number AU3630934N and check digit 7.

Update

Germany switched to the Euro in January 2002. The Euro uses a check digit that utilizes modulo 9.

Answers

AG0614294U2
DA6819403G4

The
Parallel Climbers
Puzzle

ALAN TUCKER
SUNY, Stony Brook

A Case Study in the Power of Graph Models

Two climbers start at points A and Z on the left and right sides, respectively, of the mountain range in Figure 1. We pose the following puzzle, which we call the Parallel Climbers Puzzle.

Parallel Climbers Puzzle: Is it possible for the left and right climbers to move from A and Z, respectively, along the range in Figure 1 to meet at M in a fashion so that *they are always at the same altitude every moment?*

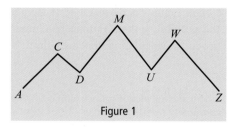
Figure 1

The objective of this paper is to show that this puzzle has a solution for any mountain range. The two assumptions we make are that: i) A and Z are at the same height, and ii) there is no point lower than A (or Z) and no point higher than M. We shall solve this problem by means of a graph-theoretic model.

Graph theory is an important field of discrete mathematics. While the term 'graph' is used most commonly in mathematics to refer to the set of points satisfying a functional relationship between two (or more) variables, the word 'graph' has another meaning as a mathematical object $G = (V,E)$, consisting of a non-empty, finite set V of **vertices** and a set E

of **edges** joining certain pairs of vertices. Figure 2 shows a graph $G = (V, E)$ with vertex set $V = \{a, b, c, d, e\}$ and edge set $E = \{a\text{-}b, a\text{-}d, b\text{-}c, c\text{-}d, c\text{-}e\}$. (Note that an edge such as $a\text{-}b$ is normally written as the ordered pair (a, b), but this standard notation turns out to be confusing in this particular modeling problem, because we need ordered pairs for another purpose.) The mountain range depicted in Figure 1 can be viewed as a graph with vertex set $V = \{A, C, D, M, U, W, Z\}$ and edge set $E = \{A\text{-}C, C\text{-}D, D\text{-}M, M\text{-}U, U\text{-}W, W\text{-}Z\}$.

Graphs are used to analyze a wide variety of mathematical problems, most commonly in operations research and computer science. For example, graphs

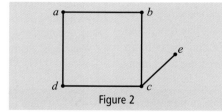
Figure 2

can represent transportation and telecommunication networks. The problem of simultaneously routing thousands of long-distance calls between various pairs of parties is a graph theory problem. The data structures of computer science that are used to organize linkages between various pieces of information are graphs.

An example of such a data structure is a search tree for looking up a word in a spell-checking dictionary. The spell

checker does not start at the beginning of the dictionary and sequentially check the unknown word against all 50,000 words in the dictionary. Even for a fast computer, that would be unnecessarily time-consuming. Rather, it compares the unknown word with the middle word in the dictionary's alphabetical list of words to see if the unknown word occurs in the first half or second half of the dictionary. Depending on the outcome of the first test, the spell checker next compares the word with the middle word in the first half or second half of the dictionary, and so on. This strategy uses each comparison to divide the set of dictionary words that need to be checked in half. The data structure organizes the comparisons, telling what comparison to do next depending on the outcome of the current comparison. Figure 3 shows the beginning of a possible data structure graph for spell checking.

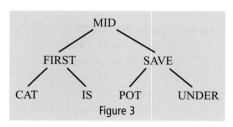
Figure 3

A related but simpler type of search problem involving graphs is the following: can one find a path—a sequence of connecting edges—from a specified starting vertex to a specified stopping vertex in a given graph. The vertices in the graph

Reprinted from November 1995, pp. 22–24

might represent all the possible positions in some puzzle and the edges legal moves between positions in the puzzle. "Solving" the puzzle reduces in the associated puzzle graph to finding a path from the starting position to the stopping, or winning, position.

only if the two people can move constantly in the same direction (both going up or both going down) from point P_L to point P'_L and from P_R to P'_R respectively.

First we redraw the mountain range in Figure 1 as shown in Figure 4 with points added parallel to peaks and valleys. For

of a vertex and make two simple observations about degrees. This information, when applied to range graphs, will then lead to a simple proof that every Parallel Climbers puzzle has a solution.

The **degree** of a vertex is the number of edges incident to that vertex.

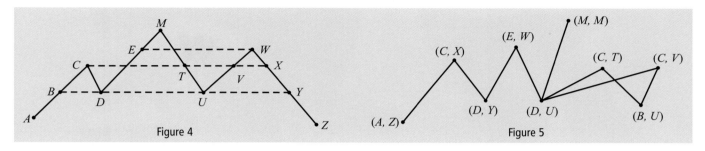

Figure 4

Figure 5

The mountain climbing problem posed above is such a puzzle with a starting position—the left and right climbers starting at locations A and Z, respectively—and a stopping position—both climbers at location M. The challenge in creating such a puzzle graph is to determine what should be the set of vertices representing possible positions of interest. There are an infinite number of points along the initial ascend segments the two climbers take, but a graph is defined to have a finite number of vertices.

Like the spell-checker problem, we are concerned in the Parallel Climbers problem with making a sequence of decisions. These decisions occur at places along the climb where the climbers will have choices. That is, in the mountain range in Figure 1, when one of the climbers arrives at a peak or valley (the labeled points in Figure 1), we must correctly decide what should be done next. Thus, the peaks and valleys are the locations of interest that should be used to define the vertices in our graph model.

A **range graph** is a graph whose vertices are pairs of points (P_L, P_R) at the same attitude with P_L on the left side of the summit and P_R on the right side, such that one of the two points is a local peak or valley (the other point might also be a peak or valley). An edge in a range graph will join two vertices (P_L, P_R) and (P'_L, P'_R) if and

example, when the left climber is at C, the right climber could be at point X or point V or point T. Thus (C, X), (C, V) and (C, T) will all be possible vertices in the range graph, since the two climbers at some stage might come to any of these three pairs of locations.

The range graph for the mountain range in Figure 4 is shown in Figure 5. Our question is now: Is there a path in the range graph from the starting vertex (A, Z) to the summit vertex (M, M)? For the graph in Figure 5, the answer is, by inspection, obviously yes.

Following the path in Figure 5 from (A, Z) to (M, M) takes us first up to (C, X), then the left climber moves lower but still towards the summit, while the right climber backs down, coming to (D, Y). Next both climbers move upwards to (E, W). Next the left climber backs down while the right climber moves lower but towards the summit, coming to (D, W). From there, the climbers can jointly ascend to the summit, (M, M).

The model we made in Figure 5 recasts the puzzle in a fashion that made it easy to solve by inspection. This is one of the great values of graph models. They can often reformulate problems in a fashion that makes the answer easy to "see." However graphs also provide a framework for proving general results. First, we shall introduce the concept of the degree

For simplicity, we shall assume that the two ends of an edge must be distinct vertices.

Theorem. *The sum of the degrees of all vertices in a graph is equal to twice the number of edges in the graph.*

Proof. The sum of all degrees counts every occurrence of an edge being incident to a vertex. Since every edge has two end vertices, twice the number of edges also counts every occurrence of an edge being incident to a vertex. The theorem follows.

Corollary. *In any graph, there must be an even number of vertices of odd degree.*

Proof. Since the sum of all degrees is an even number—namely, twice the number of edges, by the Theorem— then the number of odd terms in this sum must be even.

Now let us examine the degrees of the vertices in a range graph. We claim that vertices (A, Z) and (M, M) in any range graph have degree 1 while every other vertex in the range graph has degree 2 or 4. (A, Z) has degree 1 because when both people start climbing up the range from their respective sides, they have no choice initially but to climb upwards until one arrives at a peak. In Figure 4, the first peak encountered is C on the left, and so the one edge from (A, Z) goes to (C, X). A similar argument applies at (M, M). Next

Illustration by Loel Barr

consider a vertex (P_L, P_R) where one point is a peak and the other point is neither peak nor valley, such as (E, W). From the peak we can go down in either direction: at W, we can go down toward Z or toward U. In either direction, the people go until one (or both) climbers reach a valley. At (E, W), the two edges go to (D, Y) and (D, U). So such a vertex has degree 2. A similar argument applies if one (but not both) points are a valley. It is left as an

exercise for the reader to show that if a vertex (P_L, P_R) consists of two peaks or two valleys, such as (D, U), it will have degree 4. (A vertex consisting of a valley and a peak will have degree 0—why?)

Suppose there were no path from (A, Z) to (M, M) in the range graph. We use the fact that the starting vertex (A, Z) and the summit vertex (M, M) are the only vertices of odd degree. The part of the range graph consisting of (A, Z) and all the ver-

tices that can be reached from (A, Z) would form a new graph with just one vertex of odd degree, namely, (A, Z). This contradicts the corollary and so any range graph must have a path from (A, Z) to (M, M). Thus, we have proved:

Theorem. *The Parallel Climbers puzzle has a solution for any mountain range.* ■

A Perfectly Odd
Encounter in a
Reno Cafe

DAN KALMAN
The American University

My father and I were sitting in a cafe in Reno. I was giving him some examples from number theory, including a problem that has been unsolved since the days of ancient Greece. Unintimidated, my father came up with a solution in about a minute flat. The reasoning behind his solution, and my skeptical reaction, reveal something about a central concern of mathematics: the nature of proof.

We were killing time while others in our party were playing the slot machines. Ted, that's my father, had been talking about the mathematics in a book he was reading, Michener's *The Source*. In fact, it wasn't really mathematics at all. It was numerology, the mystical interpretation of numerical relationships for purposes of divination. It occurred to me that number theory is the closest thing in real mathematics to the numerology Ted had been talking about, and I tried to describe the subject.

As an example, I told him about perfect numbers. The number 6 is perfect, because if you add up its proper divisors, 1, 2, and 3, the total is 6. Another example is 28: the proper divisors are 1, 2, 4, 7, and 14, and these sum to 28. Are there any others? Can you find some?

It has been known for over 200 years how to find all the even perfect numbers. There is a formula: $2^{n-1}(2^n - 1)$. For $n = 2$ this gives $2^1(2^2 - 1) = 2 \cdot 3 = 6$. For $n = 3$

we get $2^2(2^3 - 1) = 4 \cdot 7 = 28$. The next possibility, $n = 4$, yields 120, and that isn't a perfect number because 60, 40, and 30 are all divisors. The trouble is that for $n = 4$, $(2^n - 1)$ isn't a prime number. Euclid showed that when $(2^n - 1)$ is prime, the formula $2^{n-1}(2^n - 1)$ always produces a perfect number. Thus, for $n = 5$, $2^n - 1 = 31$ is prime, so we can be sure that $2^{n-1}(2^n - 1) = 496$ is perfect. Some two thousand years later, Euler proved that Euclid's formula actually generates all the even perfect numbers. So today we know the complete story on even perfect numbers.

So what? Who cares? What possible use could there be in knowing about perfect numbers? Well, number theory is like that. It is bursting with curious relationships that aren't particularly good for anything, but which have fascinated amateur and professional mathematicians for centuries. The number theorist and the numerologist share this fascination with numbers, but the number theorist doesn't try to draw mystical conclusions from the number patterns. The object is simply to understand the mysteries and to back up each insight with proof.

Of course, it isn't always easy to find proof. That is why number theory abounds with conjectures: that is, statements fitting all the known data, and seeming to be valid general laws, but for which no proof has been found. Number theorists

do not despair of ever finding proofs for these conjectures. Why, Fermat's last theorem was recently proved after standing as a conjecture for 350 years. Fermat wrote in the 1640's that $x^n + y^n = z^n$ could never hold for positive integers x, y, z, and n, with $n > 2$. That is, when working with positive integers, the sum of two cubes is never a cube, the sum of two fourth powers is never a fourth power, and so on for all powers greater than 2. From Fermat's day until our own, no proof could be found for his statement. But in 1993, Andrew Wiles announced that he had discovered such a proof, and today it is generally accepted that Fermat's theorem has been established. So, number theorists continue to hold out for proofs. No matter how overwhelming the evidence, no matter how clear the insight, true understanding is not conceded until there is proof.

That is exactly the situation with odd perfect numbers. Since Euclid's time, no one has ever been able to find an odd perfect number, even though the numbers checked by computer reach into the millions and beyond. It seems like an inescapable conclusion that odd numbers simply cannot be perfect and perfect numbers simply cannot be odd. But no one has been able to prove it.

So there I was, explaining to Ted about number theory, how it is like numerology, and how it is unlike. To illustrate the ideas

of conjecture and proof, I told him about perfect numbers commenting that one of the oldest open questions in number theory is whether there are any odd perfect numbers. I told him, as I have told you, that the even case is completely solved. I described the state of affairs for the odd case: no one can find an odd perfect number, yet no one can prove that none exist.

Then, to my astonishment, Ted announced that it was completely obvious that an odd perfect number is an impossibility. He explained his reasoning this way: An even number is divisible by 2, and when you divide it by 2, you get one of its divisors. In fact, you get its largest possible proper divisor, half of the original number. For an even perfect number, adding up the remaining divisors produces the other half of the original number. But if the original number (n) is odd, the smallest factor (other than 1) is at least

3, so the largest proper factor is at most $n/3$. In that case, in order for n to be perfect, the other divisors—all of which are even less than $n/3$—have to add up to 2/3 of n, and that is impossible.

Is that a proof? Was the famous problem of odd perfect numbers solved in a Reno cafe? I was instantly skeptical. Surely this argument could have occurred to Gauss, or Euler, or even *me.* But even more compelling, just from its inherent structure, I instantly realized that Ted's argument was not a proof. Can you see why?

One of the foremost skills of the trained mathematician is to recognize what is a proof, and what is not. Yet it is not always easy to clearly explain what constitutes a proof. My father's argument is logical, it is insightful, it seems to explain things. And yet there is a huge hole, a gap in the reasoning. *Why* is it impossible for there to be enough small factors

to total 2/3 of the original number? Ted could give no further explanation.

On the surface, the nature of proof seems clear cut. There must be a logical reason for each conclusion. If any of the conclusions is questioned, the prover must be able to provide reasoning that justifies it. This additional reasoning, in turn, is open to challenge, and must likewise be defended. And so on, and so forth, the prover must be prepared to provide a justification for each conclusion that is questioned. But this process cannot be taken infinitely far. At some point, won't the prover be reduced to the same position as my father? At some point, a step will be reached that is so self-evident that no further explanation can be advanced. The prover can only insist that the skeptic must surely agree with the conclusion, just as my father insisted that no further argument was needed for his proof. His conclusion was transparently self-evident! It was obvious! This is where deciding what is a proof gets tricky. It comes down to recognizing what is obvious, and what isn't.

Well, how *does* one recognize the obvious? It reminds me of what the Supreme Court justice said about pornography. I may not know how to define it, but I know it when I see it. Ted's final assertion was definitely not obvious. His inability to explain further invalidated the proposed proof. By training and long habit of thinking, my eye is instinctively drawn to potential gaps or holes in arguments. I constantly probe the fabric of a proof. Is there a weakness here? Can I argue a point further there? It is my years of practice that qualify me to judge what is obvious. All mathematics students need to work at this same kind of skepticism. Be wary of the obvious, distrust it, be as obtuse as you possibly can. If it is overly obvious, one ought to be able to explain why. Dig deeper, push harder, however brightly lit the corner, shine an even brighter light there, until the shadows are driven out utterly. Make this your standard practice. Only then will you be qualifed to say what is obvious.

Of course, there is little satisfaction in simply denying what someone else claims is obvious—far better to demonstrate that the desired conclusion need not follow. In the case of Ted's argument, this can be accomplished by considering the example of 945. That is an odd number whose proper divisors sum to 975. Although the largest divisor is 945/3 = 315, and all of the other divisors are even smaller, there are enough of these small divisors to add up to more than 2/3 of the original number. This is just what my father's argument claimed was impossible. This example highlights the flaw in his reasoning. And since it is possible for the proper divisors of an odd number to have a sum that exceeds the number, it might also be possible for an odd number to be perfect.

At this point, you are probably asking where the 945 came from. How did I find this example? I went looking for it. I no-

ticed that Ted's argument, if valid, would not only rule out the possibility of odd perfect numbers, but would also rule out the existence of an odd number which is *exceeded* by the sum of the proper divisors. Also, I knew a useful fact from number theory: Given the prime factorization $m = p_1^{e_1} p_2^{e_2} \cdots p_n^{e_n}$, the sum of all the divisors, including m itself is equal to

$$\left(1 + p_1 + p_1^2 + \cdots + p_1^{e_1}\right) \times$$
$$\left(1 + p_2 + p_2^2 + \cdots + p_2^{e_2}\right) \times \cdots \times$$
$$\left(1 + p_n + p_n^2 + \cdots + p_n^{e_n}\right).$$

It is easy to see that this is true. Simply observe that when the product is multiplied out, the terms are all the divisors of m. Things can be simplified a little by applying the formula

$$\left(1 + p_i + p_i^2 + \cdots + p_i^{e_i}\right) = \frac{p_i^{e_i+1} - 1}{p_i - 1}.$$

This gives the sum of the divisors as

$$\frac{p_1^{e_1+1} - 1}{p_1 - 1} \frac{p_2^{e_2+1} - 1}{p_2 - 1} \cdots \frac{p_n^{e_n+1} - 1}{p_n - 1}.$$

Using this last expression, and a handheld calculator, it is nearly effortless to compute the sum of the divisors for any

number. For example, if

$$m = 3^4 \cdot 5^2 \cdot 13$$

then the sum of the divisors is

$$\frac{242}{2} \cdot \frac{124}{4} \cdot \frac{168}{12} = 121 \cdot 31 \cdot 14.$$

That's a pretty nifty way to total up the divisors of 26325!

Remember the goal is to find an odd number m which is smaller than the sum of its proper divisors, that is, all the divisors other than m itself. Put another way, we need the sum of all the divisors of m to exceed $2m$. Well, experiment a bit. Remember to use odd numbers, and consider some m's with two or three different prime factors, some with exponents greater than 1. It does not take long to stumble on the example

$$3^3 \cdot 5 \cdot 7 = 945.$$

And what happened to my father? Did he change his mind about odd perfect numbers? I am sorry to report that he did not. By the time I found my example, a day or two had gone by and he had moved on to other things. I don't think he recalled what the main thread of his argument had been, or indeed, what the entire dispute was about. Reno, after all, has many other diversions, and one's relatives can only be expected to sit still for so much mathematics. But I hope this article has contributed to your own understanding of proofs. And the next time something appears obvious, think of my father and the cafe in Reno where, from his point of view, the non-existence of odd perfect numbers was proved.

I am grateful to William Dunham of Muhlenberg College for suggesting many improvements to this paper.

In
Prime
Territory

ELLEN GETHNER
University of Colorado, Denver

A Math Question Posed by Michael Jordan

Did you know that Chicago Bulls player Michael Jordan used to be a math major at the University of North Carolina before switching over to geography? We're probably better off not discussing the moral of *that* story. In any case, during the 1994–95 basketball season, MJ renewed his interest in mathematics when he changed his new team number (45) back to his old and formerly retired team number (23). He posed the following question: 'What's in a number, anyway?' MJ was quite likely wondering about *prime numbers.*

What is a prime number anyway? Let's review. First, *a whole number,* or *counting number,* is a number such as 1, 2, 3, 4, 5, 6, 7,… and so on, forever. A *prime number* (or just prime for short) is a whole number that is divisible only by 1 and itself. For example 2, 3, 5, 7, 11, 13, 17, 19, and 23 are primes, but $4 = 2 \times 2$, $6 = 3 \times 2$, and $45 = 5 \times 3 \times 3$ are not primes. In fact, numbers that are **not** primes are called *composites.*

For later use, let's expand the set of whole numbers by including 0 and the negatives of the whole numbers, like, 0, –1, –2 –3 –4, –5, –6, –7 and so on. This new expanded set is known as the *integers.*

What Euclid Knew About Primes

Before moving on to further technicalities, it might be interesting to take a historical perspective on the prime numbers. In particular, Euclid was a great mathematician who lived in Alexandria during the 3rd Century, B.C. and who wrote a thirteen-volume work called *Elements.* The purpose of *Elements* was to compile all of the known geometry of the time into one comprehensive work.

In volume IX Euclid asked if there were infinitely many primes. To the uninitiated, the answer seems to be a straightforward and resounding '*yes*,' simply because there are infinitely many whole numbers. But, mathematicians need a more convincing (and for that matter correct!) argument, as we shall see.

Euclid did, in fact, prove that there are infinitely many primes and here's the idea behind his argument. First assume there

Euclid

are only a finite number of primes in the world. For example, suppose 2 and 3 are the only primes, and the rest of the whole numbers are composite (humor me). Now think about the number

$$2 \times 3 + 1 = 7$$

Well, 7 isn't supposed to be prime (remember, 2 and 3 are the *only* primes for the time being), which means that 7 must be divisible by smaller primes. Well, our only choices are 2 and 3. But dividing 7 by either 2 or 3 leaves a remainder of 1, which means that 7 is neither divisible by 2 nor by 3. What does this mean? Our assumption was incorrect, and therefore, 2 and 3 cannot be the only primes in the world.

The above paragraph was only a warm-up to Euclid's real argument. Here is Euclid's more general argument to prove that there are infinitely many primes. For a contradiction, assume that there are only finitely many primes. Here they are: 2, 3, 5, 7,…, p where p is the largest prime (in the warm-up argument, $p = 3$). Now multiply all of the existing primes together and add 1 to the result:

$$(2 \times 3 \times 5 \times 7 \times \cdots \times p) + 1.$$

The above gigantic number cannot be a prime (remember, p is the largest prime) and therefore must be divisible by one or some of our primes, namely 2, 3, 5, 7,…, p. But dividing the gigantic number by any of these primes leaves a remainder of 1. This is the contradiction that Euclid

Reprinted from April 1996, pp. 8–13

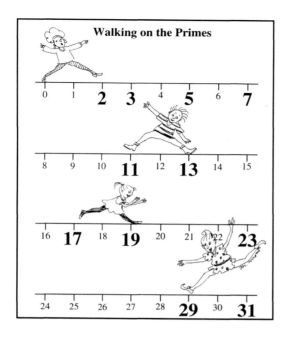

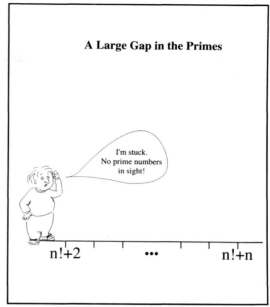

Illustrations by Marissa Moss

was seeking, namely that the gigantic number is neither prime itself nor is it divisible by any of the finitely many primes in the world. So, our assumption was wrong. In particular, there must be infinitely many primes.

Take A Hike!

Now that we've got all of these primes, what should we do with them? Try the following game. Stand on the number 0 and start walking. The only rule to this game is, as you take each step, you are **only** allowed to step on the primes. A curious question: If you keep walking forever, will you, every once in a while, have to take longer and longer steps? Or, might you be able to keep walking and take only 'small' steps. This is how the experts ask the above question: Can you walk to infinity on the primes in steps of bounded length? The answer is 'no,' and here's why.

In order to show that a prime-hiker will eventually have to take longer and longer steps, it suffices to show that we can find an arbitrary number of composites in a row. For example, if we were to find 4 composites in a row, a prime-hiker would be forced to take a step of length 5. More abstractly, if we were to find n composites in a row, then the prime-hiker would be forced to take a step of length $n + 1$. But let's start out slowly and aim for finding 4 composites in a row. *Voilá*—here they are:

$$5 \times 4 \times 3 \times 2 + 2 = 122,$$
$$5 \times 4 \times 3 \times 2 + 3 = 123,$$
$$5 \times 4 \times 3 \times 2 + 4 = 124,$$
$$5 \times 4 \times 3 \times 2 + 5 = 125.$$

The first number, 122, is divisible by 2, the second number by 3, the third by 4, and the fourth by 5, and hence the prime-hiker will be forced to take a step of length (at least) 5.

Now let's be more ambitious. Let's find 99 composites in a row to force the poor hiker to take a step of length (at least) 100. But first we need help with the notation because we'll have to multiply lots of numbers together. In particular, the number

$$100 \times 99 \times 98 \times \cdots \times 3 \times 2 \times 1$$

is more easily written as 100! and is pronounced 'one hundred factorial.' If you are unfamiliar with this notation, you might at first think that 100! means that 100 is a very exciting number, but remember, one hundred factorial means 'multiply all the whole numbers less than or

equal to 100 together.' Now we're ready. Here are 99 composites in a row:

$$100! + 2,$$
$$100! + 3,$$
$$100! + 4,$$
$$\vdots$$
$$100! + 99,$$
$$100! + 100.$$

Got it? The first number, 100! + 2 is divisible by 2, the second number 100! + 3 is divisible by 3, and so on up to the last number 100! + 100 which is divisible by 100. In all there are 99 composites in a row, which forces the prime-hiker to take a step of length (at least) 100. By now perhaps you see the way the general construction works. Specifically, for any n, you can force the prime-hiker to take a step of length (at least) n by producing the following $n - 1$ composites in a row ($n!$ means multiply all whole numbers less than or equal to n together):

$$n! + 2,$$
$$n! + 3,$$
$$n! + 4$$
$$\vdots$$
$$n! + (n - 1)$$
$$n! + n.$$

All in all we've learned that one can-

Carl Friedrich Gauss

not walk to infinity on primes in steps of bounded length.

C.F. Gauss Visits the Planet Vulcan

Carl Friedrich Gauss was a famous 18th-century German mathematician whose interests were very diverse.

In fact, mathematics on the planet Earth was not enough to keep Gauss interested (ahem, the author begs pardon), and so he visited the planet Vulcan to learn how the inhabitants count. As in any respectable math paper, we need a good example. How about Mr. Spock of Star Trek fame?

A bit of background is in order here. Spock was born on the planet Vulcan in the year 2230. His father, Sarek, was a Vulcan diplomat, and his mother, Amanda Grayson was a human scientist. So, just how *does* Spock count? Think about it. When you were a kid and were first learning how to count, what did you do? You held up your hand and pointed to your fingers. Eventually, you could very smugly count in your head.

What does Spock do when *he* counts? When Spock was a

kid, he held his hand up and, well, you know what that looked like.

Apparently, Spock counts in *pairs* of numbers, in particular, in pairs of integers.

About the Gaussian Integers (or Spock's Numbers)

Spock's numbers, known as *Gaussian integers* to mathematicians, are pairs of integers like $(7, 12), (-2, 6), (13, 4), (-206, 10027)$, and so on, forever. The picture of all Gaussian integers is too large to fit on this page (there are infinitely many Gaussian integers!), but a small portion of the picture is shown in figure 1 and is a grid made up of many small squares.

Another way of describing the Gaussian integers (and this is what Gauss himself did) is to write, for example, $(2, 3)$ as $2 + 3i$ and $(1, 4)$ as $1 + 4i$, where $i^2 = -1$. Then multiplying Gaussian integers is just as you think it should be. That is $(2, 3) \times (1, 4)$ is

$$(2+3i) \times (1+4i) = 2 + 8i + 3i + 12i^2$$
$$= -10 + 11i$$

(or $(-10, 11)$). The general method for multiplying two Gaussian integers, say (a, b) and (c, d), is the same as above:

$$(a + bi) \times (c + di)$$
$$= ac + adi + bci + bdi^2$$
$$= (ac - bd) + (ad + bc)i$$

(or $(ac - bd, ad + bc)$). Once we have

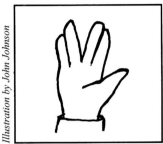

Counting in Pairs

this method, it makes sense to talk about Gaussian *primes*. That is, a Gaussian integer is a Gaussian prime exactly when it can't be written as a product of 'smaller' Gaussian integers, where 'smaller' means closer to the origin, $(0, 0)$. This amounts to the condition that (a, b) is smaller than (c, d) if $a^2 + b^2 < c^2 + d^2$. The Gaussian integers that appear in boldface in figure 1 are actually Gaussian primes.

A Quick Trick

There is a nice trick for determining exactly when a Gaussian integer is prime. Suppose someone hands you the Gaussian integer $(5, 2)$ and wants to know if this number is a Gaussian prime. What to do? You first compute $5^2 + 2^2 = 25 + 4 = 29$, and check to see if the result is a (regular) prime. In this case, 29 is prime, so you're in luck: $(5, 2)$ is a Gaussian prime. This trick works on all Gaussian integers (a, b) as long as neither one of a nor b is zero. That is, given a Gaussian integer (a, b), compute $a^2 + b^2$. The result is a (regular) prime precisely when (a, b) is a Gaussian prime.

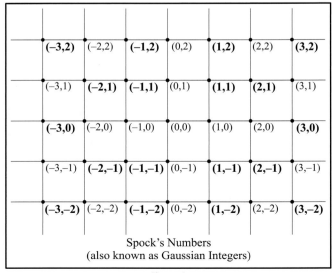

Spock's Numbers
(also known as Gaussian Integers)

Figure 1

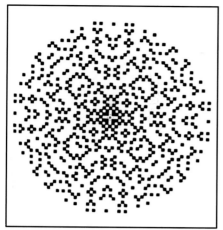

Some Gaussian Primes

Gaussian 2-moat

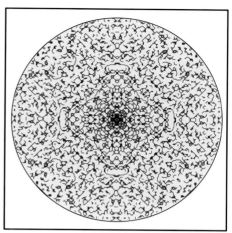

Gaussian 3-moat

To finish off this trick, we need to know what to do for the Gaussian integers $(a, 0)$ or $(0, b)$ (remember, the trick only works, so far, for (a, b) when neither a nor b is 0). The answer is that $(a, 0)$ (respectively $(0, b)$) is a Gaussian prime precisely when $|a|$ (respectively $|b|$) is a regular prime which is exactly 3 greater than a number divisible by 4. For example, $(11, 0)$, $(-11, 0)$, $(0, 11)$, and $(0, -11)$ are all Gaussian primes, whereas $(-5, 0)$, $(5, 0)$, $(0, 5)$, and $(0, -5)$ are Gaussian composites. (For the skeptic in the audience, what happens when you compute $(2, -1) \times (2, 1)$?) As usual, a picture is worth 10,000 words, and the graphic above and on the left shows all Gaussian primes within a distance $\sqrt{1000}$ from the origin, $(0, 0)$.

Take Another Hike!

Now that we've got all of these Gaussian primes, what should we do with them? Take another hike, of course. We'll try the prime-hiking game on the Gaussian primes this time. Here goes: Stand on the number $(0,0)$ and start walking. Remember, you can only step on Gaussian primes. The question is, if you keep walking forever, will you, every once in a while, have to take longer and longer steps? Or might you be able

to keep walking and taking only 'small' steps? Now that you're an expert, you would ask: Can you walk to infinity on the Gaussian primes in steps of bounded length? The answer? Nobody knows… (dramatic pause) … yet.

A GIANT Homework Problem

The Gaussian prime walking problem was posed in 1962 by the mathematician Basil Gordon from UCLA. It remains unsolved to this day. The truth is, the answer isn't in the back of any textbook, and isn't outlined in any paper. It just isn't known! (The author happily admits to having spent more than 1.5 years looking for the elusive solution, and expects to spend a great deal *more* time doing so.) This problem is a research problem, and

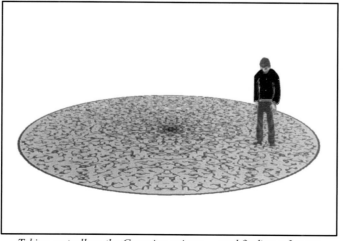

Taking a stroll on the Gaussian primes … and finding a 3-moat.

is just one of a plethora of such unsolved or open problems.

"Moativation"

The usual image that comes to mind when one thinks of a moat is a body of water surrounding a castle to keep out intruders. In our case the center of the castle is at the origin and the moat is a squiggly band (of nonconstant width) completely made up of Gaussian composites. Here is an idea for solving the Gaussian prime walking problem. If we can keep finding 'fatter and fatter' Gaussian moats surrounding the origin, then the prime-hiker will have to keep taking longer and longer steps. But, as usual, pictures are worth multitudes of words thanks to Stan Wagon (Macalester College) and Brian Wick (University of Alaska, Anchorage).

Stan Wagon, using *Mathematica,* programmed the Gaussian 2-moat shown above, by observing the following rules: connect two Gaussian primes with a black line if a) they are at most distance 2 apart, and b) either prime can be reached from the origin in steps of length at most 2. Connect two Gaussian primes with a blue line if a) they are at most distance 2 apart, and b) neither prime can

Brian Wick, using Mathematica, produced a condensed picture of a $\sqrt{18}$ - moat. His method was to look at 25 × 25 blocks of Gaussian integers and reduce each down to a single point.

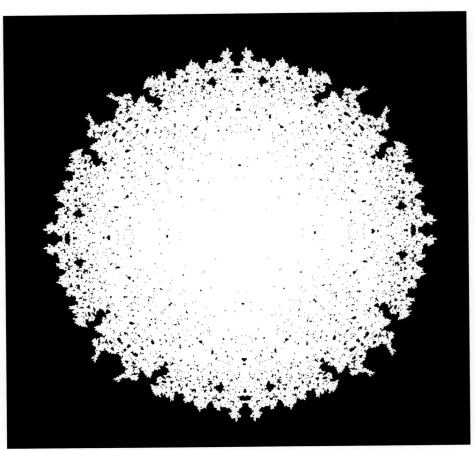

Gaussian $\sqrt{18}$-moat

be reached from the origin in steps of length at most 2.

So, for example, the unfortunate prime-walker who thought she or he could walk to infinity in steps of length at most 2 would start following the black paths and then get stuck at the moat of Gaussian composites that lies between the black lines and the green lines.

Stan Wagon used the same technique to generate a Gaussian 3-moat.

Perhaps Spock, a very clever fellow, thought he could walk to infinity in steps of length at most 3. He started walking along the network of black paths, and discovered that all of these paths were dead-ends. In other words, he ran into a Gaussian 3-moat. To get from the region with black lines to the region with blue lines, Spock would have to take a step of length strictly greater than 3.

The existence and location of both the 2-moat and 3-moat were known to the two mathematicians J.H. Jordan (no relation to Michael) and J.R. Rabung in 1970, and in fact, they were able even to find a $\sqrt{10}$-moat. Unfortunately, they ran out of money, and were unable to continue their search.

Brian Wick, using *Mathematica,* produced a condensed picture of a $\sqrt{18}$- moat. His method was to look at 25 × 25 blocks of Gaussian integers and reduce each down to a single point. If a block contains one (or more) Gaussian prime(s) which is (are) reachable from the origin in steps of length at most $\sqrt{18}$ then the corresponding point is colored white. Otherwise, it is colored black. The change from mostly white to all black dramatically outlines the $\sqrt{18}$-moat. The radius

of a circle containing the uncondensed moat is approximately 10,000.

Finally, using *Mathematica,* E. Gethner, S. Wagon, and B. Wick have a computational proof that a moat of width $\sqrt{26}$ must exist.

Final Stretch

Lewis Carroll, who is best known for having written *Alice in Wonderland,* was also a mathematician. Perhaps he was thinking about walking on prime numbers when he wrote:

"Would you tell me, please, which way I ought to walk from here?"

"That depends a good deal on where you want to get to," said the Cat.

"I don't much care where," said Alice.

"Then it doesn't matter which way you walk," said the Cat.

"—so long as I get *somewhere*," Alice added as an explanation.

"Oh, you're sure to do that," said the Cat, "if you only walk long enough!"

Now you know about some of the interesting progress that has been made on the Gaussian prime walking problem, but, really, the journey has just begun. ■

References

1. L. Carroll, *Alice in Wonderland,* Grosset and Dunlap, 1988.

2. Euclides, *The Elements of Euclid,* E.P. Dutton, New York, 1933.

3. E. Gethner, S. Wagon, and B. Wick, "A stroll through the Gaussian primes," *Amer. Math. Monthly* 105:4 (1998).

4. J.H. Jordan and J.R. Rabung, "A conjecture of Paul Erdös concerning Gaussian primes," *Math. Comp.*, 24 (1970), 221–223.

Acknowledgments

The author wishes to thank Bill Dunham, Bob Osserman, and Bill Thurston for their advice and inspiration, Marissa Moss for her illustrations, Stan Wagon for his technical help with the Gaussian 2- and 3-moats, and Brian Wick for his technical help with the Gaussian $\sqrt{18}$-moat. The author is especially grateful to Nancy Shaw for her computer wizardry and vast patience in the preparation of all of the graphics.

"Would you tell me, please, which way I ought to walk from here?"

"That depends a good deal on where you want to get to," said the Cat.

"I don't much care where," said Alice.

"Then it doesn't matter which way you walk," said the Cat.

1996 — A Triple Anniversary

WILLIAM DUNHAM
Muhlenberg College

People tend to observe anniversaries. This is true whether the observance is of happy events—like weddings—or of unhappy ones—like stock market crashes. Somehow, the very act of remembering seems irresistible.

The scientific community, too, commemorates its milestones. Earlier this year, for instance, the University of Pennsylvania held a conference to recognize the fiftieth anniversary of the development of ENIAC, the world's first electronic computer. As part of those festivities, Chess Grandmaster Garry Kasparov matched wits with the IBM "Deep Blue" computer in an intellectual clash between carbon and silicon. I take no little pride in reporting that Kasparov cleaned Deep Blue's (internal) clock.

Because we are creatures of the base-10 numeration system, certain anniversaries assume particular importance. That is why Penn went ape over the fiftieth—as opposed to, say, the 37th or 43rd. We tend to celebrate events remote from us in multiples of ten. And we remember those separated from us by centuries ($10^2 = 100$) as being especially significant.

Thus, for mathematicians, 1996 is a year rich in anniversaries. Not only is it the centennial of the first proof of the prime number theorem, and not only is it the bicentennial of the discovery of the geometric constructibility of the regular 17-gon, but it is also the tricentennial of the publication of the first calculus textbook. We have not one, not two, but *three* reasons to celebrate the current year.

The Prime Number Theorem (1896)

Primes, of course, are the multiplicative "atoms" from which all whole numbers can be built. For two and a half millennia, mathematicians have studied their fascinating but surprisingly elusive properties. It was Euclid in 300 B.C. who proved that no finite collection of primes can include them all—in other words, that primes are infinitely abundant. His argument, appearing as Proposition 20 of Book IX of the *Elements,* is regarded as a logical *tour de force,* one of the most elegant, most beautiful proofs in all of mathematics.

One reason for its fame is that Euclid established the infinitude of primes without providing any explicit formula or pattern for these numbers. Indeed, the distribution of primes seems quite haphazard. Consider the first three dozen of them:

$$2, 3, 5, 7, 11, 13, 17, 19, 23, 29, 31, 37, 41, 43,$$
$$47, 53, 59, 61, 67, 71, 73, 79, 83, 89, 97, 101,$$
$$103, 107, 109, 113, 127, 131, 137, 139, 149, 151$$

Anyone see a pattern here?

Well, all but the first are odd, but that's not terribly profound. Slightly more perceptive is that, after the first two, each is either one more or one less than a multiple of 6, but again this is of marginal value. Sometimes adjacent primes are close together, like 41 and 43; at other times they seem fairly far apart, like 113 and 127. All in all, primes appear to be distributed pretty much at random.

But they're not. To see why, we first define the arithmetic function $\pi(n)$ to be the number of primes less than or equal to n. For example, $\pi(9) = 4$ because there are four primes less than or equal to 9—namely, 2, 3, 5, and 7. Likewise, $\pi(10) = 4$, $\pi(100) = 25$, and $\pi(151) = 36$.

Instead of counting primes less than or equal to n, we can look at the *proportion* of primes among the numbers less than or equal to n. That is, we consider $\pi(n)/n$. Clearly $\pi(9)/9 = 0.4444$, $\pi(10)/10 = 0.4000$, $\pi(100)/100 = 0.2500$, and $\pi(151)/151 = 0.2384$. This can be extended to much larger numbers, yielding such proportions as $\pi(10^7)/10^7 = 0.06645790$ or $\pi(10^{10})/10^{10} = 0.04550525$.

Hidden amid these fragmentary results is a subtle asymptotic pattern known as "the prime number theorem." Like so many profound results from number theory, it was suspected long before it was conclusively proved.

One who perceived order amid the chaos was young Carl Friedrich Gauss (1777–1855). As a pastime in his mid-teens (no kidding!), Gauss compared the frequency of primes and the entries in a table of logarithms. "I soon recognized," he wrote, "that behind all of its fluctuations, this frequency is on the aver-

Reprinted from September 1996, pp. 8–13

age inversely proportional to the logarithm."[1] The 15-year-old Gauss confided to his diary the cryptic statement:

$$\text{prime numbers below } a \ (= \infty) \ \frac{a}{l(a)}.$$

Replacing $l(a)$ with the modern "$\ln(a)$" and using the function $\pi(a)$ as defined above, we can translate Gauss's jottings into:

$$\text{for large } n, \ \pi(n) \approx \frac{n}{\ln(n)}$$

$$\text{or, equivalently, } \frac{\pi(n)}{n} \approx \frac{1}{\ln(n)}.$$

For example, compare $\pi(10^{10})/10^{10} = 0.04550525$ with $1/\ln(10^{10}) = 0.04342945$. That's darn close.

With one final adjustment, the prime number theorem in modern form is stated as:

$$\lim_{n \to \infty} \frac{\pi(n)}{n/\ln(n)} = 1.$$

The young Gauss had recognized this pattern by employing both a phenomenal insight and a tremendous amount of perseverance (after all, these computations were done at a time when "digital technology" meant counting on your fingers). His discovery provides dramatic evidence—as if any were needed—of the vast difference between guesswork and Gausswork.

Although inferring the theorem, Gauss gave no proof. All he really provided was a promising conjecture. *Proving* the prime number theorem would occupy mathematicians throughout the nineteenth century. Although frustrating, this quest gave birth to the important field known as analytic number theory.

It is to Peter Gustav Lejeune-Dirichlet (1805–1859) that analytic number theorists usually trace their beginnings. In 1837 Dirichlet proved the long-suspected fact that any suitable arithmetic progression must include infinitely many primes. More precisely, if we begin with relatively prime whole numbers a and b and examine the progression

$$a, a + b, a + 2b, a + 3b, \ldots, a + nb, \ldots,$$

then there must be infinitely many primes among them. Note that if $a = 1$ and $b = 1$, the progression is simply the set of all whole numbers, so that Euclid's result on the infinitude of primes becomes a corollary of Dirichlet's stronger theorem.

What made his proof so remarkable was that it employed the analytic tools of convergence and divergence to answer a question about whole numbers. There is something unexpected about applying analysis—the science of the continuous—to something as discrete (i.e., non-continuous) as the positive integers. This surprising and wonderful interconnection is the essence of analytic number theory.

But the prime number theorem remained unproved. In the early 1850s Pafnuti Lvovich Tchebycheff (1821–1894) showed that, *if*

$$\lim_{n \to \infty} \frac{\pi(n)}{n/\ln(n)}$$

exists, then it must fall somewhere between 0.92129 and 1.10555. Unfortunately he never could establish the existence of this limit and thus could not draw the desired conclusion from all his efforts. Georg Friedrich Bernhard Riemann (1826–1866) next took up the challenge and in 1859 brilliantly advanced the frontiers of analytic number theory by examining what we now call the Riemann zeta function. Alas, his pursuit of the prime number theorem was also unsuccessful.

(By the way, it seems to me that analytic number theory—boasting such innovators as Peter Gustav Lejeune-Dirichlet, Georg Friedrich Bernhard Riemann, and Pafnuti Lvovich Tchebycheff—favors those with mellifluous, multisyllabic names. A syllabically-challenged person like "Cher" just wouldn't stand a chance.)

The prime number theorem resisted the efforts of mathematicians until the century had nearly run its course. At last, in 1896—exactly one hundred years ago—two individuals independently furnished the longsought proof. One was the aptly named Charles-Jean-Gustave-Nicholas de la Vallée-Poussin (1866–1962). The other was Jacques Hadamard (1865–1963), whose name seems insufficiently flamboyant for this crowd.

Ch. J. de la Vallée Poussin

J. Hadamard

Their simultaneous discoveries, coupled with their nearly identical lifespans, have led some to doubt that these were really two different people. Rest assured, they were.

Vallée-Poussin and Hadamard finally laid the proposition to rest. A century ago, thanks to their genius and the power of analytic number theory, the prime number conjecture became the prime number *theorem*.

Construction of the Regular Heptadecagon (1796)

In the previous section we met the 15-year-old Gauss counting primes. Here we celebrate the bicentennial of his breathtaking announcement that a regular heptadecagon (hepta = 7 and deca = 10) can be constructed using only the Euclidean tools of compass and straightedge. This he discovered at the relatively advanced age of 18.

The problem was an ancient one: which regular polygons can be drawn with compass and straightedge? Euclid, who addressed the subject in Book IV of the *Elements,* knew how to construct regular triangles, squares, pentagons, hexagons, octagons, and pentadecagons (15-gons). Euclid also knew (although he somehow neglected to mention it) that polygons formed by repeatedly doubling the number of sides of any of

these were likewise constructible. Thus, one could do a regular $2 \times 8 = 16$-gon or a regular $2 \times 16 = 32$-gon, and so on.

What Euclid left unsaid was whether these were the *only* regular polygons that were geometrically constructible. Over the centuries no one had found any others, so there was reason to believe that Euclid had netted them all in 300 B.C. It thus came as a shock when, in 1796, the young and as yet unknown Carl Friedrich Gauss informed the mathematical world that Euclid had missed some. In Gauss's words, "It seems to me then to be all the more remarkable that besides the usual polygons there is a collection of others which are constructible geometrically, e.g., the 17-gon."[2]

Needless to say, his argument cannot be squeezed into a few paragraphs; if it were that simple, someone would have stumbled upon it during the previous 21 centuries. Yet the difficulty lies more in the intricate interconnections of its steps than in the complexity of any step in particular. Let me outline the basic structure of his proof.

First, it was known since the time of Descartes in the early seventeenth century that, beginning with a unit length, one can construct any magnitude whose length is expressible in terms of integers and finitely many applications of the operations $+$, $-$, $\times$, $\div$, and $\sqrt{}$. Such expressions are called "quadratic surds."

Of course, it is obvious how, starting with a segment of length 1, we can construct a segment of length 2, or 3, or 4. Less clear is how to construct one of (say) length $\sqrt{5}$. If you haven't seen it before, here it is:

Along a straight line, mark off segment AB of length 5 and BC of unit length (as shown in Figure 1). Bisect AC at O—a familiar compass and straightedge construction—and using O as center and $OA = OC$ as radius, draw a semicircle. From B, erect a perpendicular to AC meeting the semicircle at D, another simple construction.

Now cash in on some elementary geometry. Triangle ADC is right because $\angle ADC$ is inscribed in a semicircle. Thus DB, the altitude to the hypotenuse, splits $\triangle ADC$ into two similar triangles, namely $\triangle ABD$ and $\triangle DBC$. By similarity,

$$\frac{AB}{BD} = \frac{BD}{BC}$$

Figure 1

so that

$$(BD)^2 = AB \times BC = 5 \times 1 = 5.$$

Hence BD has length $\sqrt{5}$. And here's the bottom line (pun intended): The length $\sqrt{5}$ was constructed *with compass and straightedge.*

As noted, one can also construct products and quotients of previously constructed lengths, although the interested reader will have to discover how for herself. (O.K., if you don't want to do it yourself, look at [3].)

So—to repeat—by nesting these sorts of constructions, it is possible to draw any segment whose length is a quadratic surd. For instance, it is possible to construct a segment of length

$$\frac{2+\sqrt{3+\sqrt{5-\sqrt{7}}}}{4-\sqrt{1+\sqrt{6}}},$$

although I wouldn't particularly want to.

Next, Gauss recognized that if he could construct $\cos(2\pi/n)$, then he could easily construct a regular n-gon. That is, inside a unit circle, copy the constructed length $\cos(2\pi/n)$ as segment OC in Figure 2. From C construct a perpendicular upward, meeting the circle at B. If $\theta = \angle BOC$, then clearly

$$\begin{aligned}\cos\theta &= \frac{OC}{OB} \\ &= \frac{\cos(2\pi/n)}{1} \\ &= \cos(2\pi/n),\end{aligned}$$

and so the central angle $\theta = 2\pi/n$. Copying the chord AB n times around the unit circle, we must return exactly to A and in the process shall have constructed a regular n-gon.

Summarizing to this point: we know that a regular 17-gon is constructible if $\cos(2\pi/17)$ is and, further, that $\cos(2\pi/17)$ is constructible if it is a quadratic surd. The remaining obstacle, then, was to establish the quite unexpected fact that $\cos(2\pi/17)$ really is such a surd. This is what Gauss did.

In the process, he displayed his extraordinary genius by veering off into the world of complex numbers. At first this seems bizarre. Geometric constructions, after all, occur in the real world; there's nothing imaginary about them.

C. F. Gauss

But note that $\cos(2\pi/17)$ appears as the real part of the complex number $z = \cos(2\pi/17) + i\sin(2\pi/17)$, which is itself one of the seventeenth roots of unity. This suggests a bridge into the imaginary realm, a bridge Gauss crossed with spectacular success. Once on the far side, he proved that $\cos(2\pi/17)$ is indeed a quadratic surd. To be specific,

$$\begin{aligned}\cos(2\pi/17) = &-\frac{1}{16} + \frac{1}{16}\sqrt{17} + \frac{1}{16}\sqrt{34-2\sqrt{17}} \\ &+\frac{1}{8}\sqrt{17+3\sqrt{17}-\sqrt{34-2\sqrt{17}}-2\sqrt{34+2\sqrt{17}}}.\end{aligned}$$

Although this may strike some readers as being a quadratic *ab*surd, it is perfectly correct.

Because this complicated expression is built from integers that are added, subtracted, multiplied, divided, and square-rooted, it is constructible. Hence $\cos(2\pi/17)$ is constructible, and it follows from the criterion above that the regular 17-gon is constructible as well. So goes the proof.

Wow! Like investigating primes with the techniques of analysis, the construction of geometric polygons using the properties of complex numbers links two seemingly unrelated subjects. Once again we see that unexpected mathematical interconnections can yield the most remarkable of theorems.

A striking feature of this result is its Janus-like quality. In addressing the constructibility of regular polygons, Gauss looked backward to the Greeks; in introducing complex variables, he looked forward to a subject whose importance would explode in the coming century. Of course, the *practical* significance of all this was nil. It didn't help anyone balance a ledger or powder a wig. In truth, it was quite useless. But if theorems can ever be breathtaking in their boldness and sweep, this was one.

And Gauss's proof served another purpose. It showed that Euclidean geometry runs deeper than is usually imagined. From

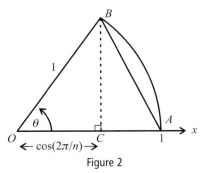
Figure 2

my own student days, I remember thinking that Euclid's geometry was simply the careful demonstration of self-evident facts. It was fun to prove that the base angles of an isosceles triangle were congruent, but surely anyone would have guessed this by drawing a few pictures. The proofs may have required insight, but the theorems as stated drew only yawns.

Not so here. Prior to Gauss, no one—*no one*—anticipated that a regular 17-gon could be constructed within the constraints of Euclid's innocent-looking system. The construction of the regular heptadecagon exactly two hundred years ago forcefully reminded mathematicians that Euclidean geometry still held some big surprises.

The First Calculus Text (1696)

If our first two anniversaries commemorate specific theorems, the third recognizes a broader achievement: the publication in 1696 of the first textbook on calculus. Its author was Guillaume François Antoine de l'Hospital (1661–1704), and the story of how he came to write it demands a brief digression.

L'Hospital was a French marquis, a minor nobleman who was an ardent, if unschooled, mathematician. Late in 1691, l'Hospital was introduced to the young Johann Bernoulli (1667–1748), then fresh from his triumphant discovery of the catenary curve. In making that discovery, Bernoulli had artfully employed the techniques of calculus as first described in the 1684 and 1686 papers of Gottfried Wilhelm Leibniz, its creator. Bernoulli later recalled, "I knew right away that he [l'Hospital] was a good geometer … but that he knew nothing at all of the differential calculus, of which he scarcely knew the name, and still less had he heard talk of the integral calculus, which was only just being born."[4]

L'Hospital wanted to learn. And, as a marquis, he had deep pockets. He thus hired Johann Bernoulli to teach him this new and powerful subject. For the better part of a year, Bernoulli worked to bring l'Hospital up to speed. In the process, he provided (or, more precisely, "sold") l'Hospital his own discoveries in calculus. Essentially, l'Hospital bought the rights to Bernoulli's theorems.

By all accounts, l'Hospital made excellent progress and soon felt himself ready to write a book on the subject. It appeared three hundred years ago, in 1696. Titled *Analyse des infiniment petits (Analysis of the infinitely small),* it was largely the recycled memoirs of Johann Bernoulli. L'Hospital acknowledged his debt to Leibniz and, especially, to Bernoulli and forthrightly observed that "I frankly return to them whatever they please to claim as their own."[5]

Later in life Johann grumbled that l'Hospital had garnered undeserved credit for this text. Initially skeptics dismissed Bernoulli's protestations as sour grapes. Johann Bernoulli was, after all, a contentious, argumentative egotist—the sort of person who gives arrogance a bad name. Actually, as documents

subsequently revealed, Bernoulli was quite justified in claiming much of the book as his own.

And so it was that, in 1696, the world saw its first calculus text. *Analyse des infiniment petit* treated differential calculus only. It was modest in size, unlike today's 14-pound monsters that require a wheelbarrow to carry to class. There were no color illustrations nor any calculator exercises indicated in the problem sets. In fact, there were no problem sets.

Yet it *was* a calculus book.

L'Hospital began with a few postulates and definitions. For instance:

Postulate I: Grant that … a quantity which is increased or decreased only by an infinitely small quantity may be considered as remaining the same.

Definition II: The infinitely small part whereby a variable quantity is continually increased or decreased is called the differential of that quantity.[6]

To such statements, the modern reader is likely to respond, "Huh?"

"Infinitely small quantity?" "Continually increased?" What do these mean? Clearly the precision and rigor of modern analysis lay far, far in the future.

Early in the book, l'Hospital noted that the differential of a constant is zero. That is, if a is constant, then $da = 0$ (where the symbol *"d"* for differential was borrowed from Leibniz's original 1684 paper). A bit later, l'Hospital gave the following proof of the product rule:

To find the differential of the product $xy,$ we increase x by an infinitely small part dx to get $x + dx.$ Similarly, y is increased by an infinitely small part to $y + dy.$ Thus the product xy will be increased to $(x+dx)(y+dy) = xy+xdy+ydx+dxdy,$ and so the differential of xy will be the difference

$$d(xy) = [xy + xdy + ydx + dxdy] - xy$$
$$= xdy + ydx + dxdy.$$

Then l'Hospital noted that, "because $dxdy$ is infinitely small with respect to the other terms,"[7] we simply can throw it away: And so—Bingo! —we have the product rule $d(xy) = xdy+ydx.$

L'Hospital then dispatched the quotient rule in short order:

To find $d(x/y),$ introduce the auxiliary variable $v = x/y$ so that $x = vy.$ Apply the just-proved product rule to this last expression to get

$$dx = d(vy) = vdy + ydv,$$

and therefore

$$dv = \frac{dx - vdy}{y}.$$

Multiplying numerator and denominator by y yields

$$dv = \frac{ydx - vydy}{y^2},$$

and we now merely recall that $v = x/y$ and $x = vy$ to deduce the famous quotient rule:

$$d\left(\frac{x}{y}\right) = \frac{y\,dx - x\,dy}{y^2}.$$

Such was calculus in 1696.

We must not conclude without mentioning the result that appeared in Section IX of the *Analyse*. There l'Hospital gave the famous rule for finding

$$\lim_{x \to a} \frac{f(x)}{g(x)}$$

when both $f(x)$ and $g(x)$ tend to 0 as x approaches a (although he didn't state it in this modern form). In his words,

> if the differential of the numerator be found, and that be divided by the differential of the denominator, after having made $x = a$..., we shall have the value ... sought.[8]

Modern textbooks state this result more concisely as

$$\lim_{x \to a} \frac{f(x)}{g(x)} = \lim_{x \to a} \frac{f'(x)}{g'(x)}.$$

It should come as no surprise that l'Hospital's Rule was actually discovered by Johann Bernoulli. It is one of those nuggets that l'Hospital had promised to return to its rightful owner. Unfortunately for Johann, the return was never made. Everyone today knows "l'Hospital's Rule" as a wonderful application of differential calculus. Not everyone knows that it should be called "Bernoulli's Rule."

After stating the result that would guarantee him immortality, l'Hospital provided his first example. It was a doozy. He asked for the limit as x approaches a of

$$\frac{\sqrt{2a^3 x - x^4} - a\sqrt[3]{a^2 x}}{a - \sqrt[4]{ax^3}}.$$

This, I repeat, was his *first* example!

My, how textbooks have changed. For the sake of comparison, I checked out some of the popular texts of today:

Stewart's first example of l'Hospital's Rule is the tame

$$\lim_{x \to 0} \frac{2^x - 1}{x}.$$

Finney, Thomas, Demana, and Waits (the authors, not the law firm) start with the easy

$$\lim_{x \to 0} \frac{3x - \sin x}{x}.$$

And Anton begins with the positively wimpy

$$\lim_{x \to 2} \frac{x^2 - 4}{x - 2}.$$

These problems are mere lightweights compared to l'Hospital's snarl of symbols. If you dare, try it for yourself—although be mindful of that old adage, "A chain rule is only as strong as its weakest link."

With this challenge, we end our anniversary celebration of three mathematical milestones—the prime number theorem, the regular heptadecagon, and the first calculus text. Readers are now free to break out the champagne, throw confetti, and cheer wildly!

But don't celebrate overlong. After all, one big question remains to be answered: a century from now, what mathematical milestone will our ancestors be celebrating from 1996?

Get working!

Acknowledgment. I would like to thank Penny Dunham of Muhlenberg College for help on this project, and so many others.

References

1. L. J. Goldstein, "A History of the Prime Number Theorem," *American Mathematical Monthly* 80 (1973), p. 612.

2. J. Fauvel and J. Gray, *The History of Mathematics: A Reader,* Macmillan (1988), p. 492.

3. R. Courant and H. Robbins, *What is Mathematics?,* Oxford (1941), pp. 121–122.

4. *Ibid.,* p. 441.

5. D. Struik (Ed.), *A Source Book in Mathematics: 1200–1800,* Princeton (1986), p. 312.

6. *Ibid.,* pp. 313–314.

7. *Ibid.,* p. 315.

8. V. Katz, *A History of Mathematics: An Introduction,* Harper Collins (1993), pp. 484–485.

A Nice Genius

DON ALBERS
Mathematical Association of America

As a boy, Ron Graham, was small for his age and was never chosen to play on school teams. He never attended the same school for more than one year and did not graduate from high school. He did not take any mathematics courses during his first three years of college at The University of Chicago and eventually got a bachelor's degree in physics from the University of Alaska.

Today at age 61 Dr. Graham stands six feet-two inches and looks very much the athlete. He is regarded as one of the top mathematicians in the world. Graham is famous for his work in combinatorics, which won him election to the National Academy of Sciences. He is past-president of the American Mathematical Society, and he has just been named Chief Scientist of AT&T Laboratories, successor to the much venerated Bell Labs. In spite of his heavy responsibilities at AT&T, he manages to find time to pursue other interests, especially juggling and gymnastics. In college he was a California state trampoline champion. He also is highly skilled in Ping-Pong, tennis, bowling, and boomerang. Bungee trampolining is his latest interest.

Harvard mathematician Persi Diaconis who has collaborated with Graham many times, describes him as "a remarkably accomplished mathematician. Ron is always willing to help a struggling student or a colleague. He never leaves you hanging. He's a genius, but a nice genius."

Diaconis remembers giving a talk about joint research that he and Graham had done. He ended his talk by saying

Ron Graham, Chief Scientist of AT&T Labs, in unicycling gear—suit, tie, and unicycle. Photograph by Ché Graham.

"This problem is still unsolved." At that point, Graham, who was in the audience, stood up and gave a solution on the spot. The audience, thoroughly impressed, burst into applause, an unusual outpouring of emotion for a group of mathematicians.

Reprinted from November 1996, pp. 18–22

Dr. Graham at work on a graph theory problem. Photograph by Carol Baxter.

Calculus At Age 11

Ron Graham remembers liking mathematics a lot as a little boy. In fifth grade, Miss Smith showed him an algorithm for calculating square roots. He soon developed his own algorithm for calculating cube roots. His natural ability for the subject resulted in a high level of confidence. "By the time I got to seventh grade," he recalls, "I knew algebra and trigonometry and thought I could solve any problem given to me. But one day, Mr. Schwab gave me one I couldn't do. The problem was to find the size of a population of mice if it was known that the death rate was proportional to the size of the population. It took a knowledge of differential equations to solve it. He then gave me a book and told me that I would be able to solve the problem by the time I finished it. That book was *Calculus* by Granville, Smith, and Longley. He was right. By the end of the semester, I finished the book and had a solution. He also chose me to be on the school chess team. I owe a lot to Mr. Schwab."

Ron was only eleven years old at the time. He had skipped a few grades as his family moved around the country, primarily in California and back and forth to Georgia. Until he entered The University

of Chicago when he was 15, he never attended any school for more than one year. Family stability was diminished in 1941 when Ron's father, an oilfield worker, enlisted in the Merchant Marine, and essentially left his wife and three children to fend for themselves. Young Ron had learned that superior academic achievement went a long way with teachers and peers. In his words, "I was a good kid and I learned to adapt."

Ron saw his father again one day six years later when he was delivering the *Berkeley Gazette.* He delivered papers both in the morning and in the evening. He remembers living in the housing projects and being poor.

Graham's tenacious pursuit of solutions has continued to this day. More often than not, he triumphs over problems, but after expending lots of time on one without significant progress, he will occasionally post a $100 reward for a solution. "Some problems are capable of driving me crazy," says Graham, "and I will pay just to be put out of my misery."

Much of Graham's work has been in graph theory in support of AT&T's communications network. Telephone networks are examples of mathematical graphs. Telephone networks are replete

with graph theory problems. The *shortest network problem* is one of the easiest of those problems to state but it is far from being easy to solve. The problem asks for the shortest network of line segments interconnecting an arbitrary set of, say, 100 points. The solution to this problem has frustrated the best mathematical minds and overwhelmed computers.

Adding Points

Sometimes adding points yields a shorter network. Take for example three points arranged at the vertices of an equilateral triangle. Any two sides of the triangle then give a solution. If, however, a point is added to the center of the triangle, then a shorter network results from joining that point to the vertices of the triangle.

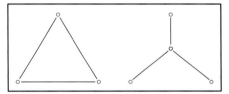

Adding a point (right) yields a shorter network.

Complete Disorder Is Impossible

It is his contribution to Ramsey theory, however, that mathematicians regard as his finest work. "Ramsey theory," according to Graham, "says that complete disorder is impossible. There is always structure somewhere." Ramsey theory shows that in any group of six or more people, three will either know one another or will all be strangers. It is somewhat harder to prove that in a group of 18 people there must be four who all know each other or are all strangers. The minimum size of the group needed to ensure that there are always five mutual friends or five mutual strangers is unknown.

In his junior year of high school, Graham's mother encouraged him to apply for a new scholarship program set up by the Ford Foundation that enabled talented students to enter certain universities before they finished high school in order to provide them more of a collegiate education before they might be drafted into military service. He scored high on the scholarship exams and was admitted to The University of Chicago. Since he had done especially well on the mathematics portion of the exam, he was exempted from further mathematics courses at Chicago. As a result, during his three years there, he took no mathematics, but he did discover the world of gymnastics and juggling. When he arrived at Chicago, he was still small for his age, but the right size for gymnastics.

Acrotheater

One of the classes that Graham enrolled in at Chicago was Acrotheater. It was a combination of dance, circus arts, and gymnastics. The teacher was E.F. (Bud) Beyer, a national gymnastics champion in 1935. Ron remembers the class with great fondness: "The class met twice a week. We would do shows in the Chicago area high schools and elsewhere; it was a recruiting device of sorts. At the end of the year we'd do a big show with music, make up—the whole nine yards. Acrotheater hooked me on gymnastics."

During his three years at Chicago, the undersized kid developed into a six-foot-two-inch eighteen-year old. Juggling and the trampoline also developed into life-long interests for him during that period. The year after he left Chicago for the University of California, Berkeley, he won the California Intercollegiate Trampoline championship. Some years later he was elected president of the International Jugglers Association.

After three years at Chicago, Graham transferred to Berkeley in 1951 as an electrical engineering major. "The problem was," he recalls, "that by then I was behind in math requirements and would not complete a degree in four years. At that time you could be drafted if you had not completed a degree in four years. So I enlisted in the Air Force and was stationed

in Alaska." He arranged his schedule so that he attended classes by day at the University of Alaska and did his Air Force job in communications at night. He took more mathematics classes at Alaska, but ended up with a degree in physics in 1958 because Alaska was not yet accredited in mathematics.

Twelve balls at once! How about thirteen, Ron? Photograph by Ché Graham.

The Berkeley System

In 1959, nine years after starting college, Graham returned to Berkeley to begin his doctoral studies—in mathematics! He remembers that his first semester at Berkeley was very challenging. "I didn't catch on to the 'system' for a while. In my first year, I walked into Chern's class on differential manifolds and on the first day things were moving pretty fast. And the next day, people in the front rows were saying, 'Come on, let's get on to the good stuff!' Chern thought he was moving too slowly for the class, so he really started ripping. It turned out that graduate students typically

audited a course once or twice before taking it for credit. That was the system and it stimulated me to work hard."

By 1962, Graham had thoroughly mastered the system and was the proud possessor of a PhD. His thesis on Egyptian fractions was supervised by the legendary number theorist D.H. Lehmer. An Egyptian fraction is a fraction having numerator 1, a so-called unit fraction. Thus 1/2, 1/7, 1/15 are Egyptian fractions. Notice that 3/5 = 1/2 + 1/10. Now 2/7 is a bit harder to write as a sum of Egyptian fractions: 2/7 = 1/5 + 1/13 + 1/115 + 1/10465. It's known that every fraction can be written as a finite sum of unit fractions. What if you restrict yourself to using fractions with odd denominators? The answer turns out to be yes, provided that fraction has an odd denominator. How about using distinct unit fractions, using only those that are only perfect squares. Graham showed that, again, the answer is yes, where it's reasonable. [1]

Graham's gymnastics interests continued unabated during his graduate student years. He and two other students formed a professional trampoline group, the Bouncing Bears, and earned money by performing at schools, supermarket openings, and the circus! Bungee trampolining is Graham's latest sport. In Bungee trampolining, a pair of Bungee cords are mounted over a trampoline and connected to a harness that is usually attached to a twisting belt, so that you can twist and somersault freely while being suspended by the two cords. "You go up about 30 feet above the surface, and if you miss, well, you get only bruised ribs. Eventually, you take off the harness and do the trick, unassisted. It's important, I think, to keep doing new things when you get into a rut, and start operating solely on reflexes."

Erdős

In 1963, Graham met the famous Hungarian mathematician Paul Erdős at a mathematics meeting in Colorado. Erdős was one of the most prolific mathematicians in history with more than 1500 published papers to his credit. For more than fifty years, Erdős traveled from university to university with one small suitcase, gathering and sharing mathematical ideas with hundreds of mathematicians. Number theory and graph theory, Graham's specialties, were also the specialties of Erdős. In short order, they collaborated on a number of papers. As Erdős aged, Graham helped him tend to some of the basics of life—paying taxes, buying clothes, etc. Graham even set up an Erdős room in his home. Up until the time of his death in September of this year, Erdős frequently stayed with Graham and his wife Fan Chung.

Graham has vivid memories of his first meeting with Erdős. "I saw this rather senior guy of 50, already quite famous, playing Ping-Pong during one of the breaks. He asked me if I wanted to play, and I agreed. He absolutely killed me! I had played casual Ping-Pong, but I couldn't believe that this old guy had beaten me. So I went back to New Jersey and I got a machine that shoots out Ping-Pong balls at you. I bought a table, joined a club, started playing at Bell Labs, and in the state league. I eventually became the Bell Labs champion at Ping-Pong, and won one of the New Jersey titles. Finally, I reached the point where I thought I could play Erdős. Erdős, of course, was getting older, and I was playing more. When I could play him, I'd do it sitting down. I would sit in a tall chair, and let him start at 20. Sitting down in a chair is really not that big a disadvantage. You'd think that's going to handicap you. You can't get up from the chair. If you get up, you lose. It turns out that isn't that big of a disadvantage against somebody who's not such a serious competitor."

"As you work at Ping-Pong, you begin to understand the subtleties of the spin and the sound of Ping-Pong, so to speak. I always liked that aspect of it. The trouble

is, it takes a certain investment to do that, and you can't do that for everything. And there's a certain carryover, but sometimes it's negative. What you do at Ping-Pong isn't necessarily good for what you do at tennis. It's like learning Chinese. If you learn Mandarin, that may or may not be helpful in learning Cantonese."

He flies through the air with the greatest of ease, the man on the bungee trampoline.
Photograph by Ché Graham.

True Mark of Teaching

Graham's passion for learning is quite remarkable. He loved school because "I could learn something new, and I enjoyed especially teaching to someone else what I had learned and taking them one level beyond. It's very satisfying teaching someone a particular skill like juggling, the trampoline, or mathematics. But it's even better to take them to the next level where you teach them how to teach."

Graham claims that in the juggling world, there is a tradition that you teach people what you know, and they will in

turn teach someone what they have been taught and what they've learned in addition. "I taught Tom Brown who in turn taught Joe Buhler how to juggle and he now is a better juggler than I am. That's a true mark of your teaching ability if you produce better students than you are."

After completing his doctorate at Berkeley in 1962, Graham joined the research staff of Bell Labs. He had been warned by academic mathematicians that he would be dead mathematically in three years if he went into industry. His elections as President of the American Mathematical Society and as a member of the National Academy of Sciences argue against the prediction. He found the "open door" policy of Bell Labs especially stimulating and very conducive to cross fertilization with physicists, chemists, engineers, economists, and other mathematicians.

"That results in a different mind set to some extent," according to Graham, "and that's important for students to know if they hope to take jobs in industry. That's not easy to teach because most people in university settings aren't accustomed to working that way." He notes that doors at many universities aren't open so much.

One of Graham's goals as Chief Scientist of AT&T Labs is to mold programs which establish better exchanges with people in universities and the Labs. During his 34 years with AT&T, Graham has taught courses seven times in universities such as UCLA, CalTech, Stanford, Princeton, and Rutgers. He finds contact with students very helpful. "It's good for someone like me to stand up in front of good students who ask questions that you take for granted. They may say: 'Well, why do you do it this way?' The answer might be: 'We always do it this way.' That forces you to think about what you've been doing and to question authority. I keep a saying on my refrigerator to remind me of staying open to new ideas. It says:

The Main Obstacle to Progress is Not Ignorance, But the Illusion of Knowledge.

Graham at home with his spouse, mathematician Fan Chung of the University of Pennsylvania. Photograph by Ché Graham.

"When you think you know, that can blind you to the fact that you don't know. Once you know you don't know, then you might be more receptive to looking at other approaches."

Research And Arrows

Graham took over as Chief Scientist of AT&T Labs in October at a transition point in the history of the company, which recently split into AT&T and Lucent Technologies. The quickening pace of technology has placed pressure on researchers to direct more of their efforts toward the business side. Graham's job, in part, is to foster the scientific excellence of the cultural careers of the researchers. He says that "I will make sure that they feel comfortable and are encouraged to pursue very fundamental ideas such as quantum computing that may not have any business impact for 20 years or more." He also notes that "pioneers are the people who have arrows in their chests, so you may not want to be out there too far ahead. At the same time, it's crucial to have the places and people who believe in fundamental research from the long-term viewpoint."

Graham is very much aware that in most businesses you often don't have years: the next model has to be out the door in a few months or the competition will blow you away. At the same time, someone somewhere has to be doing the basic research. Graham believes that most of that research will be done in universities, with some cooperation with industry. He also thinks that some large organizations such as AT&T are sufficiently forward-looking, have the resources, and understand the necessity of doing basic research.

As our interview came to a close in Graham's office at the National Academy of Sciences, one of three offices that he maintains, I remarked that he probably would be glad to get home to New Jersey and relax after flying back from Washington. He said that he planned to relax by working on a new juggling trick of the site-swap type, and perhaps listen to some classical music. He explained that a site-swap is a way of taking a numerical pattern such as 3-4-5 and converting it to a juggling trick. The 3, 4, and 5 tell you how high the balls go, and the sequence tells you which hands you're to do it with. So the right hand throws the 3 the left hand 4, the 3 crosses, and the 4 doesn't. But the next hand, which is the right, throws a 5. Then this is repeated, and you've got to do all of this in real time.

He says that it's not simply a control issue, but that you have to know where to look. "If you're juggling, say, five balls, you don't watch every ball all the way because there are too many balls and not enough eyes. You have to look at a given ball only briefly. Typically, you look at each ball at the top of the arc and get some information there. You mentally compute where it's going and put your hand there. Every throw is a little bit off, some off more than others, and juggling in large part is maintaining that radius of allowable mistakes, so to speak, in the 'region' of stability."

When juggling five balls, the typical pattern for each of the balls is essentially the same. You see a very symmetrical and beautiful pattern. They're all crossing. In a typical pattern they cross at the top, so you watch right there. The other patterns, such as a site swap, where different hands are throwing different balls to different heights can be hard to follow: There's no single place to look. Different hands are throwing balls to different heights, and there's no good place to look. You kind of back off and look globally. It's better

Cheryl Graham, Ron's daughter, is an accomplished photographer who likes juggling too. Photograph by Ché Graham.

to focus on each ball, even though briefly. Graham says, "Well, that may not be the best way to juggle, but it kind of helps your normal juggling, because it teaches you something that you did when you were first learning, and you watch, you want to keep your eye on the ball, so to speak—on the balls— and you get lazy." For Dr. Graham, it's always been the case that if he goes more deeply into a subject, he appreciates it more. He says "To understand how juggling evolves, if it's not so hard for four balls, try five, or try it underwater, or while trampolining, or with your arms crossed."

Music, Understanding, and Magic

Graham also likes music very much. Naturally, the same question comes up— would he appreciate it more by going more deeply into it? If he really tried to understand the structures of what Beethoven, Bach, or Mozart had been doing, would he appreciate it more? He worries that it might somehow lose the magic: "It's as though you like to be impressed by a magic trick, but once you know how it's performed, the magic is gone."

"For example, I can flip a coin and be very 'lucky.' I can be extremely 'lucky.' I can, in fact, be perfectly 'lucky' if necessary. But once you understand how it happens, the mystery and the magic is gone. Would that happen with music? I don't know, but I have deliberately stayed away from seeking a deep understanding of music." Musicians tell me, "No, you'll appreciate it even more when you really understand how great these composers were."

One of the complaints about mathematics is that you couldn't really understand the beauty and elegance, the power unless you're another mathematician. How do you convey that to someone else? There's a certain truth in that although you can certainly show people to some extent the power or the elegance and the surprise of some parts of mathematics.

In a few weeks, Graham will travel to the Disney Institute to participate in special mathematical programs for the general public. Dr. Graham, the mathematician, teacher, and athlete, continues to work at displaying a friendly side of mathematics to the public. He is, indeed, a nice genius. ■

Note

[1] The fraction needs to be in the union of the intervals $[0, \pi^2/6 - 1)$ and $[1, \pi^2/6)$.

An ABeCedarian History of Mathematics

STEVE KENNEDY
Carleton College

George Santayana told us that those ignorant of history are doomed to repeat it. Underwood Dudley, in this magazine, amended Santayana: we're all doomed to repeat history, those ignorant of it will merely be more surprised by life than those learned in it. I suspect they were thinking of the serious study of history. Popular, as opposed to scholarly, histories often seem no more than a mishmash of myth, rumor, anecdote, tall tale, and fairy tale. These may not lead to wisdom, but they sure can be fun. In mathematics we have Archimedes in his bathtub, Abel's alcoholic parents, the bickering Bernoulli brothers, and Lazzarini's bogus Buffonery. The list below contains some of the more enduring and endearing myths and fables in the historical folklore of mathematics. See if you can identify the mathematician associated with each bit of silliness.

A "Give me a place to stand and I will move the earth."

B Hungary issued a stamp in his honor, but as no portrait of him exists, they faked the picture.

C Ended his life in a madhouse. (That's what happens if you think about infinity long enough.)

D Sacrificed to the vanity of a queen: summoned to Stockholm to tutor Queen Christine, the early hours he was forced to keep and the severe cold combined to kill him.

E His sieve captures primes.

F Marginalia.

Reprinted from February 1997, pp. 14–15

G Antoine LeBlanc and her correspondent.

H This cricket fan, facing a stormy English Channel crossing, sent postcards reading "I have settled the Riemann Hypothesis." He was sure that God would not let him die with the credit for such an accomplishment, therefore his safe passage to France was ensured.

I "… as far as the propositions of mathematics refer to reality they are not certain; as far as they are certain, they do not refer to reality."

J In response to Fourier's claim that the purpose of mathematics is the explanation of natural phenomena he retorted, "A philosopher like Fourier should know that the glory of the human spirit is the sole aim of all science!"

K As a very young child this mathematician would sit, silently studying the walls of the nursery which were wallpapered with the pages of dad's calculus notes. Later as a teenager learning calculus the formulas would leap magically, or so it seemed to her tutor, into her mind.

L Napoleon, in jest, chided him about his *Mécanique Céleste*: "You have written this huge book on the system of the world without once mentioning the author of the universe." He replied, "Sire, I had no need of that hypothesis."

M With a strip of paper, one can find out how twisted he was.

N "If I have seen further than others it is because I have stood on the shoulders of giants."

O Fathered thirteen children and thousands of formulas; Arago described him thus: "____, who could have been called almost without metaphor, and certainly without hyperbole, analysis incarnate."

P Proved theorems about the cycloid to distract himself from a toothache; you know his triangle.

Q Designed and analyzed the first Belgian census, contributing to the founding of statistics.

R S is the set of all sentences that do not describe themselves. Is the previous sentence in S?

S Don't have a cow, man! Just approximate the definite integral.

T Real name Niccolo Fontana, this nickname refers to the stammer he acquired as the result of a dispute with a sword.

U "There is no royal road to geometry."

V He used consonants for constants and vowels for variables and established the use of letters as symbols for quantities in algebra.

W In college he majored in beer and fencing; later he introduced ε-δ proofs.

X If P is a point on a conic section, the ratio of the distance from P to the focus to the distance from P to the directrix is a constant.

Y Trained as a concert pianist, she collaborated with her husband on over 200 mathematical papers, articles, and books (all published under his name only) while home-schooling her six children (teaching each a different musical instrument) and completing the coursework for a medical degree.

Z He put the Z in ZFC.

Answers to the ABeCedarian History of Mathematics can be found below. ■

Answers to the ABeCedarian History of Mathematics

Archimedes, Bolyai, Cantor, Descartes, Erastothenes, Fermat, Germain and Gauss, Hardy, Einstein, Jacobi, Kovalevskaia, LaPlace, Möbius, Newton, Euler, Pascal, Quetelet, Russell, Simpson, Tartaglia, Euclid, Viète, Weierstrass, Eccentricity, Young, Zermelo.

Some Surprising Theorems About Rectangles in Triangles

MARTIN GARDNER

It is easy to construct the largest rectangle that will fit inside a given triangle. Simply draw lines from the midpoints of any two sides to meet the third side perpendicularly as shown in Figure 1. Rectangle *ABCD* will be the largest rectangle that will go inside the triangle. There are three such rectangles (not necessarily the same shape) in any acute triangle. Right triangles have two, and obtuse triangles have only one which rests on the side opposite the obtuse angle.

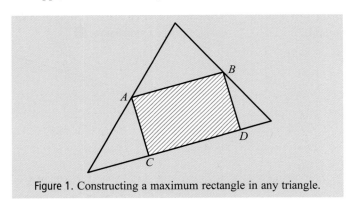

Figure 1. Constructing a maximum rectangle in any triangle.

It is not hard to show by calculus or algebra that the area of the maximum rectangle is exactly half the triangle's area, and the rectangle's base is half the triangle's base. This result goes back to Euclid's Book 6, Proposition 27. A neat way to demonstrate it is by paper folding. Cut the triangle from paper, then fold over the three corners as shown in Figure 2, folding along the rectangle's three interior sides. The flaps will fit snugly into the rectangle. This provides a simple way to construct a maximum rectangle by paper folding.

Paper folding also will demonstrate that no other inscribed rectangle can be larger. Try drawing one that is shorter and fatter, or taller and thinner. Fold over the corners as before by folding

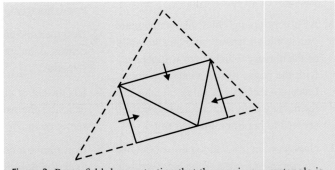

Figure 2. Paper fold demonstration that the maximum rectangle is half the triangle's area.

along the new rectangle's three sides. You'll find that the flaps either overlap each other, or overlap the rectangle, showing that its area is less than half that of the triangle.

A question now arises. Could there be a larger triangle entirely inside—that is, one that does not have a side resting on a side of the triangle? (See Figure 3.) The answer is no, although proving it is not so easy. You'll find proofs in the published references listed at the end of this article. An informal demonstration can be made by paper folding. Cut along the dotted line, then fold over the flaps by folding along the rectangle's

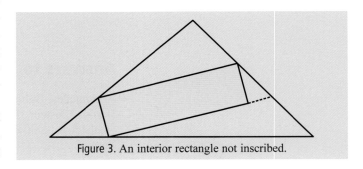

Figure 3. An interior rectangle not inscribed.

Reprinted from September 1997, pp. 18–22

sides. You'll find the flaps overlapping the rectangle, proving that their combined area exceeds the rectangle's area.

It is not possible, by the way, for any rectangle to be inscribed in any triangle without resting on one of the triangle's sides. Do you see why? A polygon is said to be "inscribed" in a larger polygon only if all its vertices are on the perimeter of the circumscribed polygon. Apply the pigeonhole principle. A rectangle has four corners but a triangle has only three sides, therefore two corners must touch the same side.

When we seek the largest square that will go inside any given triangle, the task becomes more interesting. Like maximum rectangles, such squares must rest on a side of the triangle. Call this the triangle's base. Let a represent the base and b the triangle's altitude. The formula giving the side of the maximum inscribed square is wonderfully simple:

$$\frac{ab}{a+b}$$

The late geometer Leon Bankoff, in a letter cited here as a reference, provided two algebraic proofs of this formula. The simpler one makes use of the triangle shown in Figure 4 with its largest inscribed square. Triangle Ade is similar to triangle ABC, so we can write:

$$\frac{b-x}{x} = \frac{b}{a}$$
$$xb = ab - ax$$
$$ax + xb = ab$$
$$x(a+b) = ab$$
$$x = \frac{ab}{a+b}$$

There are different ways to construct such maximum squares. One of the simplest is shown in Figure 5 with respect to right, acute, and obtuse triangles. Erect a square on the outside of the triangle's base, then draw lines from A and B to C. They will intersect the triangle's base at points that mark the base of the side of the largest square resting on that base. (Another way to construct such squares can be found in George Pólya's *How To Solve It*, 1945, Part 1, Problem 18.)

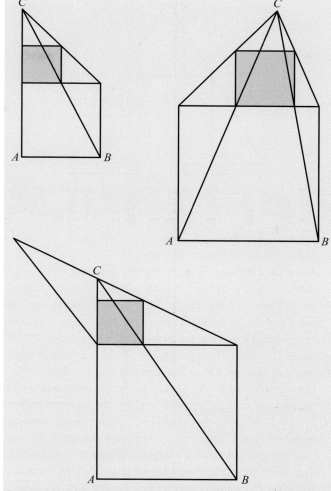

Figure 5. Constructing the maximum interior square on a triangle's side.

Unlike maximum rectangles in triangles, maximum squares that are on different sides of a triangle need not have the same area. On the right isosceles triangle in Figure 6, for example, the square shown is slightly larger than the largest square resting on the triangle's hypotenuse.

Figure 7 displays the largest square that goes inside an equilateral triangle of side 1. It is slightly smaller than the largest inscribed rectangle. On this triangle the three maximum squares

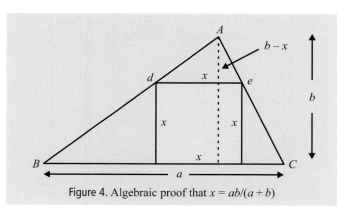

Figure 4. Algebraic proof that $x = ab/(a+b)$

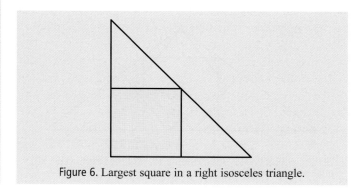

Figure 6. Largest square in a right isosceles triangle.

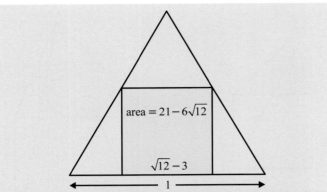

Figure 7. Dimensions of the largest square in an equilateral triangle of side 1.

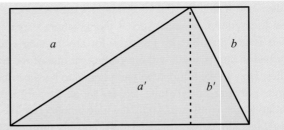

Figure 9. Proof that a maximum-area triangle inscribed in any rectangle has an area equal to half the rectangle's area.

on the triangle's three sides are, of course, equal. Is this the case only for the equilateral triangle?

Amazingly, the answer is no. There is just one other triangle for which this also is true. It is the obtuse isosceles triangle shown in Figure 8. I found it in *The Book of Numbers,* by John Conway and Richard Guy (1996), page 206, where it is credited to Eugenio Calabi, a geometer at the University of Pennsylvania. The ratio of base to side is 1.55138..., a number that is the positive value of x in the equation $2x^3 - 2x^2 - 3x + 2 = 0$.

The formula $ab/(a + b)$, applied to the largest square within the equilateral triangle of side 1, gives the square's side as $\sqrt{12} - 3 = 2\sqrt{3} - 3 = .46410161513...$. Note this number carefully. We will meet it again in a completely unexpected place.

How can one construct a triangle of largest area that will fit inside a given rectangle? Simply draw lines from the corners of the rectangle's base to any spot on the opposite side. There obviously is an uncountable infinity of such triangles. Like rectangles of the largest area in any triangle, the largest triangles in any rectangle are each half the rectangle's area. We can't demonstrate this by folding, but if you snip off the two small triangles at each side of a maximum triangle, they will fit neatly into the large triangle. Figure 9 shows how this can be proved geometrically. Triangles a and a', and triangles b and b', obviously are congruent.

Consider now the largest equilateral triangle that will fit within a unit square. Slightly larger than such a triangle resting on a square's base, it is shown in Figure 10. To be of largest

area its corners must touch all four sides of the square, but since a square has four sides, one of the triangle's corners must be at a corner of the square. An informal proof that this triangle cannot be enlarged is to imagine it rotated in either direction, keeping its lower left corner fixed and sliding the other two corners along the square's other two sides. Obviously this will make one side of the triangle shorter than the other two so the triangle will no longer be equilateral. I do not know if a formal proof by calculus would be easy or hard.

The side of the equilateral triangle is the secant of 15°, which is $\sqrt{6} - \sqrt{2} = 1.03527...$. We can now determine the triangle's area by finding its altitude ($\sqrt{6 - 3\sqrt{3}} = .8965...$) then halving the product of altitude and base, or by using Heron's formula for the area: $\sqrt{s(s-a)(s-b)(s-c)}$, where s is the semi-perimeter (half the perimeter of the triangle), and a, b, c are the triangle's sides. The area of the maximizing triangle turns out to be $2\sqrt{3} - 3$, the same as the *side* of the largest square that fits into an equilateral triangle of side 1! Has this been noticed before? I found it extremely mystifying until Princeton's John Conway came up with a beautiful proof that I will explain in a moment.

Abul Wefa (940–998), a Persian geometer, gave five different ways to construct this largest equilateral triangle with compass and straightedge. Three are reproduced in David Wells's *Penguin*

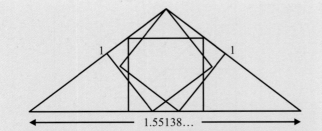

Figure 8. The only triangle, aside from the equilateral, with three equal maximum squares.

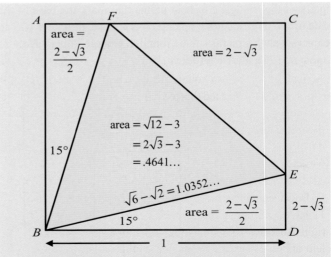

Figure 10. The largest equilateral triangle inside a unit square.

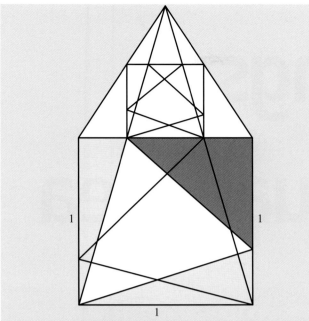

Figure 11. A pattern linking the largest square in an equilateral triangle of side 1 with the maximum equilateral triangle in a unit square.

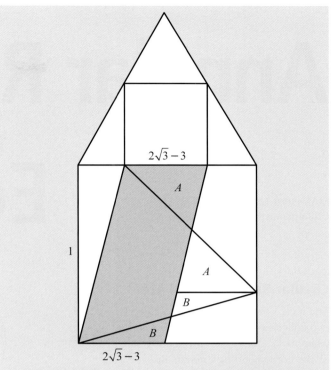

Figure 12. John Conway's dissection proof of a curious equality.

Book of Curious and Interesting Puzzles (1992, Puzzle 38). Wells refers readers to *Episodes in the Mathematics of Medieval Islam* (1986), by J. L. Berggren, a book I have not seen.

Henry Ernest Dudeney, in *Puzzles and Curious Problems* (1931, Problem 201) shows how to construct the triangle by folding a square sheet of paper.

An easy way to draw the triangle (I do not know if it is one of Wefa's methods) is shown in Figure 11. As science writer Barry Cipra discovered, the same lines that construct the largest square in the equilateral triangle also mark the points on the top of the larger square that are the top corners of the two maximum equilateral triangles that fit within the unit square. By drawing these two triangles we get a lovely bilaterally symmetrical pattern that shows how closely related the two constructions are.

I mentioned to Cipra the surprising fact that the area of the largest equilateral triangle inside a unit square exactly equals the side of the largest square inside a unit equilateral triangle. Cipra in turn mentioned this to Conway. Almost at once Conway saw how to prove the equality by a dissection. On the diagram of Figure 12 draw the parallelogram shown in blue. Its area clearly is 1 times $2\sqrt{3} - 3$, the side of the square at the top. By slicing the parallelogram into three pieces as shown, they will fit precisely into the equilateral triangle inscribed in the bottom square!

Because the narrow right triangles colored in gray in Figure 11 have a combined area exactly equal to the area of the blue right isosceles triangle, a pretty geometrical dissection problem arises. What is the smallest number of pieces into which the two gray triangles can be cut so they will fit without gaps or overlaps into the blue triangle? Clearly the number of pieces cannot be less than four because each gray triangle must be sliced at least once to fit inside the blue triangle. A solution in four pieces is given below.

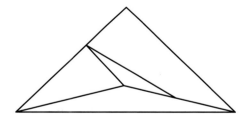

Are there other interesting properties or puzzles based on the pattern of Figure 11?

I wish to thank Don Albers and Peter Renz for many good suggestions that I have followed in writing this article. ∎

References

Bankoff, Leon. Letter in *Mathematics Teacher* 79, May 1986, 322.

Bird, M.T. "Maximum Rectangle Inscribed in a Triangle." *Mathematics Teacher* 24, May 1993, 759–760.

Embry-Wardrop, Mary. "An Old Max-Min Problem Revisited." *American Mathematical Monthly 97,* May 1990, 421–42

Lange, Lester H. "What is the Biggest Rectangle You Can Put Inside a Given Triangle?" *Mathematics Teacher* 24, May 1993, 237–240.

Niven, Ivan. *Maxima and Minima Without Calculus.* Mathematical Association of America, 1981.

Annular Rings of Equal Area

MAMIKON MNATSAKANIAN
California Institute of Technology

Circular Rings of Equal Area

The area S of a circular ring bounded by two concentric circles with radii r and R, where $0 < r < R$, is the difference in area of the corresponding circular disks: $S = \pi R^2 - \pi r^2$. In other words,

$$S = \pi a^2,$$

where $a^2 = R^2 - r^2$. Therefore, the area of the annular ring is equal to that of a circular disk of radius a, where $a = \sqrt{R^2 - r^2}$ (Figure 1).

This note reveals that equality of these areas is an intrinsic geometric property that can be deduced without using the formula for the area of a circular disk. Moreover, the property holds for many noncircular annular rings as well.

First we refer to Figure 2 to obtain a simple geometric meaning of the quantity a. Because $a^2 + r^2 = R^2$, the converse of the Pythagorean theorem tells us that the triangle in Figure 2 with edges a, r, and R is a right triangle. Therefore, a represents the length of a segment tangent to the inner circle and extending from the inner to the outer circle as shown in Figure 2.

In Figure 3a several segments of length a are shown tangent to the inner circle at different points. If we represent each tangent segment as a vector of length a and translate each of these vectors parallel to itself to a common initial point, their terminal points lie on a circle of radius a, as shown in Figure 3b.

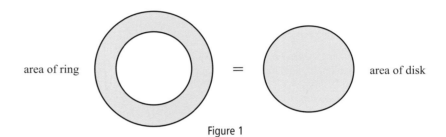

area of ring = area of disk

Figure 1

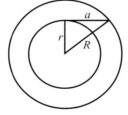

Figure 2

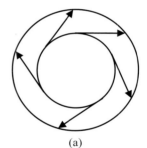

(a)

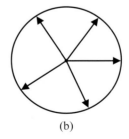

(b)

Figure 3

Now we see that as the point of tangency moves once around the original inner circle, the vectors in Figure 3b sweep out a circular disk of radius a. Figure 4 shows many circular rings swept out by tangent vectors of given length a. All these rings have equal area, each being equal to the area of the disk of radius a, which can be regarded as a degenerate case of a ring in which the radius of the inner circle shrinks to zero.

Areas of Oval Rings Generated by Smooth Closed Curves

The result of the foregoing section has a surprising generalization. Replace the inner circular disk by a convex plane region whose boundary is a simple closed curve that has a tangent line at each of its points. We refer to such a boundary as a *smooth closed curve*. Recall that a set of points is called *convex* if the line segment joining any two points of the set is also in the set.

Reprinted from November 1997, pp. 5–8

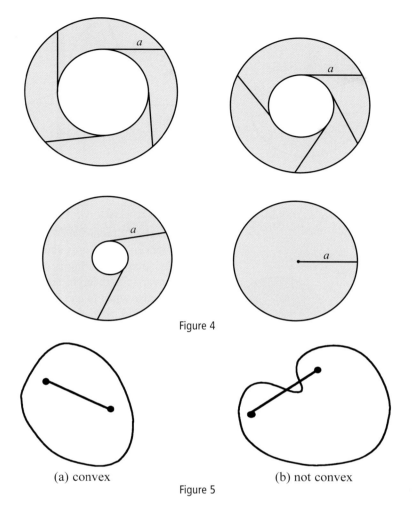

Figure 4

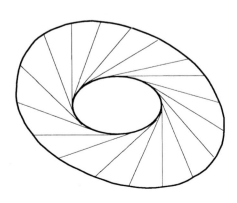

(a) convex   (b) not convex

Figure 5

Invariance theorem for areas of oval rings. *All oval rings swept out by a line segment of given length that is tangent to every point of a smooth closed curve have equal areas, regardless of the size or shape of the inner curve. Moreover, the area depends only on the length a of the tangent segment and is equal to πa^2, the area of a circular disk of radius a, as if the tangent segment was rotated about a single point.*

The proof employs the ancient Greek method of exhaustion, in which an arbitrary closed curve is approximated by inscribed polygons. Because polygons do not have well-defined tangent lines at their vertices, the foregoing definition of oval rings is not directly applicable. So we begin by formulating an appropriate definition for rings generated by polygons.

Areas of Oval Rings Generated by Polygons

Start with a triangle as shown in Figure 7a, and choose a segment of length a. Although the triangle is convex, its boundary is not smooth at the vertices and we need to formulate a special definition for the oval ring generated by this triangle. By rounding off the corner at each vertex, we can approximate the triangular boundary with a smooth curve. As the point of tangency moves in a given direction (clockwise, for example) around the 'smooth triangle' the tangent segment generates the oval ring shown in Figure 7b. As the rounded corners become sharper and sharper, the oval ring becomes more and more like that in Figure 7c, which is the union of three circular sec-

For example, the region in Figure 5a is convex, but that in Figure 5b is not convex.

Imagine a tangent segment of given length a placed at each point of a smooth closed curve. As the point of tangency moves once around the curve, the tangent segments sweep out a region that we call *an oval ring generated by the curve.* The examples in Figure 6 show that the region inside the outer boundary may or may not be convex, depending on the elongation of the inner curve.

What is the area of an oval ring generated in this manner? As in the case of circular rings, it is the area of the outer region minus the area of the inner region, but there is no simple formula for determining the areas of these more general regions. Nevertheless, we can prove the following remarkable theorem.

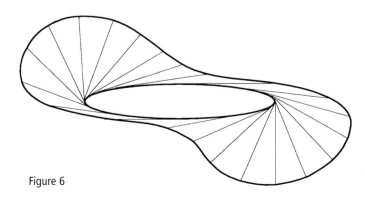

Figure 6

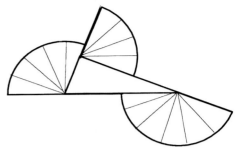

(a) triangle   (b) oval ring generated by 'smooth triangle'   (c) oval ring generated by triangle

Figure 7

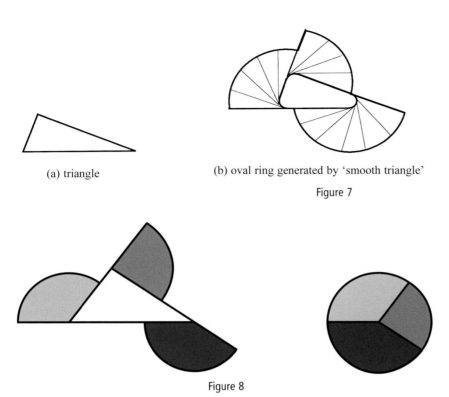

Figure 8

tors together with the boundary of the triangle. By definition, this is the oval ring generated by the triangle. A corresponding definition applies to any convex polygon.

Figure 8 shows how the three sectors in Figure 7c can be translated without rotation so that they fill out a circular disk of radius a without overlapping. Therefore the area of the ring generated by the triangle is equal to that of the circular disk, πa^2.

The same idea can be applied to a convex polygon with any number of sides. Figure 9 shows a ring generated by a six-sided polygon, and how the sectors can be pushed together without overlapping to form a circular disk of radius a. Again, the area of the ring is equal to that of the disk, πa^2.

Every smooth closed curve can be approximated to any degree of accuracy by polygons. As the number of sides increases and the length of each side decreases, the polygon looks more and more like the curve, as suggested by the example in Figure 10. The area of the oval ring generated by each approximating polygon is πa^2, so the limiting value is also πa^2. This proves the area invariance theorem for arbitrary oval rings.

A New Proof of the Pythagorean Theorem

The invariance theorem leads to a new proof of the Pythagorean theorem based on the fact that the area of a circular disk is a constant c times the square of its radius. This is a well-known property of similar figures. (We know that this constant is π, the ratio of circumference to diameter, but this fact is not needed for this proof.) Given any right triangle whose legs have lengths a and r and whose hypotenuse has length R, draw two concen-

tric circles of radii r and R, and a tangent segment of length a as shown in Figure 2. The area of the ring is $cR^2 - cr^2$, and the invariance theorem tells us that this area is also equal to ca^2. Therefore, $cR^2 - cr^2 = ca^2$, so $a^2 + r^2 = R^2$, which is the Pythagorean theorem.

Extensions of the Invariance Theorem

The invariance theorem can be generalized. First, the inner region of the ring need not be convex—its boundary can be any smooth simple closed curve. Second, the closed curve need not lie in a plane. It can be any smooth simple closed curve in 3-space. Just as in the plane case, a tangent segment of given length moving once around this curve will sweep out a surface in 3-space whose area depends only on the length of the segment and is proportional to the swept angle. Surfaces swept out by a family of line segments are called *developable* surfaces. Familiar examples are cylinders and cones. They

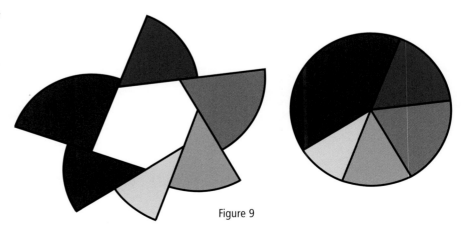

Figure 9

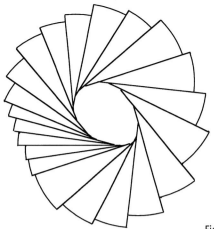 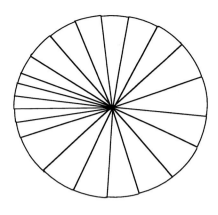

Figure 10

its points and extend it until it intersects the outermost circle, and let c denote its length. An example is shown in the diagram below.

Prove that the three segments so obtained can be arranged to form a right triangle. *Note:* The result is also true if the three circles are replaced by concentric spheres! Generalize the problem to four concentric circles (or four concentric spheres). ■

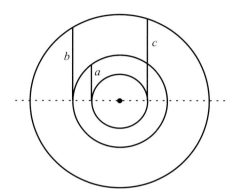

can be developed (unfolded or unrolled) onto a plane without distortion of length or angles. Consequently, internal properties of the surface such as area and arc length, are preserved during the unrolling process.

The invariance theorem on areas was discovered by the author in 1959 while he was an undergraduate at Yerevan University in Armenia. The first proof, using the methods of calculus, was published in the *Proceedings of the Armenian Academy of Sciences,* v. 73, 1981, pp. 97–102. The intuitive methods used in this note can also be applied to area problems in which the moving tangent segment changes its length as it moves along the curve. These methods form the basis of a new approach to calculus that was developed by the author during the years 1959–61. An English-language text based on these methods is currently being developed with the cooperation of Professor Tom M. Apostol under the auspices of *Project MATHEMATICS!*.

Exercise

Given three arbitrary concentric circles. Draw a segment tangent to the innermost circle at any of its points and extend it until it intersects the middle circle; let a denote the length of this tangent segment. Now draw a segment tangent to the middle circle at any of its points and extend it until it intersects the outer circle; let b denote the length of this second tangent segment. Finally, draw a segment tangent to the innermost circle at any of

Acknowledgements. This work was done under the auspices of *Project MATHEMATICS!* at the California Institute of Technology, with joint support from the Division of Physics, Mathematics and Astronomy, and National Science Foundation grant ESI 9553580. I express my deepest gratitude to Professor Tom M. Apostol, *Director of Project MATHEMATICS!*, for arranging this support and for his assistance and generous contributions in the preparation of this article. It is only at his insistence that his name is not included as co-author.

Some New Discoveries About 3×3 Magic Squares

MARTIN GARDNER

One afternoon, while sitting in a car waiting for my wife to finish supermarket shopping, I located a pencil and paper and drew on the sheet the unique *lo shu,* or magic square made with distinct digits 1 through 9.

2	7	6
9	5	1
4	3	8

Having nothing better to do, I decided to see what happens if instead of adding digits in each row and column I multiplied them. The top row's product (2 × 7× 6) is 84. The second row's product is 45, and the third row's product is 96. I added the three products to see if the sum was of any interest. It was 225, the square of the *lo shu*'s magic constant of 15. Would the sum of the column products be equally interesting? To my amazement, it also was 225. This equality obviously holds regardless of how the square is rotated or reflected. Could this be just a coincidence? What about the two main diagonals? Might their products also add to 225? No, the sum turned out to be 200.

Back home I began experimenting with other 3 × 3 magic squares having numbers not necessarily in sequence. In every case the sum of the products of rows equaled the sum of the products of columns.

Like so many discoveries in mathematics, far more significant than this, I had found a result by experimenting with numbers in a manner exactly like the way physical laws are discovered. The big difference, of course, is that mathematical conjectures, obtained experimentally, can be confirmed or falsified by rigorous proofs.

I recalled the following algebraic structure for all order-3 magic squares:

$a + b$	$a - b - c$	$a + c$
$a - b + c$	a	$a + b - c$
$a - c$	$a + b + c$	$a - b$

When distinct real values (including fractions, negative numbers, irrational roots, pi, e, and so on), even complex numbers, are substituted for *a, b,* and *c,* the result is always magic. The matrix provides a simple though tedious way to prove the conjecture that for all 3 × 3 magic squares the sum of row products must equal the sum of column products.

I thought I had stumbled on a new property of order-3 magic squares not noticed before, but Lee Sallows, a computer scientist in Holland, quickly disabused me. Hwa Suk Hahn, of West Georgia College, in Carrollton, Georgia, reported the result in "Another Property of Magic Squares" (*The Fibonacci Quarterly,* Vol. 73, 1975, pages 205–208). He called magic squares with the sum of row products equal to the sum of column products a "balanced square," and proved that all order-3 magic squares are balanced. The property was rediscovered by D. B. Eperson, who mentioned it in a brief note to *The Mathematical Gazette* (Vol. 79, July 1995, pages 182–83).

Hahn found similar squares of orders 4 and 5, but not of any higher order. Shown below are balanced squares of orders 4 and 5:

1	14	7	12
15	4	9	6
10	5	16	3
8	11	2	13

1	7	13	19	25
14	20	21	2	8
22	3	9	15	16
10	11	17	23	4
18	24	5	6	12

After reading a first draft of this article, John Robertson suggested adding here the following paragraph:

We have seen above that all order-3 magic squares are balanced while for any higher order, some squares are balanced and some are not. Hahn also found a more subtle

Reprinted from February 1998, pp. 11–13

difference between magic squares of order 4 and magic squares of higher order. He proved that if a constant is added to every element of any balanced order 4 magic square, the result is always another balanced magic square. If a constant is added to a balanced magic square of order greater than 4 the result will be a magic square, but it might or might not be balanced. If a constant is added to every element of the order 5 magic square in the diagram above, the result is always a balanced magic square. (For proof, note that you need only test five nonzero constants as the sum of the row or column products is a fifth-degree polynomial in the constant.) There is a known order 5 balanced magic square, with non-integral entries, that loses the property of being balanced when a certain constant is added. It would be of interest to find an order 5 balanced magic square with integral entries that produces a non-balanced magic square when some constant is added, or prove there are none.

A question arises. Are there order-3 magic squares for which the sum of the products of the two diagonals is also the same as the other two sums? I was unable to construct such a square.

In correspondence with Robertson, of Berwyn, PA, who is far more skilled than I in solving problems in number theory, I mentioned my discovery about the rows and columns, and wondered if diagonals could have the same magic sum. A few days later he sent a proof that there is an infinity of such squares.

Here's how to construct them. Start with any sequence of three square numbers in arithmetic progression, x^2, y^2, and z^2. For a in the algebraic matrix substitute the value $2y$. For b substitute the value of x, and for c substitute the value of z.

Let's take the simplest example: 1^2, 5^2, 7^2. The three squares, 1, 25, 49, are in arithmetic progression with a difference of 24. Within the matrix, then, a has a value of $2 \times 5 = 10$, b has a value of 1, and c has a value of 7. Lo and behold, like sorcery, the result is a magic square of the type we are seeking (shown on the right). The sum of the row products is 1,500. This is also the sum of the column products, and the sum of the diagonal products.

11	2	17
16	10	4
3	18	9

Figure 1 shows examples, provided by Robertson, of three higher squares with the same strange property. Above each square I show the arithmetic progression of square numbers which generates the square, and the values taken by a, b, c in the matrix. Below each square is its magic product-addition sum.

These magic squares—let's call them Robertson squares—are remarkable patterns. Not only are they magic in the tradional sense, they are also magic in an entirely different sense. As known to the ancients, the additive constant of any order-3 magic square is $3a$, or three times the square's central number. Robertson squares have in addition what we can call the multiplicative-additive constant. It is $a(2a^2 - b^2 - c^2)$, or more simply, $3a^2/2$.

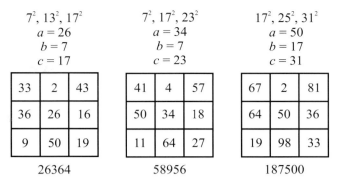

	$7^2, 13^2, 17^2$	
	$a = 26$	
	$b = 7$	
	$c = 17$	

33	2	43
36	26	16
9	50	19

26364

	$7^2, 17^2, 23^2$	
	$a = 34$	
	$b = 7$	
	$c = 23$	

41	4	57
50	34	18
11	64	27

58956

	$17^2, 25^2, 31^2$	
	$a = 50$	
	$b = 17$	
	$c = 31$	

67	2	81
64	50	36
19	98	33

187500

Figure 1. Three Robertson Squares

Here is a simplified account of how Robertson discovered the procedure for constructing such squares. From the algebraic matrix it is easy to determine that the sum of the row (or column) products is $a(3a^2 - 3b^2 - 3c^2)$. As mentioned above, the sum of the diagonal products is $a(2a^2 - b^2 - c^2)$. Setting these equal and simplifying yields $a^2 = 2(b^2 + c^2)$. Because a must be even, we can rewrite the equation as $2(a/2)^2 = b^2 + c^2$.

The last expression tells us that b^2, $(a/2)^2$, and c^2 are squares in arithmetical progression. There is an infinity of such triples, and well-known formulas for producing them. This makes it easy to construct an infinity of Robertson squares.

At the most, three squares can be in arithmetic progression. Such triples are closely related to Pythagorean triangles—right triangles with integral sides. The smallest square root of a number in the progression is the difference between the triangle's legs, the largest square root is the sum of the legs, and the middle square root is the Pythagorean triangle's hypotenuse.

Consider, for example, the familiar 3, 4, 5 Pythagorean triangle. The difference between its legs is 1, the hypotenuse is 5, and the sum of its legs is 7. The squares of these numbers, 1, 25, and 49, form the progression that generates the smallest Robertson square. The three squares shown in Figure 1 correspond to Pythagorean triangles with sides 5, 12, 13; 8, 15, 17; and 7, 24, 25. Thus from any given Pythagorean triangle you can construct a Robertson magic square.

A question remains. Are there order-3 squares other than the *lo shu* such that the sum of the row products equals the sum of the column products, and such that this sum also equals the square of the magic constant? (Diagonals are not considered.) Robertson has shown that the *lo shu* is the only 3×3 square in distinct positive integers that has this property. If zero is allowed, the square to the right meets the provisos.

14	0	10
4	8	12
6	16	2

The magic constant is 24. The sum of the row (or column) products is 576, and this equals 24^2. Similar squares exist if negative numbers are allowed. For example, to the right is such a square found by Sallows.

1	-4	3
2	0	-2
-3	4	-1

Its sum of row products, sum of column products, and sum of diagonal products in each case is zero. Zero also is the square of the magic constant. As Sallows pointed out, adding 5 to each number produces the *lo shu*.

Robertson has found a simple way to construct order-3 magic squares with distinct positive integers such that the sum of the row (or column) products is an integral multiple of the square of the magic constant. He is writing a note explaining the procedure.

An interesting property of order-3 magic squares, long known, is that the sum of the squares of the first row equals the sum of the squares of the third row. This is also true of course of the sums of the squares of the first and third columns, though the two sums are never the same.

Another curious and little known property of all order-3 magic squares, called to my attention by Monte Zerger, involves treating such squares as matrices. If such a square matrix is multiplied by itself the result is never another magic square. However, if the matrix is cubed, the result is always another magic square! For example, when the *lo shu* is multiplied twice by itself, the result is the following magic square with a constant of $15^3 = 3375$.

1053	1173	1149
1221	1125	1029
1101	1077	1197

For a proof of this theorem as well as other results based on magic matrices see N. Gauthier's paper "Singular Matrices Applied to 3×3 Magic Squares," in *The Mathematical Gazette* (Vol. 81, July 1997, pages 225–220).

Robertson adds:

For order-3 magic squares that are nonsingular when considered as matrices, it is known that all odd matrix powers are magic squares, including negative powers (Problem E3440, *American Mathematical Monthly,* proposed Vol. 98, No. 5, May 1991, page 437, solved Vol. 99, No. 10, December 1992, pages 966–967). Robertson has shown that an arbitrary matrix product of an odd number of order-3 magic squares is a magic square. It is straightforward to show this for a product of three squares, and then proceed by induction. He conjectures that both of the above results (inverse is magic, arbitrary products of odd numbers of squares are magic) are true of all magic squares of odd order. A proof or disproof would be of interest. ■

The Eccentricities of Actors

JOHN M. HARRIS and MICHAEL J. MOSSINGHOFF
Appalachian State University

Is Kevin Bacon the center of Hollywood? Apparently so, according to a popular game invented by three college students in 1993. The object of the game is to connect any actor or actress to Kevin Bacon through appearances in movies. For example, Sally Field can be linked to Kevin Bacon in just two steps: Field starred in *Forrest Gump* with Tom Hanks, who appeared with Bacon in *Apollo 13*. Field and Bacon have never appeared in a movie together, so two steps is the best we can do.

Let's define an actor's *Bacon number* as the fewest number of steps required to connect the actor to Kevin Bacon in this game. Thus, Sally Field has a Bacon number of 2. Steve Martin's Bacon number is 1, since they both appeared in *Planes, Trains, and Automobiles*, and Alfred Hitchcock can be linked in far fewer than *39 Steps*. Hitchcock appeared in *Torn Curtain* with Paul Newman, who co-starred in *The Color of Money* with Tom Cruise, who had the leading role in *A Few Good Men* with Bacon. Right now, there are no shorter paths from Hitchcock to Bacon, so Hitchcock's Bacon number is 3. Bacon himself is assigned a Bacon number of 0, and if someone cannot be linked to Bacon at all, we assign her or him a Bacon number of infinity.

In a popular version of the game called "Six Degrees of Kevin Bacon," the chains connecting actors to Bacon can have at most six links. This makes one wonder: does every actor that can be linked to Bacon at all have a Bacon number of 6 or less? Is Bacon really the center of Hollywood, or is the game easier if we pick another actor? To answer these questions, let's introduce some ideas from the field of graph theory.

Some Terms from Graph Theory

By a *graph* we mean a collection of *vertices* together with a collection of *edges*, where each edge connects two vertices. An example of a graph is *G* in Figure 1. *G* has eight vertices and ten edges.

The *distance* from a vertex v to a vertex w in a graph is the fewest number of edges required to travel from v to w. For example, the distance from e to g in *G* is 3, since three edges are required to travel from e to g: either ec, cd, dg, or ec, cb, bg. The *eccentricity* of a vertex v is the maximum distance from v to any other vertex in the graph. In the graph *G*, the eccentricity of e is 4, since the distance from e to h is 4 and the distance from e to any other vertex is 3 or less. The eccentricity of each vertex of *G* is shown in Table 1.

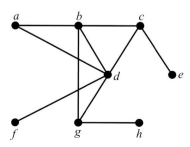

Figure 1. The graph *G*.

Vertex	Eccentricity	Farthest Vertices
a	3	e, h
b	2	e, f, h
c	3	h
d	2	e, h
e	4	h
f	3	e, h
g	3	e
h	4	e

Table 1. Eccentricities of vertices in *G*.

The *radius* of a graph is the smallest value of the eccentricity among all the vertices in the graph, and the *center* of a graph is the collection of vertices whose eccentricity is this minimal number. Among the vertices of *G*, b and d have the smallest eccentricity, 2. Thus the radius of *G* is 2 and the center of *G* is the set $\{b, d\}$.

Finally, the *diameter* of a graph is the largest value of the eccentricity among all the vertices, so the diameter of *G* is 4.

The Hollywood Graph

Let's create the Hollywood graph, *H*, by establishing a vertex for every actor that can be linked to Kevin Bacon, and connecting two vertices with an edge if the corresponding actors have appeared in a movie together. An actor's Bacon number then is just the distance from the actor's vertex to Bacon's vertex in the graph *H*. A small portion of *H* is shown in Figure 2. We have omitted the titles of the movies that make these connections—perhaps you can fill them in.

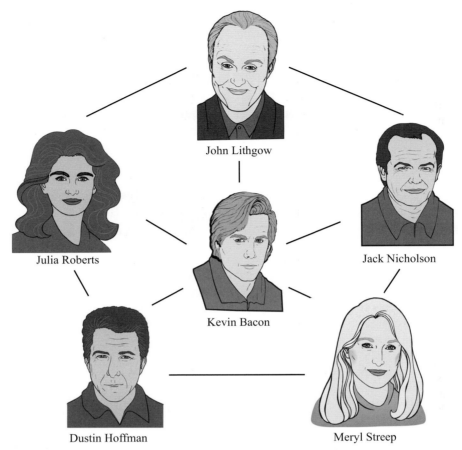

Figure 2. A portion of the Hollywood graph.

Of course, the graph H is constantly growing: every actor's debut adds a new vertex, and almost every movie released adds new edges. We will study the graph as it appeared on August 1, 1997. On this day, the Internet Movie Database [3] had records on the roles of more than 270000 actors. Of these, 240390 can be linked to Kevin Bacon. Brett Tjaden, a post-doctoral researcher in the computer science department at the University of Virginia, and Glenn Wasson, a doctoral student in computer science at Virginia, have calculated Bacon numbers for all 240390 of these actors. Their results are shown in Table 2.

We conclude that Bacon's eccentricity in the graph H is 7, and there are 57 actors that are guaranteed to stump your opponent in "Six Degrees of Kevin Bacon." What happens if we try linking to another actor, say Meryl Streep, or James Earl Jones, or Henry Fonda? Tjaden and Wasson find eccentricity 7 for each of these actors. Their data appear in Table 3.

In fact, Tjaden and Wasson find that no actors have eccentricity 6 or less, so the radius of H is 7, and Kevin Bacon, along with several thousand other actors, is indeed in the center of Hollywood. Can you determine an upper bound on the diameter of H?

It is interesting that Bacon only recently entered the center: three months earlier, on May 1, there were two actors with distance 8 from Bacon.

A web site run by Tjaden and Wasson [5] contains a wealth of information on Bacon numbers, connections between actors, and centers of Hollywood. On this site, you can request the Bacon number of any actor, and you can find a shortest path between any two actors. Tjaden and Wasson also publish a "Hall of Fame" in the Kevin

n	Kevin Bacon
0	1
1	1187
2	74075
3	133519
4	29313
5	1980
6	258
7	57

Table 2. Number of actors at distance n from Kevin Bacon.

n	Streep	Jones	Fonda
0	1	1	1
1	892	1865	2309
2	71685	102411	100700
3	142111	119344	121366
4	23955	15418	14672
5	1478	1159	1133
6	212	137	162
7	56	55	47

Table 3. Number of actors at distance n from Streep, Jones, and Fonda.

Bacon game on this web site: to be admitted, you have to find one of the actors with Bacon number 7.

Some Other Numbers

A similar construction helps us analyze other problems about connecting people. We give two examples.

First, construct a graph J where vertices are researchers and an edge joins two people if they have written a joint paper together. The Erdős number of a researcher is the distance from the person to the prolific Hungarian mathematician Paul Erdős in J. A web site [2], maintained by Jerry Grossman at Oakland University and Patrick Ion at *Mathematical Reviews*, keeps a record of all people with Erdős number 1 or 2. There are more than 470 people with Erdős number 1, and over 5000 with Erdős number 2.

Surprisingly, Paul Erdős has a finite Bacon number! Erdős was the subject of the 1993 documentary film *N is a Number*. The opening scene shows Erdős receiving an honorary doctorate from the University of Cambridge in 1991. Among the others shown receiving this award is British actor Alec Guinness, who appeared in *Lovesick* with Elizabeth McGovern, who co-starred in *She's Having a Baby* with Kevin Bacon. So Erdős's Bacon number is 3, while, to the best of our knowledge, Bacon's Erdős number is infinity.

Second, construct an "acquaintance graph," where every person is represented by a vertex and an edge joins two vertices if the corresponding people are acquainted with one another. The psychologist Stanley Milgram studied distances in this graph in some experiments in the late 1960's [4]. He gave a packet containing the name and some background information on one person (the target) to another person (the source) who lived far from the target. If the source knew the target, he or she sent the packet directly to that person. Otherwise, the source sent the packet to a personal acquaintance who seemed most likely to know the target, and the procedure repeated. Milgram found that the number of steps needed to reach the target varied between 2 and 10, with a median value of 5.

What would you guess the radius and diameter of the acquaintance graph to be? What is the distance from you to President Bill Clinton in this graph? To Michael Jordan? To Kevin Bacon?

Bacon Bits

Try your hand at the Kevin Bacon game!

Centers in Hollywood. Many great basketball players in Los Angeles have landed roles in movies. Try linking the following big men to Kevin Bacon in at most 3 steps: Wilt Chamberlain, Kareem Abdul-Jabbar, Vlade Divac, and Shaquille O'Neal.

Animals and Aliens. Connect these extraordinary stars to Kevin Bacon in 3 or fewer steps: Bugs Bunny, Lassie, Jaws, Chewbacca, Godzilla, and the entire bridge crew of the final season of *Star Trek: The Next Generation*.

Canadian Bacon. Link the following Canadian actors to Bacon (all have Bacon number 1 or 2): Raymond Burr, Jim Carrey, Margot Kidder, Martin Short, and Alex Trebek. ■

References

1. G. Chartrand and L. Lesniak, *Graphs & Digraphs*, Chapman & Hall, 1996.

2. J. W. Grossman and P. D. F. Ion, *The Erdős Number Project*, www.oakland.edu/enp.

3. *The Internet Movie Database*, www.imdb.com.

4. S. Milgram, The Small-World Problem, *Psychology Today* **22** (1967), 61–67.

5. B. C. Tjaden and G. S. Wasson, *The Oracle of Bacon at Virginia*, www.cs.virginia.edu/oracle.

Update

Since this article first appeared in 1998, the graphs described here have changed substantially. We add a few remarks on the three main graphs discussed in the article to summarize some of the changes that have occurred in the last seven years.

The Hollywood graph H has nearly tripled in size: the Internet Movie Database now tracks nearly 800,000 actors and actresses, more than 660,000 of whom can be linked to Kevin Bacon.

The radius of H has grown to 8, its diameter is presently 15, and Kevin Bacon remains in its center, together with more than 16,000 other actors.

The table to the right shows the current number of actors at distance n from Bacon in H (we thank Patrick Reynolds, who now maintains the Oracle of Bacon website, for his assistance in obtaining this data).

In addition, we remark that Alfred Hitchcock's Bacon number changed from

n	Kevin Bacon
0	1
1	1846
2	148221
3	406883
4	98030
5	7520
6	899
7	108
8	13

3 to 2 in 2000. Hitchcock's last film, *Family Plot*, included William Devane, who played a supporting role in *Hollow Man* with Kevin Bacon. Some might find this link to be a bit tenuous, since Hitchcock's requisite cameo in *Family Plot* consisted only of a view of his trademark silhouette, but we feel it is valid, since after all Kevin Bacon's character was invisible in most of *Hollow Man*.

The joint publication graph *J* has also changed considerably over the last few years. According to the Erdős Number Project, there are now 509 people with Erdős number 1, and nearly 7000 with Erdős number 2. The radius of (the large component of) *J* is presently 12, its diameter is 23, and Erdős's eccentricity is 13. MathSciNet (www.ams.org/mathscinet) now has a "Collaboration Distance" tool that allows one to determine the distance between any two researchers in *J*. In addition, it is interesting to note that Bruce Schechter, in his biography of Erdős, *My Brain is Open* (Touchstone, 1998), pointed out independently that Erdős has a finite Bacon number, though the path exhibited there does not link through Alec Guinness and has length 4.

Last, the acquaintance graph, first investigated by Stanley Milgram, has been the object of some recent studies. A recent experiment led by Columbia University sociologist Duncan Watts measured some qualities of this graph, with participants making connections by email. More than 60,000 people participated in this study, and the median number of links required varied between 5 and 7, depending on the physical distance between the source and target. See smallworld.columbia.edu for more details, or Watts' recent book, *Six Degrees of Separation: The Science of a Connected Age* (Norton, 2003). Also, another large experiment led by James Moody at Ohio State University is now underway; see smallworld.sociology.ohio-state.edu.

What's Left?

RICHARD K. GUY
University of Calgary

Your editor recently asked me: "After Fermat's Last Theorem, what's left?"

Paradoxically, even more than there was before. With most good mathematical problems, the more you solve, the more new problems are propagated.

What people are asking is: are there any more problems out there like Fermat's Last Theorem? What was it that made FLT particularly capture the imagination? Three things: it appears to involve only simple arithmetic; it held out so long against the efforts of generations of famous mathematicians; Fermat said that he had a proof. There aren't too many problems which have this third quality. Kempe thought that he'd proved the Four Color Theorem, but in that case he published his proof for all to see and it was only a dozen years before it was found to have a hole in it.

But there are plenty of problems which have been around as long as FLT and much longer and there are plenty which seem only to involve simple arithmetic. What is difficult to estimate is how much **mathematics** there is in them.

Only Simple Arithmetic?

Here are a couple of questions that probably don't contain any mathematics, and (perhaps for that reason?) seem unlikely to be answered in the foreseeable future, even though we 'know' that the answers are "No!"

$$2^{86} = 77371252455336267181195246$$

is a power of two which contains no zeros in its decimal representation. Are there any larger such powers?

If we start with a number, say 89, and reverse it, 98, and add, we get 187. If we reverse that, 781, and add, we get 968. Reverse and add again, 1837. Again, 9218. Again, 17347. Then 91718. Then 173437. After seventeen more steps, if we've both done our arithmetic correctly, we arrive at the **palindromic** number 8813200023188, i.e., a number which reads the same backwards and forwards. The same thing seems to happen with most numbers. But Sprague and Trigg and others have asked: if we start with 196, will we ever come to a palindromic number? John Walker, Tim Irvin and others have pushed this to millions of digits without finding one.

And here's a problem where we seem quite unable even to do the arithmetic! Lionel Levine suggested the array:

```
1 1
1 2
1 1 2
1 1 2 3
1 1 1 2 2 3 4
1 1 1 1 2 2 2 3 3 4 4 5 6 7
...   ...   ...   ...
```

where each row is found by reading the previous row from **right to left**, e.g., the last row is got by reading the previous row backwards: 'four 1s, three 2s, two 3s, two 4s, one 5, one 6, one 7'. Neil Sloane knows fifteen members (See update.) of the sequence of last numbers in each row:

$$1, 2, 2, 3, 4, 7, 14, 42, 213, 2837,$$
$$175450, 139759600, 6837625106787,$$
$$266437144916648607844,$$
$$508009471379488821444261986503540,$$

and wonders if anyone will ever calculate the 20th term.

Don't Try This at Home!

These problems don't seem to have much mathematics in them (though it's very difficult to tell such things), but there are others where at least one new good idea is needed. There's the infamous $3x+1$ problem: if a number is odd, multiply by 3 and add 1; if it's even, divide by two; do we always get to 1?

$$7, 22, 11, 34, 17, 52, 26, 13, 40, 20, 10, 5, 16, 8, 4, 2, 1$$

Half a century ago, Erdős said, "Mathematics may not yet be ready for such problems." He might have said the same about FLT, but mathematics was rapidly in the process of getting ready!

Here's a very similar problem, perhaps even more intriguing. Conway has looked at the permutation

$$3n \Leftrightarrow 2n \qquad 3n \pm 1 \Leftrightarrow 4n \pm 1$$

which is neater in that you can go either way. Forwards: if it divides by 3, take off a third; if it doesn't, add a third (to the nearest whole number). Backwards: if it's even, add 50%; if it's odd, take off a quarter. If we start with 1 we get 1, 1, 1, 1, 1, …; if we start with 2 or 3 we get 2, 3, 2, 3, 2, 3, …; and if we start with 4 we get 4, 5, 7, 9, 6, 4, 5, 7, 9, …. And try starting with 44. But if we start with 8:

$$8 \to 11 \to 15 \to 10 \to 13 \to 17 \to 23 \to 31 \to 41 \to 55 \to 73 \to 97 \to \ldots$$

or, if we go backwards from 8:

$$\ldots \leftarrow 57 \leftarrow 38 \leftarrow 51 \leftarrow 34 \leftarrow 45 \leftarrow 30 \leftarrow 20 \leftarrow 27 \leftarrow 18 \leftarrow 12 \leftarrow 8$$

the sequences never seem to finish. We have the curious paradox that when we go 'forwards', a third of the time we multiply by 2/3 and two-thirds of the time we multiply by (roughly) 4/3; on the average, in three steps we multiply by 32/27. But when we go 'backwards', half the time we multiply by 3/2 and half the time by (roughly) 3/4; on average two steps multiply by 9/8. We increase either way! This is certainly borne out by experiment. Note that, when going backwards, each successor to an even number is a multiple of 3 — half the numbers are multiples of three! Does the chain with 8 in it close up to form a loop or is it infinite in each direction? The first number we haven't seen is 14. Does it form a different chain? Are 40, 64, 80, 82, 104, 136, 172, 184, …, each in different chains or loops? Are there infinitely many such chains? Are there any more loops? Again, there are many more questions than answers.

If you want an old problem, then perfect numbers go back to the ancient Greeks: $6 = 1 + 2 + 3$ and $28 = 1 + 2 + 4 + 7 + 14$ are the sum of their 'aliquot parts', the divisors other than the number itself. Euclid discovered a rule for giving even perfect numbers, that $2^{p-1}(2^p - 1)$ is perfect when $2^p - 1$ is prime, and Euler showed that these are the only ones. Are there any odd perfect numbers? Are there infinitely many perfect numbers?

In fact the Greeks classified numbers as deficient, perfect or abundant, according as the sum of the aliquot parts is less than, equal to, or greater than, the number. Another problem that goes back a century and a half is that of 'aliquot sequences'. If we start with a number, and sum its aliquot parts, then sum the aliquot parts of the result, and continue, where do we get to? If you try small numbers you'll soon finish up with 1, or a perfect number, or perhaps an 'amicable pair', such as 220 and 284, each the sum of the aliquot parts of the other. (Are there infinitely many such pairs?) Catalan and Dickson thought that this always happened. But John Selfridge and I believe that infinitely many numbers, perhaps almost all even numbers, give sequences which go to infinity. The smallest number for which there is doubt is 276, which leads to the sequence:

276, 396, 696, 1104, 1872, 3770, 3790, 3050, 2716, 2772, …,

which, after the efforts of generations of calculators, Catalan, Dickson, Poulet, the Lehmers, Godwin, Selfridge, Wunderlich, Devitt, Struppeck, Wolfram, Dickerman, Peter Montgomery, te Riele, Bosma, Creyaufmüller and many others that I've forgotten, is known to have terms with more than 100 digits. The trouble with aliquot sequences is that they plunge up and down erratically, and there seems to be no way of telling where they'll be a thousand terms from now.

Addition and Multiplication Don't Mix!

For the last few questions we need to know just a bit more than simple arithmetic. We need to know what a prime number is. And when you start mixing multiplication and addition, we soon find some very hard problems indeed.

Goldbach conjectured, not so very long after Fermat's time, that every even number is the sum of two primes (he counted 1 as a prime). The nearest we have got to settling this is Vinogradov's *tour de force* that every sufficiently large odd number is the sum of three primes. Even so, 'sufficiently large' is well out of computer range until someone has a new idea. Matti Sinisalo has checked the conjecture up to 4×10^{11} and Chen Jing-Run & Wang Tian-Ze have shown that Vinogradov's theorem is true for odd numbers exceeding $e^{114} \approx 3.23274 \times 10^{49}$, under the assumption of the generalized Riemann hypothesis (and there's another fairly old, very famous, and seemingly impossibly hard problem for you).

And is every even number the difference of two primes? In infinitely many ways? Take the difference 2, for example: are there infinitely many twin primes: (3,5), (5,7), (11,13), (17,19), (29,31), …? Of course there are, but who is going to prove it? Are there infinitely many sets of primes in more general patterns, such as $p - 4$, p and $p + 2$; or $p - 2$, p and $p + 4$? (Of course, $p - 2$, p and $p + 2$ can't all three be prime, unless they're 3, 5 and 7.)

Can you find arbitrarily many primes in arithmetic progression? In 1993 Paul Pritchard coordinated an effort which discovered 22 primes in arithmetic progression, starting with $a = 11410337850553$ and having common difference $d = 4609098694200$. The last term is $a + 21d = 108201410428753$. Another example, which starts later, but finishes earlier, is given by $a = 28383220937263$, $d = 1861263814410$. (See update.)

Are there in fact arbitrarily many **consecutive** primes in arithmetic progression? In November 1997 Harvey Dubner, Tony Forbes, Nik Lygeros, Michel Mizony & Paul Zimmermann joined forces to find 8 consecutive primes in A.P., $p_0 = mn + x$ and $p_k = p_0 + 210k$, $1 \le k \le 7$, where m is the product of the primes up to 193, $n = 220162401748731$ and

$$x = 19131958978991812690857851677439676683450 - 9691048718767292658692385206295221291.$$

STOP PRESS! They have since found not only 9 but 10 such primes! It will be a while before anyone finds 11.

Sums of Two Powers

It was another of Fermat's theorems that paved the way to finding numbers which are the sum of two squares in many different ways. If you form the product of $k+1$ different primes of shape $4n+1$, it will be the sum of two squares in 2^k ways:

$$5 \cdot 13 \cdot 17 = 1105 = 4^2 + 33^2 = 9^2 + 32^2 = 12^2 + 31^2 = 23^2 + 24^2.$$

Illustration by John Johnson of Teapot Graphics

For sums of two cubes there's the famous Hardy-Ramanujan taxicab number: the smallest number expressible as the sum of two cubes in two different ways is $1729 = 1^3 + 12^3 = 9^3 + 10^3$, known to Frénicle de Bessy 340 years ago. Three hundred years later Leech found the smallest number expressible as the sum of two cubes in three different ways:

$$87539319 = 167^3 + 436^3 = 228^3 + 423^3 = 255^3 + 414^3$$

and six years ago Rosenstiel, Dardis & Rosenstiel found

$$
\begin{aligned}
6963472309248 &= 2321^3 + 19083^3 \\
&= 5436^3 + 18948^3 \\
&= 10200^3 + 18072^3 \\
&= 13322^3 + 15530^3.
\end{aligned}
$$

It's known that there are numbers expressible as the sum of two cubes in as many ways as you like, but the least such for five ways isn't yet known. Euler knew that

$$635318657 = 133^4 + 134^4 = 59^4 + 158^4$$

but no-one knows of three equal sums of fourth powers.

Are there equal sums of two fifth powers?

The abc-Conjecture

Perhaps the nearest problem to FLT is that of settling the *abc*-conjecture. An important step in the proof of FLT was Frey's elliptic curve,

$$y^2 = x(x - A)(x + B).$$

The roots of the cubic are 0, A, $-B$, and their absolute differences are A, B and $A + B = C$, say, so that the discriminant is $A^2 B^2 C^2$. Put $A = a^p$, $B = b^p$, $C = c^p$ so that

$$a^p + b^p = c^p$$

and the discriminant is $a^{2p} b^{2p} c^{2p}$. The trick was to show that you couldn't have such an elliptic curve with comparatively small prime factors in its **conductor**, a number which has the same prime factors as the discriminant.

Roughly stated, the *abc*-conjecture says that if $A + B = C$ with A, B, C having no common factor, then some of the prime factors of ABC must be 'fairly large.' Define the **radical** R as the product of all the different primes which divide A, B or C and the **power** P by $C = R^P$, then the conjecture is that for a given $\eta > 1$, there are only finitely many triples (A, B, C) with $P > \eta$. The record $P = 1.62991$ was found by Reyssat with

$$2 + 3^{10} \cdot 109 = 23^5$$

where $R = 2 \cdot 3 \cdot 23 \cdot 109$. This is closely followed by de Weger's

$$11^2 + 3^2 \cdot 5^6 \cdot 7^3 = 2^{21} \cdot 23$$

where $R = 2 \cdot 3 \cdot 5 \cdot 7 \cdot 11 \cdot 23$, the product of the primes which are one less than the divisors of 24, and for which $P = 1.62599$.

The Catalan relation, $1 + 2^3 = 3^2$, does comparatively poorly with $P = 1.22629$; although 2 and 3 are small primes, they're not very small compared with 9. The Catalan conjecture, that 8 and 9 are the only powers differing by 1, has been settled by Tijdeman for 'sufficiently large' numbers, but once again, completion of this is well beyond computer range until a new idea comes up. (See update.)

A particular case of the *abc*-conjecture is the Beal conjecture, which you read about in *Math Horizons* in the February 1998 issue.

So there's plenty left: and that's only a corner of number theory, and then there's the whole of mathematics—which is increasing every day! ■

Reference

Richard K. Guy, *Unsolved Problems in Number Theory*, 2nd edition, Springer-Verlag, 1994.

Update

The Online Encyclopedia of Integer Sequences (www.research.att.com/~njas/sequences) now lists seventeen terms of Levine's sequence. The Catalan conjecture is now settled: Preda Mihăilescu, "A class number free criterion for Catalan's conjecture," *J. Number Theory*, 2003. And Ben Green and Terence Tao recently showed that one can find arbitrarily many primes in arithmetic progression!

Egyptian Rope, Japanese Paper, and High School Math

DAVID GALE
University of California, Berkeley

Many of us are no doubt familiar with the story of how the ancient Egyptians used a knotted rope to measure right angles. The rope was divided by knots into 12 equal segments and then stretched out to form a triangle with sides 3, 4, 5. The angle between the two shorter sides gave the desired right angle which, it is said, was then used for laying out the boundaries of fields along the Nile. There is apparently some doubt among mathematical historians as to whether this story is fact or myth. Either way, it is the only case I know of where the 3:4:5 triangle shows up in an application.

Until now, this is, for we wish to report that, after some four thousand years, Egyptian triangles are back in the news and in a rather surprising way.

In 1994 Kazuo Haga, former professor of Tsukuba University in Tokyo proposed a subject he called Origamics. The name is derived from the conjunction of Origami (the well-known art of making figures by folding paper) and Mathematics. The idea was to use paper folding not for making birds and flowers but rather as a tool for teaching mathematical thinking in general and geometry in particular. Here is how Professor Haga himself puts it.

"Origamics study is valuable for high school students. It helps develop mathematical and scientific thinking, and it provides opportunities for them to use a computer with a sense of purpose.

"When we fold a right-angled piece of paper, square or rectangle, we find intensely interesting mathematical phenomena resulting from the creases and the locations of intersection points of the folded piece. These phenomena give students feelings of wonder and interest which bring out their interest in solving associated equations. Pieces of paper are abundant and easily obtained at a low cost. Students get experimental data or hints to resolve problems hidden in the phenomena. When they try to fold a narrow part of paper, students often face a mechanical limit of accurate folding by hand, and in these cases, they make use of a personal computer with mathematical software such as

Cabri Geometry, The Geometer's Sketchpad, and *Turtle Graphics* to acquire further data. Through Origamics study, students are able to grow in scientific literacy and develop a pleasurable attitude toward mathematics."

The idea of doing geometric constructions via paper folding is not original with Haga. Indeed the subject has been formalized and it has been shown that not only are all the Euclidean (ruler and compass) constructions possible but also those constructions which require solving cubic and quartic equations, such as angle trisection and constructing a regular heptagon. Nevertheless, Origamics seems to have made a novel contribution by coming up with very simple examples requiring only one fold of the paper which produce unexpected results. We will present here what origamicists call Haga's first, second, and third theorems and pay particular attention first, to the frequent and unexpected appearance of Egyptian triangles. Equally striking is the relevance of this material for high school mathematics for, as we shall see, the ideas needed for "solving questions" that arise in the folding activity turn out to be exactly those that students have been drilled on over the years in traditional first-year algebra and geometry courses.

To fix notation, a sheet of origami paper will be the unit square with vertices labeled *A, B, C, D* as shown in Figure 1, and *M* will be the midpoint of side *AB*.

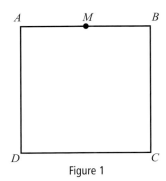

Figure 1

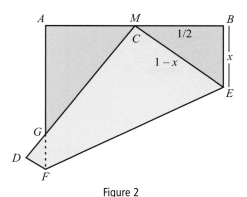

Figure 2

Haga's First Theorem

Instructions: Make a fold in the paper so that vertex C coincides with midpoint M. See Figure 2.

One way of stating the theorem is to say that all three of the triangles $\triangle MAG$, $\triangle EBM$, and $\triangle FDG$ are Egyptian. Perhaps a nicer way to give the result is as a problem:

Given that $AM = MB = 1/2$, find the lengths of all of the boundary segments, BE, EC, CG, GD, DF, FG, GA.

It would not be appropriate here to give the solution of this quite elementary problem, (that is, to give a proof of Haga's first theorem) but, as mentioned, it has an interesting feature relating to its use as a teaching tool for high school students. Indeed, it is an example of a high school geometry problem which is solved by high school algebra. It is in fact a word problem, where one analyzes the given information and then decides to let x equal something. In this case the appropriate something is the length of the interval BE, as shown in the figure because then CE has length $1 - x$. The Pythagorean Theorem then allows one to find the sides of $\triangle MBE$ by solving a (high school) quadratic equation. The other two triangles are similar to $\triangle MBE$, because (high school geometry) their corresponding sides are perpendicular, which allows one to complete all of the calculations.

Haga's Second Theorem

Here the instructions are even simpler: fold the paper along the line MC. If one does this and extends the segments CB and MB as shown in Figure 3 one again gets three triangles, $\triangle EDC$, $\triangle MAF$, and $\triangle EBF$, all of which turn out to be Egyptian.

Again the problem is to find the lengths of the boundary segments AE, EF, FD as well as BE and BF, given that $BC = 1$ and $BM = 1/2$.

The trick here is to find the coordinates of the point B which leads one to draw the perpendicular through B to sides AM, DA, and DC as shown in Figure 4.

The figure now contains seven Egyptian triangles! Furthermore, all seven are of different sizes, proportional to $3:4:5:6:10:12:15$. To prove all this we go back to high school and let x equal BK, so BG is $1-x$. Compute KC by Pythagoras

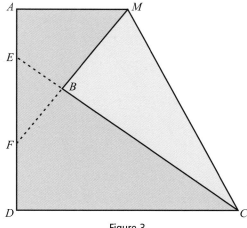

Figure 3

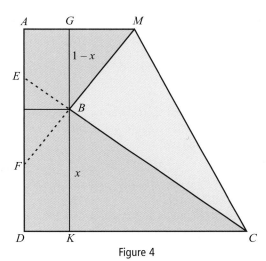

Figure 4

and observe that $\triangle BKC$ is twice $\triangle MGB$ (their hypotenuses are 1 and 1/2 respectively), solve the resulting quadratic equation, and everything drops out by similar triangles.

Finally we come to,

Haga's Third Theorem

This one requires a little more digital dexterity. Move vertex C to side DA so that side BC passes through the point M as shown in Figure 5.

Once again the triangles $\triangle MAG$, $\triangle GDF$, and $\triangle MBE$ are Egyptian, and this time we will go through the proof which, to exaggerate slightly, seems to encapsulate all of the traditional 9th and 10th grade mathematics in a single problem. For explicitness we use italics for each of the key concepts as they occur.

Let DF equal x and note that FC equals $1-x$.

Let DG equal y and note that GA equals $1-y$.

From the *Pythagorean Theorem* $x^2 + y^2 = (1 - x)^2$, and *simplifying* gives,

(1) $$y^2 = 1 - 2x$$

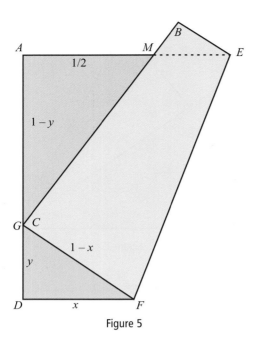

Figure 5

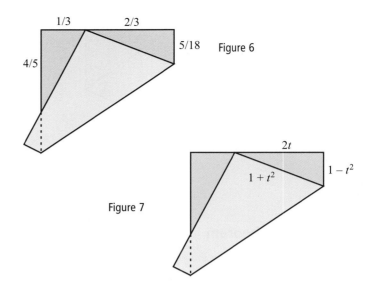

Figure 6

Figure 7

Next, angles $\angle DFG$ and $\angle AGM$ are equal because both are complementary to $\angle DGF$, so $\triangle AGM$ is similar to $\triangle DFG$. Therefore $y/x = (1/2)/(1-y)$, so

$$(2) \qquad x = 2y - 2y^2$$

and we must solve the *simultaneous quadratic equations* (1) and (2) giving, $y = 1/3$, $x = 4/9$.

Finally, since $\triangle MAG$ is Egyptian, $MG = 5/6$ so $\triangle MBE$ is Egyptian with sides $1/6$, $2/9$, $5/18$.

Note that the sizes of the three triangles are proportional to $1:2:3$.

The Mystery Resolved

All this is, of course, extremely elementary and I doubt whether many readers will have taken the trouble to do the simple calculations involved. Nevertheless, there remains a mystery. In the course of the three "theorems" we have encountered no less than thirteen Egyptian triangles. The triangle with sides $1/2$, $2/3$, $5/6$ occurs in all three but the other eleven triangles are all different. Is there some mathematical explanation of why these triangles keep appearing or is this just a mathematical "coincidence"? A little thought provides the explanation. We observe that our folding operation corresponds to drawing lines which are determined by points with rational coordinates. If we take vertex A as origin the fold in Haga's second theorem is the line through $(1/2, 0)$ and $(1, 1)$.

In Haga's first theorem it is the perpendicular bisector of this line. The new points we locate are the intersections of these lines with the boundaries of the square and are therefore points with rational coordinates. This means that if the hypotenuse of a triangle is part of the boundary of the square then it must have a rational length, as will the vertical and horizontal sides. Thus,

we will always get triangles with rational sides, and the Egyptian triangle is the simplest of these, but one can also easily get others. For example, to get $5:12:13$ triangles one folds so that vertex C falls on the point $(1/3,0)$ as shown in Figure 6.

In fact if the $2/3$ above is replaced by an arbitrary length $2t$ then we get the general triangle with sides proportional to $2t$, $1 - t^2$, $1 + t^2$. Thus, if t is rational we can divide out common factors and get all of the primitive Pythagorean triples. More precisely, from the standard theory of Pythagorean triples, choosing $t = m/n$ where $(m, n) = 1$ and m and n have opposite parities gives all the primitive triples.

Perhaps the most interesting feature of this whole exercise, aside from the tie in with high school math, is the way in which a result usually thought of as part of pure number theory gets translated into geometry. It would certainly seem rather pointless to try to draw the triangles corresponding to the primitive triples, but now, thanks to Haga and company, if one really wants to see them one can get an instant picture with a single fold of a sheet of paper.* ∎

---

* Of course, we must also be able to construct the points m/n, but for this there is an easy inductive procedure. The reader is asked to prove that if $AC = 1/n$ then $AB = 1/(n+1)$. Of course, all the lines in the figure can be obtained by folding.

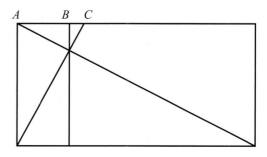

Art Benjamin—

Mathemagician

DONALD J. ALBERS
Mathematical Association of America

Mathematics professor Arthur (Art) Benjamin, 37, is a friendly person, a popular teacher, happily married, and soon to be a father. But as a child growing up in Cleveland, Ohio, all bets were off about his future. Little Art Benjamin was so rambunctious in his first nursery school that he was thrown out! Ditto for his second nursery school, and the third, ... "At nap time, all of the students except me would lie down on their blankets," he recalls. "I would be running around the room. I was hyperactive. A lot of the problem may have stemmed from the fact that I was bored with many things. I usually understood things the first time they were explained, and by the time something was being taught the fourth or fifth time, I'd rather get silly than pay attention." During the first five years of life, his hyperactivity resulted in several periods of hospitalization, often for weeks at a time. In today's language (and less understood during Art's childhood), his problem would be called ADHD, attention deficit hyperactivity disorder. He was prescribed the medication Valium, which he took for the next ten years. When Art was finally taken off Valium in the eighth grade, he promptly went out for the track team. He had always liked sports, but the Valium had for years slowed him down.

As a child, Benjamin remembers wanting a lot of attention and doing lots of things to get it. "I took up magic as a hobby, I sang, I danced, I got good at different games, and I learned to calculate quickly. When I was five, I memorized the states and their capitals in alphabetical order. I even learned the presidents in order."

Today, Professor Benjamin certainly is not lacking for attention. For starters, he gets lots of it from his students at Harvey Mudd College in Claremont, California. He is a popular teacher with a knack for involving students in his subject. He has written several research papers, including many co-authored with his students, and he is the Editor of the Spectrum Book series that is published by the Mathematical Association of America. During his spare time, he performs on numerous stages as Art Benjamin—Mathemagician. He is a regular at Hollywood's Magic Castle, the leading club for magicians in the world. He has appeared on numerous television shows in the US, England,

Could you kick this sweet little boy out of nursery school?

Canada, and Japan, including The Today Show, Evening Magazine, Square One, Live! Dick Clark Presents, and CNN Headline News. He also has given hundreds of performances for school children, high school and college students, adults, and even mathematicians.

As a premier "mathemagician," he regularly dazzles audiences with his calculating feats. He begins a typical performance by asking members of the audience with calculators to assist him. In order to test the accuracy of the calculators, members of the audience will call out two numbers for him to multiply, for example, 68 and 92. "Make sure you get 6,256," Benjamin says, before his assistants have even entered the numbers. The race between Benjamin and the machines is on. He then challenges them in squaring two-digit numbers, for example, 47, 59, 63, 89. He beats the machines every time with ease. Benjamin then moves on to squaring three-digit and four-digit numbers, and again the calculators lose. By this time, most members of the audience are thoroughly awed by his feats of lightning calculation.

He Likes All Numbers

Benjamin's interest in numbers, calculation, and number patterns goes back to his early childhood. He says that he's never met a number that he hasn't liked. One of his first toys resembled a slot machine. When he pressed the lever of the machine, four arithmetic problems appeared in each of four windows — one in addition, one in subtraction, one in multiplication, and one in division. He recalls learning his multiplication tables with that toy. By the time he was in third grade, he was discovering numerical properties resulting from multiplying two-digit numbers. "One day in school my classmates and I were in the hallway ready to go to lunch. I remember standing up against the wall daydreaming about numbers. I was marveling over the fact that 2,520 is the smallest number that all numbers between one and ten divide into perfectly. I tried a smaller number, like 1,260, and realized that eight did not divide into it. I then wondered what would be the smallest number that one through eleven would go into, and figured out that would probably be 2,520 times 11."

A few years later he hung a velcro dartboard on the wall at the end of his bed. The darts were actually velcro balls and the object of the game was to throw them at the numbered regions, 1 through 10, on the dartboard. "Sometimes I would lie awake at night staring at the dartboard. I would see the numbers, and I would multiply them together, getting various combinations as large as ten factorial (3,628,800)."

One day he took a bus into downtown Cleveland to visit his father at work. To pass the time, he started thinking about numbers that add up to 20. He started at the center with 10 and 10, and then formed the product 10×10 to get 100; 9×11 was 99; 8×12 was 96; 7×13 was 91. He then noticed that 9×11 was one less than 100; 8×12 was 4 less; 7×13 was 9 less. The pattern was 1, 4, 9; he guessed that 16 and 25 would come next. "Oh wow," he said to himself, "I wonder if this works for other numbers, and then I tried it for big numbers, small numbers, fractions. It worked!" With elementary algebra, it is easy to verify his discovery, but for a young boy with no knowledge of algebra, it was a big accomplishment.

In middle school, Benjamin's interest in games blossomed, and he started to play both chess and backgammon in a serious way. When he started college, he gave up chess in favor of backgammon because it

The Raisin

Once upon a day quite cheery,
Far from lusterless Lake Erie,
On a beach somewhere southwest of the City of Singapore,
'Twas a grape that had a notion.
If he'd rest close to the ocean,
He would tan without his lotion,
Just by resting near the shore,
And he'd meet up with adventures that he never dreamt before,
All of this and much, much more.

All of this had happened one day.
It was on a sunny Monday
As he soaked up every sun ray going into every pore.
Not a gust of wind was breezing,
And the warmth was, oh, so easing
And so beautifully pleasing.
He was filled with joy galore.
"Oh, how I wish," he said, "that I could stay here on the shore,
Remaining here forever more."

Well, that grape who had that notion,
Slowly motioned to the ocean.
He then discovered something which filled him with much gore.
For he saw by his reflection,
Nature made a small correction.
He was further from perfection,
Meaning worse off than before.
He had changed into a raisin while he rested on the shore,
And that he'd be forever more.

Well, one day a man in yellow
Gazed down at the little fellow
And looked at him and other raisins resting on the shore.
For you see his occupation
Was to go to this location,
Meaning that it was his station
To pick raisins off the shore.
So he knelt down towards that raisin on the beach near Singapore,
Picked him up with many more.

All of them were squished together
In a packet made of leather
In a factory southwest of that city, Singapore.
And that raisin was a snooper
Then he saw a spacious, super,
Massive, mammoth, monstrous scooper,
Scooping raisins by the score.
Packaged them in Raisin Bran and sold them to the store,
Scooped him up with many more.

And that raisin now is well aware
He's in a bowl in Delaware
Ready to be eaten by a child not yet four.
And the milk was slowly dropping
And the raisin heard it plopping,
Then the raisin heard it stopping
For the milk had ceased to pour.
Au revoir sweet life, he cried, life which I truly do adore,
Quoth the raisin never more.

Benjamin wrote "The Raisin" in his freshman year of high school. It is a parody of "The Raven" by Edgar Allen Poe.

Mr. Benjamin, dressed for success.

was faster and easier to play. He soon got to be very good at backgammon. In 1997, he won the American Backgammon Tour (the ABT is a series of tournaments played around the country throughout the year), and he is the all-time point leader in the history of the ABT. Benjamin says his early interest in games and calculation ties directly to his research areas of combinatorics ("clever ways of counting") and operations research ("the math of doing things efficiently").

Dreams of Broadway

After finishing high school, Benjamin attended Carnegie Mellon University (CMU). He majored in applied mathematics and statistics, and graduated in three and one-half years. During his freshman year, however, his direction was less than clear, for he spent much of his time writing song lyrics. He claims that, "if you had asked me in ninth grade and throughout much of high school, what I wanted to be when I grew up, I would have said a Broadway lyricist."

It's not surprising that young Benjamin entertained Broadway ambitions. His brother, also a teacher, was an actor for several years and remains active in Cleveland community theater as an actor and director. His sister is also active in local

theater. "My father was an accountant by day and an actor and director by night. I think he wished that he had taken a shot at a career in the theater. He passed on to all of his children his love of the stage."

Each year the Scotch and Soda Society (a theater group) of CMU produced an original musical, and over the space of a few months, Benjamin and another freshman wrote the lyrics and music for the 19 songs of "Kije" (based on the story "Lieutenant Kije Suite"). According to Benjamin, it was the biggest moneymaker that Scotch and Soda ever had. "It was a big hit. I had delusions of Broadway." As he sat telling me of that experience, he suddenly broke into song, remembering perfectly a song he had written 19 years ago (the song is performed by the villain as he attempts to seduce the princess in the second act of "Kije").

The Great Benjamini

During his middle school and high school days, Benjamin was known primarily for his magic and he performed under the name of The Great Benjamini. Initially, he performed at birthday parties for children, mostly between the ages of five and seven. Those shows involved only magic, no calculations. "I was almost a clown. I was the Great Benjamini. I would fall down and get hit over the head with

wands, and make them laugh. That was my goal."

When he started doing shows for adults, he felt the need for more sophisticated material. He added card tricks and mentalism—sort of fake ESP. After he got into mentalism, his father suggested that he put some of his mental calculations into his act. Young Benjamin was skeptical of his father's suggestion and told him, "But that's not magic, that's real, Dad!" Nonetheless, he took his father's advice, and the rapid calculations part of his act got the best response. "I decided that this must be the future of my act."

"I was something of a celebrity in high school, and not always in a positive sense. Everybody knew me as the magician." He decided that when he started college, he wanted people to know him for himself. But fate intervened twice during that year to frustrate his goal. In his first month at CMU, a magic convention was held in Pittsburgh. He attended, impressed some well-known magicians with his mental calculations, and was hired to perform at the Dove and Rabbit, a new magic nightclub in Pittsburgh. He did not inform his classmates about his off-campus job, for he was determined to have them see the "other" Art Benjamin. So on campus, he threw himself into writing lyrics for "Kije" and keeping up with classes.

High school days: "The Great Benjamini" in action!

The newly minted Dr. Benjamin with his parents, Larry and Lenore Benjamin, on the occasion of his graduation from Johns Hopkins in 1989.

But in the Spring fate struck again. Benjamin was taking a course in cognitive psychology from Marcel Just and the topic of the day was "lightning calculators." Professor Just knew a few tricks himself that were designed to give the impression of mental talent. He then asked the class, "We have lots of math and science students in this room. Does anyone here know any tricks?"

"I really shouldn't have," Benjamin recalls, "I was trying to keep my magic under wraps but when somebody asked that question... I couldn't resist. I did my act from the Dove and Rabbit. The last thing I did was to square a four-digit number, something I had just introduced in my nightclub act. While I was concentrating on the calculation and had already given out the first few digits of the answer, it was so quiet, you could hear a pin drop. I then gave the rest of the answer. It was right, and there was an audible gasp. My impromptu performance dazzled them." After class, Professor Just introduced Benjamin to Professor William Chase, who was doing research on skilled memory. For the next two years, Benjamin worked with Chase, first as a subject and later as his research assistant.

My Turtle Pancho

In spite of all his amazing mental calculating ability, Benjamin insists that his memory is only a little above average. How then can he hold so many digits in his memory when multiplying two distinct four-digit numbers in his head? Most of us can hold about seven unrelated digits in our short-term memory. When Professor Chase first tested Benjamin, he found that he could only recall 8–10 digits, just a bit above average. Chase had studied other subjects, including one man who could recall strings of 80 digits. Chase knew that such feats of recall involved the use of the grouping of digits and mnemonics (memory tricks). We can routinely recall strings of seven digits, such as when we remember telephone numbers. If we see the string 3876254, and try to commit it to memory, it's much easier to do so if we break it up into groups: 387-6254. Benjamin's calculating feats involved remembering far more than 10 digits. Chase knew that he must be using some mnemonic system.

Art can square three-digit numbers in less than a second; four-digit numbers take him about 15 seconds. Chase asked Art to explain every step of what he was doing as he squared a four-digit number. He couldn't do it, because as he explains, "The steps were so rapid that I couldn't follow them in detail." In order to slow him down, Chase asked him to square a five-digit number, something he had never done before. And that slowed him down enough that he could describe what he was doing.

For starters, Art does his calculations from left to right instead of right to left. Asked to square 46,872, Art pinches his eyes shut, clenches his fists, and paces rapidly back and forth on the stage. He utters what seem like disconnected words (his mnemonics) — *fuzz, nunnery, mover* — and then says, "two billion one hundred ninety six million." After saying these digits out loud, he never thinks about them again, depending on his audience to remember them, and freeing up space in his memory to continue with the calculation. After more pacing, hand wringing, and strings of bizarre words, he gives the answer: "2,196,984,384."

When Art started college, he did not think that multiplying distinct three-digit numbers in his head was something that he could ever do. "I could square them, but multiplying two different ones seemed beyond me." By the time he graduated from CMU, he was multiplying distinct six-digit numbers in his head! The problem would have to be called out slowly to him so that he could create the mnemonic to remember the 12 digits of the problem. The phonetic code that he employs can be found in his book *Mathemagics: How to Look Like a Genius Without Really Trying*: 0 is the "s" or "z" sound, 1 is the "t" or "d" sound, 2 is the "n" sound, 3 is the "m" sound, etc. Next he converts numbers into words by placing vowels around or between the consonant sounds (spelling doesn't matter). For example, the number 32 can become any of the following words: *man, men, mine, moon, money, menu,* etc. Here's the code he used to translate the first 24 digits of pi:

3 1415 926   5   3   58   97
"My turtle Pancho will, my love, pick

9   3   2   384   6264
up my new mover, Ginger."

With just a little bit of practice, using the phonetic code, you can memorize over a hundred digits of pi!

Choosing a Career: Math or Magic?

Benjamin's reputation as a mathemagician, researcher, and teacher continued to grow during his undergraduate days at CMU and graduate days at Johns Hopkins University. While at Hopkins, Benjamin worked on his dissertation ("Turnpike Structures for Optimal Maneuvers") under Alan Goldman, research that later earned him the prestigious Nicholson Prize from the Operations Research Society of America. Hopkins is also where he met Deena, a fellow mathematician, whom he married 4 years later at the Magic Castle in Hollywood. Before graduate school, Benjamin had already performed on several national television shows. It's clear that he could have been a full-time magician if he had chosen to do so. But he chose mathematics and teaching to be the center of his life. He

> *I wanted to be someone who would bring math to others, popularize it, and use my entertaining talents to get more people excited about it.*

says, "I wanted to be someone who would bring math to others, popularize it, and use my entertaining talents to get more people excited about it. On top of that, I had had some teaching experience as an undergraduate. Teaching was fun; it was just like performing, but better because your repertoire changed with every lecture. I sensed that if I were to become a full-time performer, it would become a bit repetitive after a while. I love applause, no question about that, but eventually I would say to myself, 'Yes, I can do this

and I can do that. But what else?' I wanted something more substantial and intellectually satisfying. I also felt that if I were going to teach mathematics professionally, I would need to go on and earn my Ph.D." It appears that he made a very wise decision, for today he is teaching, doing research, performing, and receiving applause for all of those activities.

Secrets of Benjamin

A particularly delightful aspect of Professor Benjamin is that he is eager to share his secrets of rapid mental calculations with others. Most magicians jealously guard their secrets, but not him. Several years ago he distributed within the magic community his booklet, "The Secrets of Benjamin." In 1993, he and Michael Shermer wrote *Mathemagics*, which lays out his secrets in detail. He clearly wants others to learn the art of rapid mental calculation and to come to love mathematics in the process. Can they get as good as Art Benjamin — Mathemagician? Perhaps, but that will require enormous practice and prodigious motivation. You can count on it! ■

Art and Deena Benjamin relaxing at home.

The PhD of Comedy

DEANNA HAUNSPERGER and STEVE KENNEDY
Carleton College

It's practically a tradition for math teachers to subject their students to hackneyed old jokes. Such math jokes are part of the cultural heritage of mathematics handed down from teacher to student for generations: What's purple and commutes? An Abelian grape. Did you hear about the dog in the Complex plane? He left a residue at every pole. Sometimes they can even be funny. Students in Dr. Lew Lefton's mathematics classes at the University of New Orleans, however, can be sure that not only will they learn some beautiful math, but also that the man can tell a joke. After all, he's a professional.

Growing up Geeky

Lefton grew up the youngest of four children, he has three older sisters and two parents ("one of each gender, amazingly enough; both human") in Albuquerque, New Mexico. "I guess I was a little bit of a smart-aleck at home. My dad would say, 'You're grounded!' I'd reply, 'OK, no carrying an electric charge for a month.' Or a radio station would advertise an all-request, no-commercial weekend, so I'd call up and request a commercial. It forced them to go off the air trying to resolve the paradox.

"I was pretty much the shy class nerd, the math geek, all the way through school. A real turning point in my early years was in sixth grade. I had a really great teacher, Mr. Norman Schide, who not only rode a Harley, but he told me that I could actually be a mathematician for a living! Before then, I thought that mathematician was an obsolete 19th-century job like blacksmith, cotton gin repairman, or pow-

dered wig cleaner. I didn't realize it was something that people still did. But Mr. Schide assured me that there were people who really did mathematics, which I thought was a really cool thing. So I decided in sixth grade I wanted to be a mathematician. I didn't really know what that meant of course, so I forgot about it for a while. In high school I didn't think about my career too much, because I figured I would end up changing my mind in college anyway."

Lefton double-majored in math and computer science at New Mexico Institute of Mining and Technology in Socorro. ("Ninety miles south of Albuquerque, perfect for coming home on weekends and getting Mom to do the laundry.") "I began as a computer science major because that's what the aptitude tests said I should do. But I really loved my mathematics classes the most, especially real and complex analysis. I took as many math electives as I could." After attending a school near his childhood home for four years, Lefton decided to apply outside of New Mexico to graduate schools, two in math and two in computer science. He accepted an offer from the University of Illinois at Champaign-Urbana in math.

"In graduate school I was really outgoing, no longer the shy guy. After all, I was among fellow math geeks now! Back then, the new graduate students were put in a place called the Arcade which was a huge room with 40 or 50 cubicles in it. I was told that every year, the new TAs traditionally hosted an Arcade party, so I stepped up and organized it my first year

at Illinois. Now, I grew up in a Jewish family and when we had a party, we threw a big party. So, I got a keg of beer, I put out fliers, I baked a turkey, and we had a huge potluck for the whole department. Later I found out that previous Arcade parties were more like: one Friday afternoon somebody would bring over a bag of chips. But people really liked it, and everybody had fun, for a while at least, until someone threw the turkey carcass out the second floor window."

Becoming a Comedian

Lefton was a teenager when he realized he could make people laugh. "In high school, I'd go to parties where people were telling jokes, and I realized I could remember them and tell them better than a lot of people. I learned how to tell jokes and switch them around to make them flow better. I was really intrigued by the

Lew with his sisters Linda, Charna, and Irene, 1964.

Reprinted from February 1999, pp. 5–9

Not many people know Lew started his career doing tractor jokes.

way that the same joke could be very funny if told in a certain way and fall totally flat if told only slightly differently. I discovered that telling a joke well isn't easy, you need to use the right words, the right intonation, and of course … good timing." Later in graduate school, Lefton would hang out on Friday afternoons and have a beer with professors and the other students. When he heard someone tell a new joke or "even if I just thought of something funny, I'd whip out a little pad and write it down. But for brevity, I'd only write down two or three words to remind me of the basic elements, like 'strawberry, nun, avalanche' or 'martian landing, Reagan, canoe'. Of course, after several years went by, I couldn't remember what half of the jokes were! Now I have dozens of scraps of paper with phrases like 'air-traffic controller, food, face.'"

In school, Lefton was never the class clown. "My comedy always came from personal things, not trying to get attention." He first began doing stand-up when he entered graduate school in 1982 at Champaign-Urbana. "Fortunately for me, that was a period when stand-up comedy was really growing fast, so there were lots of nightclubs opening and plenty of stage time available for aspiring comics. I had the opportunity to work with and learn from a lot of other energetic young com-

ics. In particular, I spent several of my early years in comedy doing gigs with Mark Roberts, who continued in show business with a successful career as a comic and actor." Roberts plays the character Dave on the ABC sitcom *The Naked Truth* starring Téa Leoni.

As Lefton polished his act, he included many jokes about teaching because that material played well in the college town of Champaign-Urbana. "A lot of the teaching bits are still usable today because everyone's had a math teacher."

- I asked my geometry class 'What do you get when two planes intersect?' You get fired from your job as an air-traffic controller.

- I use the standard state-approved grading policy: 90–100 is an A, 80–90 thousand dollars buys a B, and so on.

- When you die, at the very end of your life, God gives you all the answers … to your odd-numbered problems.

Becoming a Mathematician

Upon finishing his bachelor's degree in 1982, Lefton received several lucrative job offers, but he decided he'd rather go to graduate school. "I realized that money was not the most important thing. My professors seemed to be happy people with a lot of other job benefits including the security of tenure and the freedom to be a scholar. Plus, I really thought I would enjoy teaching; I believe it's one of the noblest professions." Lefton finished his PhD and when he began working as a university professor he discovered that "teaching was indeed a great job for me. It turns out that my teaching and my comedy have many of the same elements. As a stand-up comic, I'm trying to entertain the audience, and I'm most successful if I can connect with them on a basic human level. As a teacher, I'm trying to educate my students, and again I am most effective if we can both communicate on a basic human level." Teaching jokes occasionally work their way into Lefton's act, and, of course, comedy sometimes makes its way into his lectures, too.

Lefton's mathematical interests have evolved over time. After receiving his PhD in 1987 under Robert Kaufman for work on a problem in nonlinear ordinary differential equations, Lefton spent two years at the University of California Riv-

Lefton working his day job.

erside working with Victor Shapiro. At Riverside, he began writing papers and learning more about nonlinear partial differential equations (PDEs). In 1989, he came to the University of New Orleans where he began to apply his computer science background by working on the analysis and implementation of computational methods for nonlinear PDEs. "I saw some applied scientists and engineers 'solving' nonlinear PDEs numerically by naively plugging them into a poorly understood computer code with no real thought toward the mathematics or the numerics. It seemed that if the code terminated without error and the output looked 'reasonable' then they would accept that the quantities and pictures were all correct. My experience with nonlinear PDEs suggested that one has to be more careful. I began to think about issues of accuracy, stability, and whether or not the results of the computation were a valid representation of the mathematics." One of Lefton's current research interests is an interdisciplinary project with a group of geologists. "They are modeling the deformation of the Earth's crust (e.g., the India plate is indenting and deforming the Asia plate, producing the Himalayas). They view rock as a very viscous, non-Newtonian fluid modeled with the same class of nonlinear PDEs which I find interesting. I find it a nice combination of science, computation, and mathematical theory."

Mixing Math and Comedy

Lefton began doing a mathematical stand-up routine in graduate school at the annual math department Christmas party talent show. He realized "there's a whole vein of comedy that's not getting mined out there." These jokes would not work for a typical audience; "most people just don't understand what mathematicians do. When you tell someone you're a mathematician, they think you do what they did when they were in high school math, only more so. They imagine you sit down with a *really, really* big polynomial, and factor it. Or they think you take a huge,

Lefton practicing his delivery.

ten-page expression, solve for x, circle your answer, and call it a day. Of course, we do occasionally do those things, but generally not as an end in itself. Rather, we do them to gain a better understanding of some mathematical structure." Will math jokes help the public understand any better? "Probably not," says Lefton. "They're too arcane. And I don't think that the idea of requiring prerequisites for a non-mathematical audience would be well-received: 'The following three jokes require advanced calculus, linear algebra, and some exposure to the geometry of Banach Spaces.'" However, with mathematically hip audiences, he can really cut loose. Lefton has done full one-hour colloquia of original math/science/teaching jokes. He's played at a few conferences and hopes to do an AMS-MAA meeting some day. His mathematical comedy has developed since his graduate school days, but unfortunately, there aren't many places he can do the routine, so he's not quitting his day job.

• I went to a math conference, and they had booths set up for all the different branches of mathematics so students could learn about them. Unfortunately, the topology booth was closed, the algebra booth was way out in a field, and

although I could get arbitrarily close to the analysis booth, I couldn't touch it. Probably 18.4% of the booths were about statistics, but I didn't test that hypothesis because I was busy counting the combinatorics booths in three different ways. I was disappointed that the logic booth wasn't constructed, and the applied math booth just blew up!

• I failed my algebra prelim. They asked me to prove that every PID is a UFD, and I misunderstood them and proved that every IUD is a UFO, which is also a theorem, but not what they wanted me to prove.

• At Illinois, Appel and Haken solved the four-color problem, but there's still an important open question. We know that four colors suffice to color any map in the plane, but it was not shown in their paper *which four colors*. It's a non-constructive proof. I'm considering applying for a grant from Crayola on this one.

• Next I'd like to perform a simple song, which, of course, means it has no proper normal sub-songs.

• Mom always told me, 'You either believe in the Law of the Excluded Middle, or you don't.'

• 'Let P be a polynomial, brothers and sisters, amen! And the good Lord has bestowed the gift of complex coefficients on this P, hallelujah, ah-ha. Now, if P is not constant, my friends, then as sure as sinners will roast in hell and the righteous people will rise to the kingdom of heaven, P has at least one complex root.' This is, of course, The Fundamentalist Theorem of Algebra.

• I have a compact car, which means it has only finitely many open windows.

• I handed back some papers in class on which I had asked the students to do a calculation, and one student says 'How come I got zero points for this?' I said 'The answer was $\pi/2$ and you got $\sqrt{179,643}$.' The student looked me straight in the eye and said 'Well, I'm only off by a constant!'

Cartoon by Loel Barr

Robert Kennedy teaching a math class, "Some people look at this axis and they say 'y'. I distinguish a single point on the axis, and I say 'y-nought'."

- What do you call a bar napkin covered with ε's and δ's? The aftermath of mathematical small talk.

- Always remember: No man is an island, except in the discrete topology.

Real Life

Lefton is married to fellow mathematician Enid Steinbart, whom he met in graduate school. They are both tenured at the University of New Orleans. "Enid is an incredible person. She's a great wife and mother and she's very smart. She's my best friend and lover, but she's also a shrewd manager, a supportive colleague, and my most honest critic. She's beautiful. In fact, she's absolutely beautiful which means that even if you rearrange her, she remains beautiful. When we were married, we exchanged rings. I gave her the integers, and she gave me the continuous functions on the interval [0,1]." They have three daughters, the outspoken five-year-old Hannah and twin three-year-olds, enthusiastic Natalie and mischievous Monica. "We're not having any more kids. We decided to stop because after one kid, then twins, we figured we

had some kind of monotone increasing reproductive gene. After two terms, we weren't sure if it was an arithmetic or geometric sequence, but either way we didn't like the trend." His girls all think he's funny, but "reducing my humor to the pre-school level isn't that big a step for me."

With three kids, two jobs, one house, and no nearby family, Lefton says he doesn't really have such a thing as free time. "We play games, go to the zoo, take walks, chase children around the house. Usually they're our children, but sometimes other kids come over because they like to be chased, too. We have a little garden, and I sometimes sing songs with my girls so they can better appreciate the tragedy of tone-deafness."

Asked whether he had any funny teachers, Lefton replied, "I've had some great teachers and my thesis advisor has a great sense of humor, but none of my teachers were big joke-tellers. I actually learned a lot about comedy from Bruce Reznick, a mathematician whose father is a well known professional comedy writer in California." Lefton thinks that the best comedy comes from truth. "When you base your humor on yourself and your experiences, it will be funniest. Of course there are some tricks of the trade: juxtaposition, funny-sounding words, exaggeration, threes are funnier than twos or fours. It's one thing to have an idea that's funny, it's another to actually encapsulate that idea in a joke that still maintains the humor. I find it very challenging." Lefton doesn't claim to understand why people laugh, but "I think laughter's a great thing

Hannah, Monica, Natalie, and Enid

and we don't do it enough. I'm not trying to understand or analyze it, but I just like to get involved and make people experience it."

Lefton has performed comedy regularly since 1982. He's been the opening act in concert for Bobcat Goldthwait and for Three Dog Night. ("There's only two of them left. The harmonies just don't quite cut it. 'Wasn't there somebody singing a melody on this one? I just hear two harmonies.'") He regularly does shows in clubs in New Orleans. A few years ago, he and some friends started an improv troupe, Brown!, which now has five regular members and holds weekly sketch comedy shows.

Deciding between mathematics and comedy was easy for Lefton: "I had spent 20 years of my life becoming a mathematician. And mathematics is one of those fields where, if you walk away from an active research area for a period of time, you may have a huge amount to learn when you come back. I want to pursue mathematics first. But, I may one day fade out my math career and fade up the comedy. If I'm funny now, hopefully I'll still be funny in fifteen years. And that's a good thing about show business. If you're fifty and you get on stage and you have a killer act, you can get work. They don't care how old you are, as long as you're funny."

For now, Lefton is happy as a day-time professor and night-time comedian. "I don't get teaching and comedy mixed up *too* often, but it can happen. Maybe the audience is noisy, and I'll just blurt out, 'All right, put your drinks under the table. Pop quiz right now. What was the second joke I told?'" ■

The Mathematician's Blues

By Lew Lefton

I've been working for months on a theorem,
A result I believe to be true,
But my four-year-old daughter just showed me
It follows from $1+1=2$.

Oh no, I've got the mathematician's blues.
I've been working with an advanced degree,
And nothing but my sanity to lose.

I understand higher mathematics,
I do it all day and night,
But when I sit down to balance my checkbook,
The numbers don't come out right.

Oh no, I've got the mathematician's blues.
It seems I've been using epsilon,
When I should have been using epsilon over two.

To color a map in the plane,
We know that four colors will do.
To color a mathematician
You only need one and the color is blue.

Oh no, I've got the mathematician's blues.
My paychecks all say NSF,
And I can't find the cash to pay my dues.

Legislating Pi

UNDERWOOD DUDLEY
DePauw University

Everyone knows that the legislature of the state of Indiana once passed a law setting the value of π. Well, perhaps not everyone knows it, but many people with only a passing acquaintance with mathematics have it firmly in their heads that one of our sovereign states once tried to impose its will on a constant of nature. Everyone may not remember all of the details precisely: the date is often recalled only vaguely, and it might have been the legislature of Iowa, or Idaho, or maybe Illinois but no one has any doubt that π was the subject of legislation.

That is an error. The state of Indiana (or Iowa, or Idaho) never passed a law about the value of π. It never even tried to legislate the value of π. However, the idea that it did has spread widely and is as definitely fixed in the collective mind as is the knowledge that in 1492 everyone except Christopher Columbus thought that the world was flat. It will probably never be dislodged, but all of us who revere reason have a sacred duty to try to put error down whenever we can. The battle against falsehood and delusion will never be won, but it must continue to be fought.

Although the true story of Indiana and π has been told over and over again ([2], [3], [4]), it has not been told often enough to drive the false story out, and that is the reason for telling it one more time. The purpose of the bill that the Indiana House of Representatives passed was not to set the value of π by law and make the use of other values illegal, it was to give the state the privilege of using the proper value of π for free. The preamble of House Bill No. 246, Indiana State Legislature, 1897 is

A bill for an act introducing a new mathematical truth and offered as a contribution to education to be used only by the state of Indiana free of cost by paying any royalties whatever on the same, provided it is accepted by the official action of the legislature of 1897.

Why not pass such a bill? Passing it would be much the same as having the President sign a proclamation naming April as National Tool and Die Month. There was nothing possible to be lost. Gift horses need not have their mouths inspected. Then as now, legislators were busy people with no time to master all of the details of all of the bills presented to them. And when a bill contained language like

SECTION 2. It is impossible to compute the area of a circle on the diameter as a linear unit without trespassing upon the area outside of the circle to the extent of including one-fifth more area than is contained within the circle's circumference, because the square of the diameter produces the side of a square which equals nine when the arc of ninety degrees equals eight.

even a legislator with some spare time could be excused for spending it on some other piece of pending legislation. The last section of the bill, however, was reassuring:

SECTION 3. In further proof of the value of the author's proposed contribution, and offered as a gift to the State of Indiana, is the fact of his solutions of the trisection of the angle, duplication of the cube and

quadrature of the circle having been already accepted as contributions to science by the *American Mathematical Monthly,* the leading exponent of mathematical thought in this country.

The author of the bill, Edward Johnston Godwin, a medical doctor resident in southwestern Indiana, was being a little devious there. When Godwin's "Quadrature of the circle" was published in 1894 in volume 1 of the *American Mathematical Monthly* (pages 246–247) it was not an article, but part of the "Queries and Information" section where all sorts of miscellanea could be found. ("Quadrature of the circle" or "squaring the circle" means solving the problem of constructing, with straightedge and compass alone, a square whose area is equal to that of a circle. If π were rational, like 22/7, or even algebraic, like $\sqrt{10}$, this could be done, but it isn't, so it can't.) The quadrature had the subhead "Published by the request of the author" thus putting it in the category of unpaid advertisement. The editors of the *Monthly* probably needed material to fill space and thought that Godwin's piece, less than two pages long, would fill it nicely and might provide entertainment for readers. So Godwin could assert that the paper containing his quadrature had been accepted by the *Monthly,* but when he said that his quadrature had been accepted, and as a contribution to science, he was stretching the truth more than a little. (His angle trisection and cube duplication appeared in the 1895 *Monthly* (page 337) and were not distinguished. To trisect an angle, he said, trisect its chord—try that on a 150° angle and

MORNING, FEBRUARY 21, 1897—SIXTEEN PAGES.

SQUARING THE CIRCLE

DR. GOODWIN'S DEMONSTRATION ACCEPTED BY MATHEMATICIANS.

Supposed Impossible Problem the Solution of Which Changes the Multiple Pi from 3.1416 to 3.2.

Official recognition by one branch of the Indiana Legislature has been given Dr. Edward Johnston Goodwin for solving three geometrical problems which have puzzled the brains of mathematicians since the erection of the pyramids of Egypt, and which the French Academy of Science, in 1775, and the Royal Society of Great Britain, in 1776, both declared impossible of solution. The first and most important of these problems is what has been popularly

even offers $500 to any one showing an error in his demonstration.

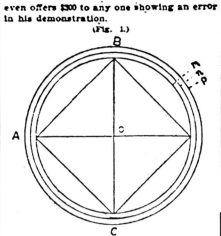

(Fig. 1)

Diameter inner circle equals 5.
Area of inner circle equals 16.
Arc A B equals 4.
Chord A B equals 3½.

Following are the simple rules embracing the points in Dr. Goodwin's solution, which are to be introduced in the text-books:

"Doubling the dimensions of plane geometrical forms quadruples their areas. The Circle—To find the circumference of a circle: Multiply the diameter (according to the ratio of 5-4:4) by 3.2. To find the diameter of a circle: Divide the circumference by 3.2. Or divide the circumference by 4 and multiply the quotient by 5-4.

"To find the area of a circle: Divide the circumference by 4 and square the quotient.

"Measurement of Volumes—The Sphere—Doubling the dimension of a sphere quadruples its spherical area and octuples its solid contents.

"To find the surface of a sphere, divide its circumference by 4 and square the quotient and multiply by 4.

"To find the solid contents of a sphere: Divide its circumference by 4 and cube the quotient."

From the Indianapolis Daily Journal.

see what it looks like—and to double the volume of a cube, increase its side by 26%—since $\sqrt[3]{2} = 1.25592\ldots$ this is close, but not exact.) In any event, House Bill 246 seemed harmless, and it passed with no votes against.

Bills passed by legislatures are of interest to newspapers, and one whose subject was mathematics would stand out. One Indianapolis daily gave a history of π, mentioning even Lindemann who had shown not many years before that π was transcendental, but since the paper was *Der Taglische Telegraph,* printed entirely in German, many legislators were unenlightened.

After passing the House, the bill went to the Senate for consideration. Even though senators were and remained ignorant about what the value of π was, enough fuss was made in the press that they, being politicians, sensed that the bill should not pass. *The Indianapolis Journal* said,

> Although the bill was not acted upon favorably, no one who spoke against it intimated that there was anything wrong with the theories it advances. All of the senators who spoke on the bill admitted that they were ignorant of the merits of the

proposition. It was simply regarded as not being a subject for legislation.

> Action on the bill by the Senate was postponed indefinitely, and so it has remained for more than one hundred years.

Godwin had, as is common with circle-squarers, been active in trying to get his discovery recognized. Actually, "discovery" is not quite the word, since he had written earlier

> During the first week in March, 1888, the author was supernaturally taught the exact measure of the circle. ... All knowledge is revealed directly or indirectly, and the truths hereby presented are direct revelations and are due in confirmation of scriptural promises.

An advantage of discovery by revelation is that proofs are not necessary.

Godwin evidently had some influence with the editor of the *Indianapolis Journal,* because nine days after the report on the bill, that paper had an editorial:

> Some newspapers have been airing their supposed wit over a bill introduced in the Legislature to recognize a new mathematical discovery or solution to the problem of squar-

ing the circle, made by Dr. Godwin, of Posey County. It may not be the function of a Legislature to endorse such discoveries, but the average editor will not gain much by trying to make fun of a discovery that has been endorsed by the *American Mathematical Journal*; approved by the professors of the National Astronomical Observatory of Washington, including Professor Hall, who discovered the moons of Mars; declared absolutely perfect by professors at Ann Arbor and Johns Hopkins Universities; and copyrighted as original in seven countries of Europe. The average editor is hardly well enough versed in high mathematics to attempt to down such an array of authorities as that. Dr. Godwin's discovery is as genuine as that of Newton or Galileo, and it will endure, whether the Legislature endorses it or not.

The editor of the *Journal* was not well enough versed in higher mathematics to know that Godwin's quadrature was nonsensical, and he was in addition gullible enough to swallow whole all that Godwin must have told him. Of course, no professor of mathematics at Johns Hopkins

A MAN OF 'GENIUS'

SOME FOLKS DON'T CALL IT THAT, HOWEVER.

Dr. Goodwin Pays no Attention to Their Taunts,

BUT KEEPS CLOSE TO HIS CRO-CHET WORK

And Hopes to See the Day When the World Wil Believe His Theory of a Round Square or Whatever You Call It.

"If I live ten years, and I hope I shall, you watch out for Goodwin. My discovery will revolutionize mathematics. The astronomers have all been wrong. There's about 40,000,000 square miles on the surface of this earth that isn't here. Watch out for Goodwin if you live ten years."

These were the parting words which Dr. Goodwin gave a SUN reporter who called to ask him about the mathematical discovery presented to the state of Indiana in the form of a bill passed in the legislature, Friday afternoon.

Dr. Goodwin practiced medicine 42 years after he had graduated from a medical college in Philadelphia. 'Twas just the other day that he decided to quit the practice entirely and devote the rest of his life to mathematical discoveries. He came up to Indianapolis on the day that Gov. Mount was inaugurated, and at once began to set pins to have his discovery passed on by the legislature: Mr. Record introduced the bill. The complexity of the terminology set the law-makers to laughing, and Speaker Pettit carried the joke a little farther by referring the bill to the committee on swamp lands. Truth triumphs in the end, Dr. Goodwin says. The bill finally got before the educational committee, and there the doctor showed a series of charts explanatory of his discovery.

Dr. Goodwin is a tall, angular man, and is stoop-shouldered. His hair is iron-gray and his mustache likewise. He wears a neglige shirt without necktie and buttons a long Prince Albert close up to his neck. He is nervous. While slipping off his tongue's end an avalanche of mathematical terms in explanation of his discovery.

The Indianapolis Sun reports on Dr. Godwin's discovery.

or the University of Michigan had declared that Godwin's quadrature was absolutely perfect. You might wonder how Godwin, a man of probity and standing, could lie through his teeth and tell the editor of the *Journal* that they had. Probably Godwin did not think that he was lying. People who square circles find it easy to believe that they have convinced authorities of their correctness. Here is evidence of that, from Augustus De Morgan's *Budget of Paradoxes* [1, v. 1, pp. 13–14]:

> An elderly man came to me to show me how the universe was created. There was one molecule, which by vibration became—Heaven knows how!—the Sun. Further vibration produced Mercury, and so on. Some modifications of vibration gave heat, electricity, etc. I listened until my informant ceased to vibrate—which is always the shortest way—and then said, "Our knowledge of electric fluids is imperfect." "Sir!" said he, "I see you perceive the truth of what I said, and I will reward your attention by telling you what I seldom disclose, never, except to those who can receive my theory—the little molecule whose vibrations have given rise to our solar system is the Logos of St. John's Gospel!" He went away to Dr. Lardner, who would not go into the solar system at all—the first molecule settled the question. On leaving, he said, "Sir, Mr. De Morgan received me in quite a different way! He heard me attentively and I left him perfectly satisfied of the truth of my system." I have much reason to think that many discoverers, of all classes, believe they have convinced every one who is not peremptory to the verge of incivility.

De Morgan wrote in 1865, but circle-squarers, angle-trisectors, and other cranks have not changed since then, nor has the way to deal with them. Do not try reason. It will not work.

All of Dr. Godwin's efforts were in vain. His sad obituary (June, 1902) was quoted in [3]:

End of a Man Who Wanted to Benefit the World

Dr. E. J. Godwin dies at his home in Springfield Sunday, aged 77 years. He had been in feeble health for some time, and death came at the end of a long season of illness. Dr. Godwin was no ordinary man, and those meeting him never failed to be inspired by this fact. He was of distinguished appearance and came from Virginia where he received an excellent education. He has devoted the last years of his life in an endeavor to have the government recognize and include in its schools at West Point and Annapolis his method of squaring the circle. He wrote a book on his system and it was commented on largely and received many favorable notices from professors of mathematics.

He felt that he had a great invention and wished the world to have the benefit of it. In years to come Dr. Godwin's plan for measuring the heavens may receive the approbation which was untiringly sought by its originator.

As years went on and he saw the child of his genius still unreceived by the scientific world, he became broken with disappointment, although he never lost hope and trusted that before his end came he would see the world awakened to the greatness of his plan and taste for a moment the sweetness of success. He was doomed to disappointment, and in the peaceful confines of village life the tragedy of a fruitless ambition was enacted.

Doomed, indeed. This is the fate of all circle-squarers, angle-trisectors, and duplicators of the cube. Let us all give thanks, loudly and daily, that we are not cranks devoting all of our energies to convincing the world that π is exactly 3.125, 22/7, or $\sqrt{10}$, (all popular values). The probability that our lives will be rich, delightful, and full of meaning may be far from 1, but it is much greater than the near-zero of the circle-squarer.

Godwin's value of π has not been mentioned. The reason is that his writing was so confused and confusing that it is impossible to determine exactly what he thought π was. His *Monthly* article starts with

> A circular area is equal to the square on a line equal to the quadrant of the circumference; ...

This clearly says that if you take a quarter of the circumference of a circle, $\pi r/2$ if the circle has radius r, and square it, you will get the area of the circle:

$$\left(\frac{\pi r}{2}\right)^2 = \pi r^2$$

Solving that equation for π gives $\pi = 4$. Later on, Godwin says

> This new measure of the circle has happily brought to light the ratio of the chord and arc of 90°, which is as 7:8.

This says, just as clearly (see figure), that

$$\frac{\sqrt{2}r}{\frac{\pi r}{2}} = \frac{7}{8}$$

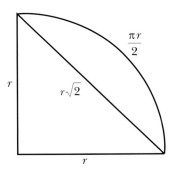

so

$$\pi = \frac{16\sqrt{2}}{7} = 3.2324888\ldots.$$

In [4], David Singmaster succeeds in finding nine different values of π implied in one place or another of Godwin's article. Godwin was giving the people of Indiana a lot. Actually, he no doubt thought that he had only one value of π, most likely 3.2, but his level of algebraic skill was so low that he could not see the contradictions. Many circle-squarers know very little algebra, and some will not let you use it to try and show them the error of their ways. Squaring the circle is a geometrical problem, they will say, and hence algebra cannot be used.

Godwin succeeded in getting the Indiana legislature to consider his value of π, but that is all. There was never an attempt by the lawmakers to set the ratio of the circumference of a circle to its diameter, there was only a refusal by them to take Godwin's value, or values, as a gift. I hope that this has driven the myth that Indiana passed a law about π out of your head. I would encourage you to devote your life to convincing the world that it was wrong in thinking the myth was true, but doing that would be just as hard as squaring the circle. ■

References

1. Augustus De Morgan, *A Budget of Paradoxes,* first edition 1872, reprint of the second edition (1915) Dover, 1954.

2. Will E. Edington, House Bill no. 246, Indiana State Legislature, 1897, *Proceedings of the Indiana Academy of Sciences* 45 (1935) 206–210.

3. Arthur E. Hallerberg, Indiana's squared circle, *Mathematics Magazine* 50 (1977) 136–140.

4. David Singmaster, The legal values of pi, *Mathematical Intelligencer* 7 (1985) no. 2, 69–72.

The
Ultimate Flat Tire

STAN WAGON
Macalester College

The Flattest Wheel

How flat can a tire be and still roll? Can something as straight as a straight line be used as a wheel? Sure, if, as explained below, one uses some care in defining its center. The insight is due to G. B. Robison in 1960 [1]; he also realized that by suitably truncating a doubly infinite straight line one could form a square wheel, which would indeed roll on a properly shaped road.

To set the stage we must be precise about what "roll" means. A round circle rolls on a straight line in the sense that the center of the wheel stays horizontal. So for a non-circular wheel, we will say that it rolls on a curvy road if the center of the wheel moves in a horizontal line as the wheel moves without slipping along the road. Here I explain why a square rolls on a sequence of inverted catenaries. (Recall that a catenary is the curve made by a flexible chain allowed to hang with both ends held at the same height; its equation is simply $y = \cosh x = (e^x + e^{-x})/2$.)

As the Wheel Turns

Suppose the road is given as a function $y = f(x)$ and the wheel is described in polar coordinates, $r = r(\theta)$. An example is given in Figure 1, where $f(x)$ is $\cos(x) - \sqrt{17}$ and

$$r(\theta) = \cos\left[2\arctan\left[\frac{1}{4}\left(\sqrt{17} - 1\right)\tan[2\theta]\right]\right] - \sqrt{17}.$$

As the wheel rolls, the distance from its center to the road must match the depth of the road: this means that the two dashed lines have equal length. And the road surface and wheel circumference must match, so the two thick curves must have equal arc length. These conditions will allow us to relate the polar equation $r = r(\theta)$ of the wheel to the Cartesian equation $y = f(x)$ of the road on which it rolls smoothly. The graph on the right shows the important function $\theta(x)$, which tells us the polar angle of the straight-down radius when the wheel has rolled enough so that its center is above the point x on the x-axis. That is, the segment joining the center of the wheel to the point on the wheel touching the road at x made the angle θ with the positive x-axis before the wheel started rolling. We are using standard polar coordinates, so $\theta(0) = -\pi/2$.

Now, the condition arising from the equality of the dashed lines leads to the radius condition

$$r\left[\theta(x)\right] = -f(x). \tag{1}$$

The negative sign is included because r should be positive but $f(x)$, which defines the road, will be negative. Note that the initial condition becomes $r[\theta(0)] = -f(0)$, or $r(-\pi/2) = -f(0)$.

Next we match the arc lengths. The road length is given by the familiar formula

$$\int_0^x \sqrt{1 + \left[f'(t)\right]^2}\, dt,$$

while the wheel circumference is the slightly less familiar

$$\int_{-\frac{\pi}{2}}^{\theta(x)} \sqrt{r(\theta)^2 + \left(\frac{dr}{d\theta}\right)^2}\, d\theta.$$

The author taking a ride. Photos by Deanna Haunsperger.

Equating these integrals, differentiating both sides with respect to x, and squaring yields:

$$1 + \left[f'(x) \right]^2 = \left(\frac{d\theta}{dx} \right)^2 \left[r(\theta)^2 + \left(\frac{dr}{d\theta} \right)^2 \right]. \qquad (2)$$

But the radius condition (1) can be differentiated with respect to x to yield $(dr/d\theta)(d\theta/dx) = -f'(x)$. Substituting into (2) yields:

$$1 + \left[f'(x) \right]^2 = \left(\frac{d\theta}{dx} \right)^2 r(\theta)^2 + \left[f'(x) \right]^2$$

which simplifies to $d\theta/dx = 1/r(\theta)$. This is what we want: a differential equation, quite simple as it turns out, that relates the rolling function $\theta(x)$ and the shape of the wheel $r(\theta)$. The variables separate into $r(\theta)\,d\theta = dx$ so integration and the initial conditions can be used to get x in terms of θ as follows:

$$x = \int_{-\frac{\pi}{2}}^{\theta} r(\theta)\,d\theta. \qquad (3)$$

If we can solve this for $\theta(x)$, we will know the shape of the road, since, for any x, $f(x) = -r[\theta(x)]$. (It is an interesting exercise to show that if we match tangent slopes instead of matching arc lengths we get the same fundamental relationship (3), in a way that avoids the arc length integrals.)

Why a Catenary?

Now consider the straight-line wheel with polar equation $r = -\csc\theta$, $-\pi < \theta < 0$. We will take the origin as center of the wheel and perform the analysis described above.

To get the rolling relationship we evaluate the integral of (3):

$$x = \int -\frac{1}{\sin\theta}\,d\theta. \qquad (4)$$

Standard calculus techniques, using the facts that the θ-domain is $(-\pi, 0)$ and the initial condition is $x(-\pi/2) = 0$, tell us that $x = -\log(-\tan(\theta/2))$.

This inverts to $\theta(x) = 2\arctan(-e^{-x})$. It follows that the road we seek is the graph of $y = r[\theta(x)] = -\csc[2\arctan(-e^{-x})]$. This simplifies to just $y = -(e^x + e^{-x})/2$, which is the familiar catenary arch, $-\cosh x$. Thus our straight line will roll on a catenary, as shown in Figure 2.

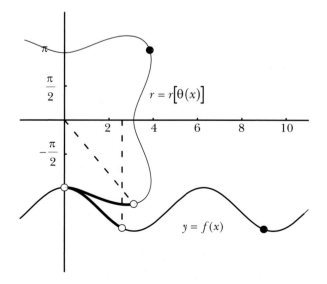

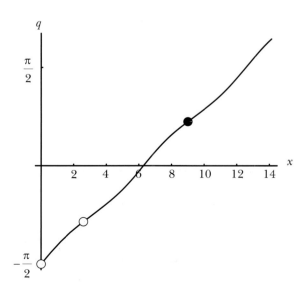

Figure 1. The left diagram shows a wheel about to begin rolling on a cosine-shaped road. The two dashed lines must have equal length, and the two thick curves must also have equal length. The image on the right shows the θ versus x relationship.

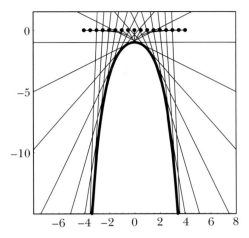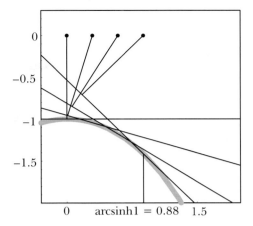

Figure 2. A wheel that is a straight line rolls on a catenary so that the hub, the point that started at $(0,0)$, stays horizontal. The close-up view shows that when the center is at $\text{arcsinh}(1)$, the line makes a $45°$ angle with the horizontal.

A Square Wheel

Now that we understand why the straight line rolls on the catenary it is easy to see how to handle a square wheel. Just truncate the catenary at the point at which the rolling straight line will make a $45°$ angle with the horizontal. Then when an identical truncated catenary is placed beside the first, the cusp will have a $90°$ angle; when a second straight line is placed perpendicular to the first one, we will have a rolling right angle. Do the same for the other angles of the square and, presto, a square wheel. Figure 3 shows how the wheel rolls; note that the locus of one of the corners occasionally goes backward, reminiscent of the classic puzzle about the locus of the point at the bottom of the flanged wheel of a train.

   Many people, on seeing the rolling square, wonder if a rolling pentagon or hexagon is possible. Indeed, essentially the same argument shows that any regular n-gon will roll on a catenary road for any $n \geq 3$. The triangular case is actually impossible in practice, the vertex of the triangle crashes into the next bump before it can settle into the cusp. It also crashes into the previous bump as it climbs out of a cusp. So, in theory you could just lay the road as you need it and rip it up as you pass over it, but that would be slow going.

Is the Ride Smooth?

There is a subtle difference between a square wheel and a round one. For a normal bike, x is proportional to θ, where x is the distance traveled and θ is the angle pedaled. If you pedal more, you travel farther, and the correspondence is linear: pedal twice as much and you travel twice as far. This is almost true of the square wheel, but not quite. Figure 4 shows this relationship for the square wheel, with a straight line shown for comparison. The discrepancy is so small that it cannot be felt by a rider. (To see more clearly the difference from linearity, examine the Maclaurin series of $\theta(x)$, which is $2\arctan(-e^{-x})$.)

A Working Model

Inspired by various models I had seen (a small one at San Francisco's Exploratorium and a larger one built by the Center of Science and Industry in Columbus, Ohio), I asked Loren Kellen, a neighbor who knows carpentry and bicycles, if he could build a working model of a square-wheel bike. He was enthusiastic and six months later it was done; the full-size model (the road is 23 feet long) is on permanent display in the science center at Macalester and is open for public riding. We decided

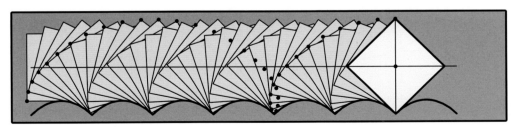

Figure 3. When a square rolls on a sequence of appropriately truncated catenaries, the ride is smooth in the sense that the hub of the wheels experiences no up or down motion. The dots show the locus of a corner.

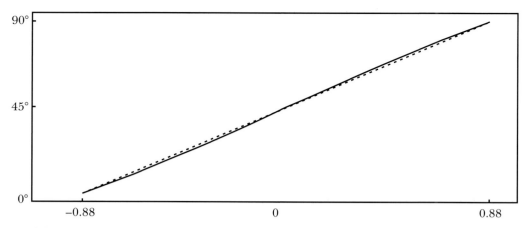

Figure 4. The θ vs. x relationship for a square wheel. The graph is very close to a straight line (dashed) so that even though horizontal speed is not a linear function of pedaling speed, the difference is very small, and not detectable on a full-size square-wheel bicycle.

on a three-wheel design for stability. Friction is a big concern: there must be enough friction between the tire and the road to prevent slippage, or "creep" as it is called by professionals in catenary road-building. Also the bike frame had to be sawn in two and rewelded so that the frame would fit the road, whose spacing is in turn decided by the size of the square wheels. I thought steering would be a problem, but in fact one can steer the thing provided one does so at the top of each arch!

More Surprises

Thanks to the power of modern software, an investigation such as this often leads to new insights. Leon Hall and I, after seeing the Exploratorium model, made an extensive investigation into the shapes of various road-wheel pairs. One surprising discov-

ery is related to the age-old definition of a cycloid as the locus of a point on a round wheel rolling on a straight road. We found that the locus of a point of a limaçon as it rolls smoothly on a trochoid (itself a type of cycloid) is also an exact cycloid! Thus the cycloid we know and love can be viewed as one of a matched pair (see Figure 5). See [2] for more such relationships. Here is a final puzzle, due to Robison: Find the unique road-wheel pair for which the road and wheel have *identical shapes*. ■

References

1. G. B. Robison, Rockers and rollers, *Mathematics Magazine,* 33 (1960) 139–144.

2. L. Hall and S. Wagon, Roads and wheels, *Mathematics Magazine*, 65 (1992) 283–301.

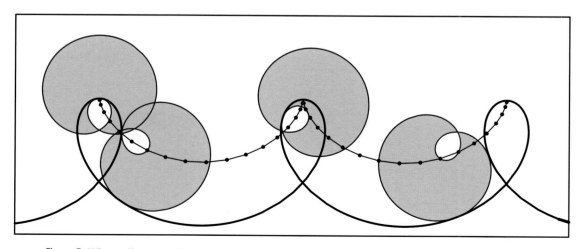

Figure 5. When a limaçon rolls on a trochoid, the locus of one of the wheel's points is an exact cycloid.

The
Roots
of the
Branches
of
Mathematics

RHETA RUBENSTEIN
University of Michigan, Dearborn

RANDY SCHWARTZ
Schoolcraft College

Mathematics is justifiably called the Queen of the Sciences. Our word **mathematics** comes from the ancient Greek term *mathematikos* (μαθηματικός). Ask a friend sometime to guess what the word meant originally: you are likely to get some narrow response like "numbers," "calculation," "equation." Actually, the root meant "mental discipline" or "learning." Plato, the Greek philosopher who flourished around 380 BCE, believed that no one could be considered educated who was not trained in such mental discipline, such *mathematikos*. He went so far as to say, "He is unworthy of the name of man who is ignorant of the fact that the diagonal of a square is incommensurable with its side." The etymology of the word "mathematics" is a concentrated residue of the sweeping way in which this activity was conceptualized in the ancient world.

The words of mathematics, like all words, have concrete origins. In this article, we explore the origins of those terms that we use to designate the various branches of our science. The tree of mathematics has many such branches, and each of them has deep and interesting etymological roots.

In ancient Egypt, when the Nile River overflowed its banks every spring, farmers had to measure the earth again to re-estab-lish their fields. The Egyptians had to invent the science of **geometry,** which comes from the Greek words for "earth" (*geo*) and "measure" (*metron*). Many English words share these same roots:

geology — the study of the earth
geography — drawing about the earth
geophysics — physics of the earth
meter — the basic unit of measure in the metric system, in poetry, and in parking!
metronome — a device to measure time in music.

It's no accident that "geometry" is a Greek word: this was the most respected branch of mathematics in ancient Greece. It is said that the door of Plato's academy bore the inscription, ἀγεωμέτρητος μηδεὶς εἰσίτω which means, "Let no one who is not geometrically minded enter." Euclid and other great geometers of the time lived in Greek colonies in Egypt, and mathematical historians trace some of the roots of Greek geometry to the "earth-measuring" and pyramid-building techniques of the Nile valley.

The same root, *meter*, can be seen in the name of another branch of the mathematical tree, **trigonometry**. In Greek, *tri* means "three," and *gonia* means "angle," so the word refers to the measure of "trigons" or triangles. The term was coined in a book by Bartholomeo Pitiscus, *Trigonometry: or The Doctrine of Triangles* (1595). While trigonometry has evolved beyond this initial idea, its source is the measurement of triangles. The Greek word *gonia* grew out of an old Indo-European root *genu* meaning "angle or knee," the basis for words like "genuflect," or "knee" itself (in Old English the 'k' was not silent).

Fitting Things Together

Arithmetic comes from the Greek word *arithmos* or "number." It stems from the Indo-European root *ar-*, "to fit together," also seen in words like "harmony." The basic operations of arithmetic—addition and multiplication—can be thought of as fitting numbers together. A surprisingly related word is "aristocrat," a person or ruler in whom the best qualities are (supposedly) fitted together. Consider these other words that stem from the same Indo-European root as does "arithmetic":

art — fitting things together creatively
arm — the limb that fits into the torso
armament — a weapon with which to outfit a soldier or army
armada — naval forces fit together in a single fleet
arthritis — an inflammation of the joints or "fit-hinges"

Reprinted from February 1999, pp. 18–20

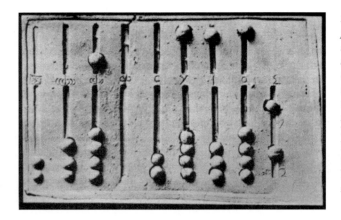

Roman hand abacus with pebbles, or calculi. Plaster cast of the specimen in the Cabinet des Médailles, Paris. Note the numerical labels (I, X, C, ...) on the grooves. From Karl Menninger, Number Words and Number Symbols: A Cultural History of Numbers, Dover Publications, 1969.

arthropod — a creature whose legs have many joints

order — to fit together in a natural sequence

ordinate, coordinate — a numeral indicating where an object fits within an ordered set

ratio — a quotient that indicates how two things would fit alongside one another.

Our term **algebra** comes from the title of an important Arabic text written around 825 by the Baghdad mathematician Muhammad ibn Musa al-Khowarizmi. His book was entitled *Hisab al-jabr w'al-muqabala*. The term *al-jabr* that he used in his title was an Arabic noun meaning "recombining broken parts"—a reference to one of his most common operations, the combining of like terms in an equation. Europeans shortened the title and altered the pronunciation to create "algeber." In England in the 14th Century, algeber meant "bone setting" or setting broken things aright. By the 16th Century it took on its current meaning, and the spelling was changed. (In the next WordWise, we'll delve into mathematical words from Arabic.)

Of Teeth and Tabulation

In ancient Rome and many other cultures, people used marbles or other small stones to help them count and do arithmetic. From the Latin word *calx* (meaning "limestone," the root of our words "calcium" and "chalk") and the ending *-ulus* ("little") came the noun *calculus* (a marble) and the verb *calculare* (to use marbles; to do arithmetic), the basis for the English word "calculate."

In the 1660s, the noun **calculus** was adopted for the newly invented branch of math that calculates the rate at which any quantity varies. But in dentistry, the term calculus is still used for the calcium-rich deposits that build up on our teeth. When you go for your yearly cleaning, and the dentist gripes about all the tartar that has to be chiseled off, at least you can agree that "calculus is the hardest thing."

In ancient Israel/Palestine, pebble calculations were performed in trays of dust or sand. The Hebrew word *abhaq* (dust) entered Latin as the term *abacus*, a calculating device using stones or beads. Nowadays, when you buy something at a store, you lay it on a table that is called a "counter" because in Medieval England, it was always equipped with an abacus. By contrast, the English words "calculator" and "computer" originally referred not to machines, but to *people* who specialized in doing computations. We might think of today's handheld calculators as

electronic pebble pushers. Even people who "lose their marbles" can use a calculator.

Breaking Things Apart Again

Analysis refers to the entire subfield of mathematics that investigates continuous functions, of which the calculus is a kind of starting point. The terms "analysis" and "analyze" stem from the Greek root *luein*, meaning "to loosen, break up, untie." Consider these related words:

lose — to be separated from something

forlorn — with hope lost

solution — the 'loosening up' of an equation or other problem

absolute value — an expression loosened of any qualification as to sign.

In what sense is mathematical analysis a matter of loosening or decomposing something? The idea is that analysis breaks complex things into their component parts, which are easier to "digest." Just as Descartes succeeded in breaking the position of any point along a curve into a pair of coordinates (hence *analytic geometry*), so Newton and Leibniz succeeded in breaking the curvature into a series of horizontal and vertical displacements, and the area under the curve into a series of inscribed rectangles. Similar approaches in other disciplines are referred to as "analytic philosophy," "analytic psychology" and the like. In chemistry, we also have these related words from the same root, *luein*:

*Egyptian geometers (earth-measurers) carry a knotted rope, useful for laying out a right angle via a 3-4-5 triangle. Detail from a mural in the tomb of Menena c. 1420 BCE. *

soluble — able to be broken apart by another substance

catalytic — aiding in the breaking down or other transformation of a chemical

hydrolysis — the breaking of a large molecule through the addition of water

lysine — an amino acid resulting from the breakdown of certain proteins.

Interestingly, the use of the word "analysis" for an entire branch of mathematics was proposed five years before Descartes's birth, when his countryman Francois Viète published *In artem analyticam isagoge* (1591). Viète lobbied for "analysis" to replace the word "algebra," on the grounds that "algebra" does not have European roots. His efforts helped popularize "analysis" to the extent that we use the term widely today. While Viète failed to eradicate "algebra" altogether, one wonders what other multicultural roots of mathematics might have been lost or buried along the path of western history.

Where We Stand

Probability is the study of the likelihood of events or situations. The name comes from the Latin adjective *probus* (upright, honest) and the derived word *probare* (to try, to test, to judge). We see these roots in words like:

probe — to test by a searching investigation

probation — a test of a convicted person's character.

The idea behind "probability," then, is "capable of being made good; able to be proven." Interestingly, the word "proof" stems from the same root as "probable," even though in modern usage, "probable" means "likely but not certain." Mathematically, what is proven and what is probable are very different, but surprisingly their roots are the same!

A Wordlover's Dream

A wonderful resource for exploring word origins is Steven Schwartzman's *The Words of Mathematics* (1994). With over 1,500 entries, this intriguing dictionary delves into the etymology of our most important mathematical terminology. From *a posteriori* (reasoning after the fact) to *zonohedron* (a convex polyhedron bounded by parallelograms), the book gives both the meaning of each mathematical term and its roots, so that connections can be made with other math words and with ordinary English words sharing the same origins. The entries are extensively cross-referenced and illustrated. An Introduction gives an overview of the interconnection between Greek, Latin, French, English and other languages, while an Appendix groups the entries of the dictionary according to their earliest known roots. With the help of his extensive training in both mathematics and linguistics, Schwartzman is able to trace the origin of a term back in time all the way to its roots in Indo-European or other precursor languages. The result makes for a fascinating read as well as an indispensable reference tool.

Steven Schwartzman, *The Words of Mathematics: An Etymological Dictionary of Mathematical Terms Used in English*, MAA, 262 pp., Paperbound, 1994. ISBN 0-88385-511-9.

In a court trial, evidence is called *probative* if it helps prove the innocence or guilt of the accused. But unless and until guilt is proven by a preponderance of evidence (or, in criminal cases, beyond any reasonable doubt), the *status* of the accused is that of an innocent person. *Status*, meaning a condition or a standing, comes from the noun form of the Latin verb *stare* (to stand), as do the following words:

stationary — standing still

statue — a model that stands still

station — a place at which to stand

statics — the physics of bodies at rest

state — the set of conditions, or the government, of a certain area, as in State of the Union Address, or State of Michigan.

The word **statistics** was first used by the Germans (*Statistik*) to mean political science, the study of (political) states, which includes data about those states. Today, "statistics" means the collection, organization, analysis, and interpretation of quantitative data. Going back to the original meaning, statistics tell you how things "stand."

Recognizing word origins can be helpful in making sense of mathematical terms and ideas, and the etymologies also provide a window onto the very lively history of our science and its connections with other subjects. ∎

References

1. David H. Fowler, *The Mathematics of Plato's Academy: A New Reconstruction.* Clarendon Press, 1987.

2. Steven Schwartzman, *The Words of Mathematics.* Mathematical Association of America, 1994.

3. David Eugene Smith and Jekuthiel Ginsburg, "From Numbers to Numerals and From Numerals to Computation", in James R. Newman, *The World of Mathematics.* Microsoft Press, 1988.

4. The MacTutor History of Mathematics archive, `www-groups.dcs.st-and.ac.uk/history/`.

5. Earliest Known Uses of Some of the Words of Mathematics, `members.aol.com/jeff570/mathword.html`.

The Magician of Budapest

PETER SCHUMER
Middlebury College

As the millennium draws to a close, many people are busy creating and debating "top ten" lists — greatest movies of all time, best fiction books of the decade, most successful musicians of the century, most significant inventions, and so on. One name which would appear on almost all such rankings of the top ten mathematicians of this century would be that of Paul Erdős (pronounced, approximately, Air Dish). In his lifetime, Erdős wrote or co-wrote nearly 1500 mathematical articles (the equivalent of a research paper every two weeks for 60 years!) He did significant work in number theory, geometry, graph theory, combinatorics, Ramsey theory, set theory, and function theory. He helped create probabilistic number theory, extremal graph theory, the probabilistic method, and much of what is now referred to as discrete mathematics. His influence on fellow mathematicians and on mathematics as a whole is bound to last for centuries to come. His reputation for being published in so many journals and in so many languages led to the following limerick:

A conjecture thought to be sound
Was that every circle was round
In a paper of Erdős
Written in Kurdish
A counterexample is found!

Epsilon Years

Paul Erdős was born in Budapest, Hungary on March 26, 1913, the son of math and physics teachers, Anna and Lajos Erdős. Paul's two older sisters died of scarlet fever while his mother remained in the hospital following his birth. This family tragedy resulted in a very close but over-protective home environment where Paul was home-schooled and his mathematical genius was able to flourish. Unfortunately, it also resulted in a socially awkward and eccentric individual who depended heavily on the care and goodwill of his many friends sprinkled across the globe.

Like the great German mathematician Carl Friedrich Gauss, Erdős's mathematical talents blossomed early. At age three Paul discovered negative numbers when he correctly subtracted 250 from 100. By age four he could multiply three-digit numbers in his head. He often entertained family friends by asking them for their birthday and then telling them how many seconds they had been alive.

At a young age Paul's father taught him two theorems about primes (i) that there are infinitely many primes, while at the same time (ii) there are arbitrarily large gaps between successive primes. To Paul, the results seemed almost paradoxical, but they led to a deep fascination with prime numbers and to a quest for a better understanding of their complicated arrangement. Again, like Gauss, an early

Paul Erdős delivering his "Sixty Years of Mathematics" lecture at Trinity College, University of Cambridge in June, 1991, the day before receiving Cambridge's prestigious honorary doctorate. Photo by George Csicsery from his documentary film N is a Number: A Portrait of Paul Erdős.

Reprinted from April 1999, pp. 5–9

fascination with number theory was the impetus for Erdős's lifetime dedication to the world of mathematics.

Another early mathematical influence was the *Hungarian Mathematical Journal for Secondary Schools*, commonly referred to as *KoMal*. The most popular part of the journal was a regular problem section where student solutions were published and name credit given for correct solutions. At the end of the year the pictures of the most prolific problem solvers were included in the journal. In this way, the best mathematics and science students in Hungary were introduced to one another. In this sense, Paul's "publications" began when he was barely thirteen years old.

The Magician of Budapest

Paul's first significant result was a new elementary proof of Bertrand's Postulate, which he discovered as an 18-year-old university student. The French mathematician, J.L.F. Bertrand, a child prodigy himself, conjectured that for any natural number $n > 1$, there was always a prime between n and $2n$. Bertrand verified the result up to $n = 3,000,000$, but was unable to prove it. Five years later, in 1850, the Russian P.L. Chebyshev proved what has become known as Bertrand's Postulate. The proof was difficult and relied heavily on analytic methods (function theory and advanced calculus). Erdős created a new proof which, though quite intricate, was elementary in the sense that no calculus or other seemingly superfluous analytical methods were used. This proof combined with other related results on primes in various arithmetic progressions constituted his doctoral dissertation.

Another early success was a generalization of an interesting observation by one of his friends. Esther Klein noticed and proved that for any five points, no three collinear, in the plane, it is always the case that four of them can be chosen which form the vertices of a convex quadrilateral. What Erdős and George Szekeres were able to show was that for any n there is a number N, depending only on n, so that any N points in the plane (with no

Photo by George Csicsery from his documentary film N is a Number: A Portrait of Paul Erdős.

three collinear) have a subset of n points forming a convex n-gon. Since Esther and George became romantically involved during this period and later married, the result was dubbed the Happy End Problem. Furthermore, Erdős and Szekeres conjectured that in fact the smallest such N will always be $N = 2^{n-2}+1$. Interestingly, the more general conjecture still has not been proven. However, the Happy End Problem was a harbinger of much of Erdős's later work — fruitful collaborations and fascinating conjectures. A good theorem often creates more questions than it answers.

Erdős also proved a neat result about abundant numbers. Let $s(n)$ represent the sum of the proper divisors of n (i.e., all positive divisors except n itself). Then n is called *deficient, perfect,* or *abundant* if $s(n)$ is less than, equal to, or greater than n, respectively. Such numbers have been studied since the time of Pythagoras. The German mathematician Issai Schur conjectured that the set of abundant numbers had positive density. That is, let $A(x)$ be the number of abundant numbers less than or equal to x, then Schur's conjecture states that $\lim_{x \to \infty} A(x)/x$ exists and is strictly greater than 0. Erdős's brilliant proof of this result led Schur to dub Erdős "the Magician of Budapest."

With life in Hungary ever worsening for Jewish intellectuals, Erdős obtained a fellowship to the University of Manchester in England. In 1934 Erdős traveled there via Cambridge University and began his mathematical travels and worldwide collaborations which never let up until his death. It was fourteen years before Erdős was able to return to Budapest where his mother had miraculously survived the war. Unfortunately, Paul's father died during that period and four out of five of his aunts and uncles perished in the Holocaust. Paul's beloved mother spent much of the rest of her life traveling with her son and using her apartment as a repository for his ever-increasing mountain of reprints.

Though Erdős's relationship with the American government was generally harmonious, it wasn't so during the McCarthy era. In 1954 while on a temporary faculty position at Notre Dame, Erdős wished to attend the International Congress of Mathematicians in Amsterdam. Knowing he came from a Communist country, an agent from the Immigration and Naturalization Service interviewed Erdős and asked him what he thought of Karl Marx. Erdős replied, "I'm not competent to judge. But no doubt he was a great man." Perhaps due to this, Erdős was denied a re-entry visa after attending the Mathematical Congress. Strong support and letters to state senators from many American mathematicians finally resulted in Erdős being allowed to return to the U.S. in 1959. From that point on, he could come and go freely.

His Brain is Open

To all who knew Erdős, it appeared that he spent ninety-nine percent of his wakeful hours obsessed with mathematics (though he somehow developed a deceptive skill at both table tennis and the game of Go). Twenty hours of work a day was not at all unusual. Upon arriving at a meeting, he would announce, in his thick Hungarian accent, "my brain is open." At par-

ties, he would often stand alone oblivious to all else, deep in thought pondering some difficult argument. When being introduced to a math graduate student, it was not unusual for him to ask, "What's your problem?" One would normally be taken aback by such a remark if it were uttered by a stranger in less than friendly surroundings, but with Erdős it was clearly meant as a friendly "hello." He was taking you seriously as a fellow dweller in the world of mathematics.

One of Erdős's greatest triumphs was his elementary proof of the Prime Number Theorem (PNT). The PNT describes the asymptotic distribution of the prime numbers and variants of it were conjectured by both Gauss and Legendre in the late 1700's. Specifically, let $\pi(x)$ be the number of primes less than or equal to x and let $Li(x) = \int_2^x dt/\log t$. The PNT states that $\lim_{x \to \infty} \pi(x)/Li(x) = 1$, that is, the number of primes less than x is asymptotic to $Li(x)$.

Significant progress towards a proof of the PNT was made by Chebyshev in the 1850's and by G. B. Riemann in 1859. Riemann's contribution was based on a deep and careful study of the complex-valued zeta function. Finally, in 1896 the

> Mathematicians are often divided into two camps: theory builders and problem solvers. Erdős was definitely a problem solver. Here are a few of Erdős's many theorems.
>
> 1. There are infinitely many odd integers that are not of the form $p + 2^k$ where p is a prime.
>
> 2. The product of two or more consecutive positive integers is never a square or any other higher power.
>
> 3. A connected graph with minimum degree d and at least $2d+1$ vertices has a path of length at least $2d+1$.
>
> 4. Let p_n be the nth prime number. Then the set of limit points of the set $\{(p_{n+1} - p_n)/\log n\}$ has positive density.
>
> 5. Let $d(n)$ be the number of positive divisors of n. Then $\sum_{n=1}^{\infty} d(n)/2^n$ is irrational.

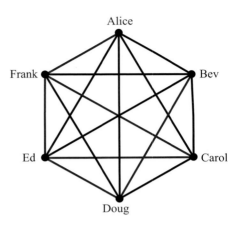

The party problem: if blue edges connect friends and black edges connect strangers, then Alice, Bev, and Ed are mutual strangers. There is no way to color the edges without forming either a blue or black triangle.

French mathematician J. Hadamard and the Belgian mathematician C.J. de la Vallée Poussin each proved the result using delicate arguments from complex function theory. Many feel that the proof of the PNT was the mathematical capstone of the nineteenth century.

In the first half of the twentieth century, the search for an "elementary" proof of the PNT seemed hopeless. However, in 1949 Paul Erdős and the Norwegian mathematician Atle Selberg, working in tandem but not together, found such proofs. Selberg won a Fields Medal for his work while Erdős won the prestigious Cole Prize in algebra and number theory for his contribution.

Another area which fascinated Erdős was classical Ramsey theory which describes the number of ways of partitioning a given set into a given number of subsets. The party problem is the classic example: at a party with six individuals, prove that there must always be at least three mutual friends or three mutual strangers. More generally define $r(u, v)$ to be the smallest integer r such that if the edges of a complete graph on r vertices are colored one of two colors, then there must be a complete subgraph on u vertices of one color or a complete subgraph on v vertices of the other color. Thus, the solution to the party problem amounts to proving that $r(3, 3) = 6$. The values for $r(3, n)$ are also known for $n = 4$,

5, 6, 7, and 9. In addition, it has been shown with significantly more work that $r(4,4) = 18$. Erdős showed $r(k, l) \le C(k+l-2, k-l)$, the number of combinations of $k+l-2$ objects chosen $k-l$ at a time. Erdős offered $250 to anyone who could prove that $\lim_{n \to \infty} r(n, n)^{1/n}$ exists. If the limit does exist, it's known to be between $\sqrt{2}$ and 4. Interestingly, the value of $r(5, 5)$ is still unknown, though it must be at least 43 and at most 49. Erdős was fond of saying that if an evil spirit was going to destroy the world unless we could determine $r(5,5)$, then it would be prudent for all of humanity to devote all of its resources to this problem. On the other hand, if the evil spirit insisted on knowing $r(6,6)$, then it would be best for all of us to try to destroy the evil spirit!

Erdős not only loved working on difficult problems and proving theorems, but always strived for the most elegant and direct proof. He had unique religious views and referred to the Almighty as the *SF* (or *Supreme Fascist*). Erdős felt that he was forever in the midst of an ongoing personal battle with the SF. However, one positive aspect was that the SF kept a secret book, *The Book*, which had all the theorems that would ever or could ever be discovered along with the simplest and most elegant proofs for each one. The highest compliment Erdős would give someone was that their proof was "one from The Book."

Nothing was more exciting to Erdős than to discover a mathematically talented child and to excite him or her about mathematics. In 1959 Erdős arranged to have lunch with a very precocious 11-year old, Lajos Posá. Erdős challenged the youngster to show why if $n+1$ integers are chosen from the set $\{1, 2, ..., 2n\}$, then there must be two chosen numbers which are relatively prime. Clearly the set of even numbers less than or equal to $2n$ does not have this property and shows that just choosing n such numbers is not sufficient. According to Erdős, within half a minute Posá solved the problem by making the striking observation that two consecutive integers must always be chosen. Erdős commented that rather than eating soup, perhaps champagne would have been more appropriate for this occasion!

Erdős's greatest influence on fellow mathematicians, young and old alike, was his continual outpouring of new conjectures coupled with various monetary rewards. Some problems had a price tag of just a few dollars while others went for several thousand dollars. To solve an Erdős problem is considered a great accomplishment, and the larger the reward the more difficult Erdős considered the problem to be. Erdős was often asked what would happen if all his problems were solved simultaneously. Could he possibly pay up? His answer was, "Of course not. But what would happen if all the depositors went to every bank and demanded all their money. Of course the banks could not pay — and besides, it's more likely that everyone will simultaneously ask for their money than that all of my problems will be solved suddenly."

Poor Great Old Man

Erdős was almost as well-known for his eccentricities as he was for his brilliant mind. By his own admission, Erdős never attempted to butter his own toast until he was already an adult. "It turned out not to be too difficult," he admitted. Many mathematicians have had the experience of either walking or driving Erdős to his next commitment, only to learn after some time that he assumed you knew where he was supposed to be going.

Erdős had an aversion to old age and infirmities and an obsession with death. When breaking off work for the night he would say, "We'll continue tomorrow — if I live." At age 60, Erdős started appending acronyms to his name. The letters

---

Here are a few outstanding conjectures and open problems that Erdős left us:

1. Do there exist infinitely many primes p such that every even number less than or equal to $p - 3$ can be expressed as the difference between two primes each at most p? For example, 13 is such a prime since $10 = 13 - 3, 8 = 11 - 3, 6 = 11 - 5, 4 = 7 - 3$, and $2 = 5 - 3$. The smallest prime not satisfying this condition is $p = 97$.

2. For every integer n are there n distinct integers for which the sum of any pair is a square? For example, for $n = 5$, the numbers: $-4878, 4978, 6903, 12978$, and 31122 have this property.

3. Can you find a polynomial $P(x)$ for which all sums $P(a) + P(b)$ are distinct with $0 \le a < b$? For example, $P(x) = x^3$ doesn't work since $10^3 + 9^3 = 12^3 + 1^3$. However, $P(x) = x^5$ is considered a likely candidate.

4. Are there infinitely many primes p for which $p - n!$ is composite for all n such that $1 \le n! < p$? For example, when $p = 101, 101 - n!$ is composite for $n = 1, 2, 3$, and 4.

5. A natural number n is *pseudoperfect* if it is abundant and it is also the sum of some of its proper divisors.

For example, $n = 66$ is pseudoperfect since $66 = 11 + 22 + 33$. A number is *weird* if it is abundant but not pseudoperfect. Try verifying that $n = 70$ is a weird number (sorry Mark McGwire). For ten dollars, are there any odd weird numbers?

6. A system of congruences a_i (mod n_i) where $n_1 < n_2 < \cdots < n_k$ is a *covering system* if every integer satisfies at least one of the congruences. For example, 0 (mod 2), 0 (mod 3), 1 (mod 4), 5 (mod 6), and 7 (mod 12) form a covering system. Erdős conjectured that for every c, there is a covering system with $n_1 \ge c$. (Currently $c = 24$ is the largest value that has been constructed.) An open question of Erdős and J. Selfridge asks whether there is a covering system with all moduli odd.

7. A $5000 conjecture: Let $A = \{a_i\}$ be any sequence of natural numbers for which $\sum_{i=1}^{\infty} 1/a_i$ diverges. Is it true that A must contain arbitrarily long arithmetic progressions? If so, one consequence would be that the set of primes contained arbitrarily long arithmetic progressions. Currently the record is 22 consecutive primes.

pgom stood for "poor great old man." Five years later he added *ld* for "living dead" and so on. Eventually he got to *cd* denoting "counts dead." Erdős explained this as follows: The Hungarian Academy of Sciences has a strict limit on the total number of members that it can have at one time. However, once you reach the age of 75, while you are still a member, you are not counted against the total. Therefore, at that point, you're counted as if you were dead.

When asked "how old are you?" Erdős would answer that he was two-and-a-half billion years old. After all, when he was a child he was taught that the earth was two billion years old and now they say it's four-and-a-half billion years old! Some say that Erdős was the Bob Hope of mathematicians. Not only did he share humorous stories, but he spent a great deal of time traveling and, by lecturing, which he called *preaching*, raising the morale of the mathematical troops.

Erdős was well-known for a shorthand language many call Erdősese. An *epsilon* was a child, *poison* meant alcoholic drink (which he carefully avoided), *noise* meant music, *boss* was wife, and *slave* husband. If someone were *captured* that meant they got married, while *liberated* stood for divorced. If a mathematician stopped publishing he *died*, while actually dying was referred to as *having left*. Nothing bothered Erdős more than political strictures which did not allow for complete freedom of expression and the ability to travel freely. Erdős referred to the Soviet Union as *Joe* (for Joseph Stalin) and the United States as *Sam*.

No account of Paul Erdős would be complete without mentioning the concept of Erdős numbers. Erdős himself had Erdős number zero. Anyone who co-authored a paper with him (there are at least 472 such people) has Erdős number one. Those who do not have Erdős number one, but co-authored a paper with someone who does, are assigned Erdős number two (there are at least 5000 such people), and so on. The largest Erdős number believed to exist is seven. Erdős himself added an interesting wrinkle for those having Erdős number one. He claimed that one should instead be assigned Erdős number $1/n$ if you co-authored n papers with him. The lowest Erdős number in this case would be held by Andras Sarkozy with number $1/57$ — just edging out Andras Hajnal with number $1/54$. (For more on Erdős numbers, see [3].)

Having Left

Paul Erdős received countless honorary degrees and his work was and continues to be the focus of many international conferences. In 1984, Erdős received the prestigious Wolf Prize for his lifetime's contributions to the world of mathematics. Of the $50,000 awarded, he immediately donated $49,280 to an Israeli scholarship named in his mother's honor. On other occasions, he donated money to Srinivasa Ramanujan's widow, to a student who needed money to attend graduate school, to a classical music station, and to several Native American causes. Always traveling with a single shabby suitcase which doubled as a briefcase, he had little need or interest in the material world. He had no home and nearly no possessions. Without hesitation, he once asked the versatile Canadian mathematician Richard Guy for a $100 adding, "You're a rich man." Richard Guy gladly gave him the money. Later Guy poignantly noted, "Yes, I was. I knew Paul Erdős."

Paul Erdős was a very versatile and creative mathematician. The vast quantity of his research output alone qualifies him as the Euler of our age. He far surpassed Einstein's ultimate litmus test for success which was to be highly esteemed by one's colleagues. Once during a lecture by the late number theorist Daniel Shanks, a long computer-generated computation resulted in a 16-digit number. Shanks, who was not known for sprinkling praise lightly said, "I don't know if anyone really understands numbers like these — well, maybe Erdős."

Paul Erdős's fantasy was that he would die in the midst of a lecture. After proving an interesting result, a voice from the audience would pipe up, "but what about the general case?" Erdős would reply, "I leave that to the next generation" and then immediately drop dead. In fact, Erdős left us on September 20, 1996 while attending a conference in Warsaw. Let's hope he is now realizing his dream of collaborating with Euclid and Archimedes simultaneously. They'd probably relish having an Erdős number of one.

Was he one of the great mathematicians of the century? Echoing Erdős himself, I'm not competent to judge. But no doubt he was a great man. ∎

For More Information

1. George Csicsery, Director, *To Prove and Conjecture: Excerpts from Three Lectures by Paul Erdős*, Film, MAA.
2. ———, Director, *N is a Number: A Portrait of Paul Erdős*, Documentary Film, A. K. Peters Ltd, 1993.
3. John M. Harris and Michael J. Mossinghoff, The Eccentricities of Actors, *Math Horizons*, February 1998, 23–25. Reprinted in this volume, pp. 77–80.
4. Paul Hoffman, *The Man Who Loved Only Numbers*, Hyperion, 1998.
5. Bruce Schechter, *My Brain Is Open: The Mathematical Journey of Paul Erdős*, Simon and Schuster, 1998.

Turning Theorems into Plays

STEPHEN D. ABBOTT
Middlebury College

After overhearing the gossip of the house staff, the young Thomasina Coverly interrupts her algebra lesson on Fermat's Last Theorem to ask her tutor, Septimus Hodge, the meaning of "carnal embrace." Septimus's first explanation, "Throwing one's arms around a side of beef," proves unacceptable as Thomasina points out that it was a certain Mrs. Chater who was discovered in carnal embrace in the garden gazebo. "I don't think you have been candid with me Septimus," Thomasina insists. "A gazebo is not, after all, a meat larder." "Ah yes, I am ashamed," Septimus finally concedes. With blunt, almost medical precision, he then provides his talented pupil with a terse description of intercourse, explaining that carnal embrace is actually sexual congress between males and females...

SEPTIMUS: ... for purposes of procreation and pleasure. Fermat's last theorem, by contrast, asserts that when x, y, and z are whole numbers each raised to the power of n, the sum of the first two can never equal the third when n is greater than 2.

THOMASINA: Eurghhh!

SEPTIMUS: Nevertheless, that is the theorem.

THOMASINA: It is disgusting and incomprehensible.

This delightful exchange from the opening moments of Tom Stoppard's *Arcadia* offers a vivid example of why the playwright himself would wonder aloud if his writing is best described as "seriousness compromised by my frivolity, or ... frivolity redeemed by my seriousness." The frivolous charm of this scene is easy to appreciate. What is not so obvious is that this banter between Thomasina Coverly and her tutor is our first clue to the ensuing debate between what can loosely be categorized as the classical and the romantic. *Arcadia* opens in the Coverly family's stately home. The year is 1809 and Thomasina, along with the rest of western culture, is emerging from the logical rigors of the Enlightenment to discover the allure of the Romantic era. Lord Byron is a house guest (though we never meet him); the pastoral gardens are being turned under in favor of gothic ruins; and Thomasina is perplexed by the fact that no matter what she does with her spoon, the red jam in her rice pudding eventually turns the entire dish an even shade of pink. "Do you think this is odd," she muses, "you cannot stir things apart." And so we have our second hint that, like nearly all of Stoppard's writing, this comedy is really a play of ideas. Throughout his work, theorems of mathematics and laws of physics habitually appear unannounced in the most unlikely places, and it is with *Arcadia* that Stoppard's affinity for the mathematical sciences reaches its pinnacle. By creating a visual image of entropy in rice pudding, and by pushing Thomasina's romantic education incongruously up against her lesson in number theory, Stoppard is setting the stage for what is to be a provocative exploration of the human implications inherent in the confrontation of classical Euclidean geometry and Newtonian physics with chaos theory and the second law of thermodynamics.

Arcadia is just one of over 40 scripts that Tom Stoppard has written for the stage, radio, television and film. His recent Academy Award for *Shakespeare in Love* has no doubt boosted his public recognition, but critics of modern theater have for several decades regarded Stoppard among the leading active playwrights writing in English. Ironically, this extremely British author was born Tomas Straussler to Czechoslovakian parents in 1937. The family lived for a while in Singapore, and then evacuated to India before World War II, although Tom's father stayed behind and was killed in the ensuing Japanese invasion. In 1946, Tom's mother married Major Kenneth Stoppard, a British army officer on duty in India, and the new family eventually found its way to Bristol, England around 1950. Adding to the confusion is that this writer of high ideas actually quit school at age seventeen to become a journalist, and eventually theater reviewer, for the *Bristol Evening World*. When asked, Stoppard unequivocally asserts that he is an Englishman, but this has not prevented proliferation of the theory that only someone with an external perspective on the culture could write dialogue with such a mastery of British language and nuance. In a similar way, being self-taught has meant Stoppard's understanding of

Photo by Peg Skorpinski.

Tom Stoppard

mathematical and philosophical concepts has come without the bias and clichés of any standard approach. The results are a refreshing and often hilarious interpretation by the artist of some of science's great achievements.

"It's poetry to me."

To be sure, the themes explored in Stoppard's plays are by no means limited to those in the sciences, nor do they respect the usual boundaries between academic disciplines. The characters in *Travesties* (1974) include Lenin, James Joyce, and Tristan Tzara, one of the founders of the artistic movement Dada. But mathematical allusions are ubiquitous, even in Stoppard's earliest work. *Albert's Bridge* is a radio play first performed in 1967 about a philosophy graduate (Albert) whose chronic disinterest in the world leads him to take a job working with three working class types painting the Clufton Bay Bridge. (It should be pointed out now that academics and other victims of higher education never fare very well in Stoppard's plays.) The particular paint they use requires re-painting every two years which is precisely the length of time it takes the four of them to complete the job. Thus, whenever the team finishes the last steel girder in the span, they return the next morning to the other side and begin all over again. What has been a twenty-year Möbius nightmare for Dad, the oldest of the painters, is actually a great relief to Albert who finds his only solace in the concreteness of his work high above the city. Consequently, it is Albert who gladly volunteers for the lonely duty required in a money-saving plan hatched by Fitch, the "clipped, confident, rimlessly-eyeglassed" town supervisor.

FITCH: You see, to date we have achieved your optimum efficiency by employing four men. It takes them two years to paint the bridge, which is the length of time the paint lasts. This new paint will last eight years, and so we only need one painter to paint the bridge by himself. After eight years, the end he started at will be just ready for re-painting. The savings to the ratepayers would be £3,529 15s. 9d. per annum.

The mathematics is not simply incidental; nothing is incidental in Stoppard's scripts. The rigidity of the algebra of this eighth-grade word problem is accented by the backdrop of the steel beams of the "fourth biggest single-span double-track shore-to-shore railway bridge in the world bar none—". Albert, meanwhile, is an amorphous soul who is utterly seduced by the definitive, angled, visible grandeur of the bridge.

FITCH: I'm the same. It's poetry to me—a perfect equation of space, time and energy—

ALBERT: Yes—

FITCH: It's not just slapping paint on a girder—

ALBERT: No—

FITCH: It's continuity-control-mathematics.

ALBERT: Poetry.

FITCH: Yes, I should have known it was a job for a university man... You'll stick to it for eight years will you?

ALBERT: Oh, I'll paint it more than once.

Some of the play's most comic moments occur when Albert is joined by Fraser, a would-be suicidal personality who, every time he climbs the bridge to jump is so calmed by the perspective that he loses the desire to kill himself. Albert is profoundly annoyed by this habit, and it is during their conversations that the flaw in Fitch's algebra is slowly revealed. After two years, Albert has painted only a quarter of the bridge, leaving three quarters of the bridge exposed under cracking two-year paint. As the town revolts at the sight of the decaying girders, Fitch's panicked solution is to send an army of painters out to finish the job in a single day. Forgetting to break stride as they march onto the top of the bridge, the resonating frequencies of the 1800 collective footsteps brings the bridge, Albert, and the play to a crashing end.

"It must be indicative of something besides the redistribution of wealth."

Stoppard's first major success as a playwright, and probably still his best known work, is *Rosencrantz and Guildenstern Are Dead*, first performed in its present form in 1967. The play tells

Alex Cranmer and Richard Price as Rosencrantz and Guildenstern in a Middlebury College production. Photo by Alex Fuller.

the story of *Hamlet* from the point of view of Shakespeare's two minor characters charged with investigating the source of Hamlet's lunacy and ultimately responsible for delivering the prince to England to be killed. The curtain rises (in Stoppard's script) to find the two misplaced Elizabethans betting on the flip of a coin; heads and the coin goes to Rosencrantz, tails and it belongs to Guildenstern. It is immediately clear that something is amiss. Each flip we witness turns up heads, and Rosencrantz's heavy bag of coins indicates that this has been happening for quite some time. Rosencrantz is embarrassed to be taking so much money from his friend, but seems uninterested in considering the matter much further. Guildenstern could care less about the money but is clearly disturbed by the implications.

GUIL: This must be indicative of something, besides the redistribution of wealth. List of possible explanations. One: I am willing it. Inside where nothing shows, I am the essence of a man spinning double-headed coins, and betting against himself in private atonement for an unremembered past.

ROS: Heads.

GUIL: Two: Time has stopped dead, and the single experience of one coin being spun has been repeated ninety times...(He flips a coin and tosses it to Ros.) On the whole doubtful. Three: divine intervention... Four: a spectacular vindication of the principle that each individual coin spun individually (he spins one) is as likely to come down heads as tails and therefore should cause no surprise each individual time it does. (It does. He tosses it to Ros.)

The mathematical aura of this scene runs deep. Beyond the explicit references to probability, Guildenstern (who is acting like the analytical "heads" to Rosencrantz's obtuse "tails," at least at this point in the play) proposes that the "scientific approach to the examination of phenomena is a defense against the pure emotion of fear," and begins organizing his arguments into logical syllogisms.

GUIL: One, probability is a factor which operates within natural forces. Two, probability is not operating as a factor. Three, we are now within un-, sub- or supernatural forces. Discuss.

But the heady Guildenstern is not done yet. Moments later, in true Lewis Carroll fashion, he attempts to turn his own logic back on itself.

GUIL: ...If we postulate, and we just have, that within un-, sub- or supernatural forces the probability is that the law of probability will not operate as a factor, then we must accept that the probability of the first part will not operate as a factor, in which case the law of probability will operate as a factor within un-, sub or supernatural forces. And since it obviously hasn't been doing so, we can take it that we are not held within un-, sub- or supernatural forces after all; in all probability, that is.

The coin is a multifaceted device. In addition to setting the existential tone in its refusal to obey the law of large numbers, it also points toward the symbiotic relationship between Stoppard's script and the script of *Hamlet*. As the Player, who is also present in both plays, says, "[It] is a kind of integrity, if you look on every exit being an entrance somewhere else." In fact, the coin does eventually come up tails—at precisely the moment when Hamlet and Ophelia swoon on-stage and the action is taken over by *Hamlet*, Act 2, Scene 1.

The Player's line in context is explicitly about his own life as a traveling actor, but it is also clearly intended as a comment on the *R&G–Hamlet* relationship. This is just one of the many ways Stoppard's play finds to talk about itself. Advertising his troop's talents, the Player asserts, "It costs little to watch, and little more if you happen to get caught up in the action." These self-referential moments abound and suggest a sympathy on the part of Stoppard for the type of recursive thinking beloved by mathematicians and crucial to such logical delights as Russell's paradox and Gödel's incompleteness theorems. At one point, Rosencrantz and Guildenstern stumble upon the troop of players rehearsing the production of *The Murder of Gonzago* requested by Hamlet to "catch the conscience of the king." Their play, however, is no less than a mimed *Hamlet* which ends with a provocative scene of Rosencrantz and Guildenstern unsuspectingly witnessing a portrayal of their own murders at the hands of the English king. In the film version of *R&G*, Stoppard even adds a puppet show of Gonzago as part of the troops' rendition of *Hamlet* happening within Stoppard's *R&G* which is inside Shakespeare's *Hamlet*. All of this is, of course, taking place in front of an audience, who, judging from the fact that Rosencrantz and Guildenstern are caught unaware in this strange loop, should not necessarily assume that the recursive levels end with the theater in which they sit.

"Saint Sebastian died of fright."

Jumpers, which followed *R&G* in 1972, is the staging of a debate, both serious and farcical, over the question of whether virtue is a social convention that has evolved as a means to keep society running smoothly (and so is open to arbitrary changes) or if goodness is absolute, existing unaltered outside of the human frame of reference. Making the case for the existence of moral absolutes is George Moore, a professor of philosophy and so, consequently, a bumbling fellow. Ultimately, though, his views are probably closest to Stoppard's own and so he is given moments of sense and sympathy during the intermittent occasions throughout the play when he dictates his speech to be given at the evening's symposium, "Man—Good, Bad or Indifferent." Meanwhile, providing con-

The Metropolitan Museum of Art, Gwynne Andrews, Rogers and Harris Brisbane Dick Funds, 1948.

Saint Sebastian, dead of fright. By Francesco Botticini

text for the academic discussion, George's debating partner's mysteriously murdered body is hanging on the back of his bedroom door. It is being concealed by his potentially guilty wife who is possibly having an affair with Sir Archibald Jumper, George's philosophical adversary and coach of a gymnastics team consisting roughly of the members of the philosophy department.

The comic twists of George's long dictation keep the audience entertained, but some underlying substance is required to keep the argument, and the play, from collapsing into triviality. What is fascinating is to see how heavily the untrained Stoppard relies on mathematics to help George analyze his opening question, "Is God?" The main attraction seems to be mathematicians' extensive experience working with the infinite. Making a loose analogy between the open interval from zero to one and the infinitude of time and space, George surprises himself by noticing, "But the fact is, the first term of the series is not an infinite fraction but zero. It exists. God, so to speak, is naught. Interesting." He (somewhat cryptically) cites Georg Cantor and Bertrand Russell, and then aggressively sets his sights on the classical paradoxes of Zeno. One of Zeno's paradoxes states that "a tortoise, given a head start in a race with, say, a hare, could never be overtaken." This is so, Zeno argued, because every time the hare reaches the place where the tortoise was, the tortoise has meanwhile moved ahead ever so slightly. For effect, George has brought a tortoise and a hare to assist with a demonstration but Thumper, the hare, has escaped from his box. George has also brought his bow and arrow to illustrate another paradox of Zeno who, in George's summary, said that "an arrow shot towards a target first had to cover half the distance, and then half the remainder, and then half the remainder after that, and so on ad infinitum, the result was, as I will now demonstrate, that though an arrow is always approaching its target, it never quite gets there, and Saint Sebastian died of fright."

George's argument for God as the "First Cause" is threatened by Zeno's notion of an infinity without an end or a beginning. To dramatically lay to rest any doubt that an infinite number of events can occur within a finite amount of time and space, he confidently notches the arrow in his bow but, startled by his wife's cry for help, fires it over the wardrobe where—as we sadly discover at the end of Act 2—it fatally impales poor Thumper.

"You get what you interrogate for."

Stoppard has said that the easiest part of creating plays is writing dialogue and the hardest part is finding material about which to write. As we have seen, Stoppard often piggy-backs off of other texts, and several times he has used historical figures in his writing. Another recurring device in Stoppard's work is the development of extended metaphors. In 1984, Stoppard wrote *Squaring the Circle*, a drama-documentary for television about Lech Walesa and the Solidarity movement in Poland. In the opening scene the narrator explains, "Between August 1980 and

December 1981, an attempt was made in Poland to put together two ideas that wouldn't fit, the idea of freedom as it is understood in the West, and the idea of socialism as it is understood in the Soviet empire. The attempt failed because it was impossible, in the same sense as it is impossible in geometry to turn a circle into a square with the same area—not because no one has found out how to do it but because there is no way in which it can be done." Now in this case it is clear that the subject matter of the play came first and that the analogy to geometry occurred as an afterthought. However, in the case of *Hapgood* (1988), Stoppard's cold-war espionage thriller, the situation is certainly reversed.

The central metaphor of Hapgood is the wave/particle duality of light described in quantum theory. A Russian scientist named Kerner who is supposedly working for British intelligence as a double agent is suspected of leaking his top secret research back to Moscow. When confronted for the truth, he responds with a poetic description of the famous double-slit experiment where light from a laser passing through the two slits forms either an interference pattern (unique to waves) or a double-peaked distribution (indicative of particles) depending on how the experiment is conducted. He then concludes with the famous adage, "The act of observing determines the reality... There is no explanation in classical physics. Somehow light is both wave and particle. The experimenter makes the choice. You get what you interrogate for." But this lesson in nuclear physics is not just to help us understand the nature of double agents; it is meant to be applied to human personality in general. Elizabeth Hapgood is a tough and classy agent who plays chess without a board and runs an otherwise all-male British intelligence office. Kerner was "turned" by Hapgood, and we learn that not only did she fall in love with her scientist but that he is also the father of her twelve-year-old son, Joe. When the goods from an intricately choreographed drop-off get mysteriously switched, Hapgood suspects an abrasive, somewhat crude agent named Ridley. Her suspicions gain momentum when, in a moment of what might be termed gratuitous mathematics, she learns that there are actually two of them.

KERNER: An ancient amusement of the people of Königsberg was to try to cross all the seven bridges without crossing any of them twice. It looked possible but nobody had solved it... Leonard Euler took up the problem of the seven bridges and he presented his solution in the form of a general principle based on vertices... When I looked at Wates's diagram (of Ridley's path during the opening drop-off) I saw that Euler had already done the proof. It was the bridges of Königsberg, only simpler.

HAPGOOD: What did Euler prove?

KERNER: It can't be done, you need two walkers.

Thus, like an electron whose location is not uniquely specified by the laws of nuclear physics, Ridley has for years been serving as his own alibi. Ridley's twin, however, is not as interesting as Hapgood's. In a plan to snare Ridley, Hapgood creates her own foul-mouthed twin-sister named Celia who conspires with Ridley to help rescue Hapgood's son from (fabricated) Russian kidnappers. The irony is deep though, as it is the "bad guy" Ridley's fondness for Hapgood that makes him willing to risk himself to save young Joe while Hapgood's superior shows no scruples about using the child as bait. Hapgood begins to realize this and—through the voice of Celia—attempts to warn Ridley. Unable to comprehend the paradox of what is happening, Ridley is confused to the point of shouting "Who the hell are you!" "I am your dreamgirl Ernie," she replies. "Hapgood, without the brains or the taste."

In the play's denouement, Hapgood is forced to shoot Ridley and there is a brief but poignant moment after the stretcher is taken away when she hopelessly asserts, "It was the shoulder." The point is that in creating Celia, Hapgood has "made up the truth." As Kerner says early in the play, "The one who puts on the clothes in the morning is the working majority, but at night—perhaps in the moment before unconsciousness—we meet our sleeper—the priest is visited by the doubter, the Marxist sees the civilizing force of the bourgeoisie, the captain of industry admits the justice of common ownership."

"Armed thus, God could only make a cabinet."

The science in Stoppard's spy-thriller/melodrama had a somewhat intimidating effect on audiences, and *Hapgood* has been significantly revised since its first production. This was not the case at all for *Arcadia* which has met with glowing reviews since its debut in 1993. One of the major influences cited by Stoppard for *Arcadia* is James Gleick's *Chaos*, the first chapter of which describes the frustrating attempts of Edward Lorenz to build a computer model capable of predicting the weather. The non-linearity of Lorenz's equations produced a phenomenon now popularly called the "butterfly effect" whereby two runs of the computer model starting with only tiny differences in the initial input produced unrecognizably different output.

To portray the butterfly effect on stage, Stoppard sets *Arcadia* across two time periods. The play opens in 1809 in the Coverly family's large country house where the daughter Thomasina is being tutored by Septimus who, when he is not teaching his talented student the classics, is engaging in carnal embrace in the gazebo with Mrs. Chater, the wife of a poet whose book he has just panned. The second scene of the play takes place in the same room, but in the present, where three academics are each trying to recreate what has taken place in this house 180 years earlier. The most laughable of the three is Bernard, an historian whose tiny errors in initial conditions lead him to conclude that Lord Byron fought a duel and killed the poet Chater before fleeing England. Hannah is also an historian, but is interested in the history of the gardens which during Thomasina's time were being transformed from smooth, undulating meadows to the "picturesque style" characterized by gloomy forests, craggy ruins,

Fractal leaf. Courtesy of Daniel Scharstein, Middlebury College.

and a hermitage in place of the infamous gazebo. As Hannah says, "It's a perfect symbol [of] the whole Romantic shame... A century of intellectual rigor turned in on itself... The decline from thinking to feeling, you see?" Central to Hannah's thesis is the hermit, "a savant among idiots, a sage of lunacy," who when he died left the hermitage stuffed with thousands and thousands of pages of mathematical proofs attempting to "save the world through good English algebra."

The third academic is Valentine, a mathematician and member of the Coverly family who is using the data in the game books to model the grouse populations as a chaotic system. In addition to explaining the essential nature of Chaos theory to Hannah (and the audience), his main service is to serve as translator for Thomasina's lesson books. No ordinary student, Thomasina grows restless with Septimus's classical geometry of plotting "xs against ys in all manner of algebraic reason. Armed thus," she frets, "God could only make a cabinet." A century and a half before Benoit Mandlebrot coins the term "fractal," Thomasina plucks a leaf from an apple and declares that she "will plot this leaf and deduce its equation. You will be famous for being my tutor when Lord Byron is dead and forgotten." One hundred and eighty years later, Valentine runs Thomasina's leaf equation through his laptop while Hannah reads the Fermatian passage from Thomasina's primer: "I, Thomasina Coverly, have found a truly wonderful method whereby all the forms of nature must give up their numerical secrets and draw themselves through number alone. This margin being too mean for my purpose, the reader must look elsewhere for the New Geometry of Irregular Forms, discovered by Thomasina Coverly."

Thomasina is as passionate as she is prodigious, at one point crumbling at the thought of the losses incurred when the ancient library of Alexandria was burned. Septimus consoles her with the argument that "we shed as we pick up, like travelers who must carry everything in their arms, and what we let fall will be picked up by those behind." Septimus believes in the perpetual mechanical wheels of the Newtonian universe, and it is clear that Stoppard is fascinated by the challenges Newton's deterministic system has sustained over the centuries. In *Hapgood* he explores the loss of "objective reality" via the uncertainty inherent in quantum theory. In *Arcadia*, determinism is discussed repeatedly, once even by Valentine's lascivious sister Chloe.

CHLOE: The future is all programmed like a computer— that's a proper theory isn't it?

VALENTINE: Yes, there was someone, forget his name, 1820s [Pierre Simon de Laplace], who pointed out that from Newton's laws you could predict everything to come—I mean, you'd need a computer as big as the universe but the formula would exist.

CHLOE: But it doesn't work, does it?

VALENTINE: No, it turns out the maths is different.

CHLOE: No, it's all because of sex.

"Ah," Valentine concedes. "The attraction that Newton left out." Mirroring this exchange is a similar one between Thomasina and her tutor.

THOMASINA: Well! Just as I said! Newton's machine which would knock our atoms from cradle to grave by the laws of motion is incomplete. Determinism leaves the road at every corner, as I knew all along, and the cause is very likely ... the action of bodies in heat.

The pun is certainly intended, but the 16-year-old genius is actually referring to the yet-to-be-formulated second law of thermodynamics. Contradicting everyone's attempts to look backwards in time, the second law of thermodynamics states that the universe is a "one-way street." In every physical process there is an inherent inefficiency—energy lost in the form of heat—and the disorder in the system increases. Without the mathematical vocabulary to express herself, Thomasina draws a diagram illustrating the principle of increasing entropy which, through the magic of the stage, is studied by both Septimus and Valentine doubled in time.

SEPTIMUS: So the improved Newtonian Universe must cease and grow cold. Dear me.

VALENTINE: The heat goes into the mix.

THOMASINA: Yes, we must hurry if we are going to dance.

VALENTINE: And everything is mixing the same way, all the time, irreversibly...

SEPTIMUS: Oh, we have time, I think.

VALENTINE: ...till there's no time left. That's what time means.

SEPTIMUS: When we have found all the mysteries and lost all the meaning, we will be alone, on an empty shore.

THOMASINA: Then we will dance. Is this a waltz?

Amid the mathematical and scientific fireworks, *Arcadia* works as storytelling because, in the end, it is heart-breaking. Thomasina and Septimus do finally dance, gracefully and to their mutual delight, passionately. But the audience has learned midway through the play that, like the library in Alexandria, Thomasina dies in a fire that night, on the eve of her seventeenth birthday. Watching the lovers dance as the curtain falls, it is finally clear that it is the devastated Septimus who is Hannah's hermit, living out his years in the hermitage, insanely trying to rescue his pupil through her own discoveries. Septimus's plight is still relevant. The question of how complex structures arise in the face of the downhill current caused by the second law of thermodynamics is considered by many to be the fundamental issue presently sitting at the forefront of science. For some, the most encouraging clues are hidden in the equations of chaos theory. Pondering over Thomasina's leaf equation on his computer, Valentine explains to Hannah,

VALENTINE: See? In an ocean of ashes, islands of order. Patterns making themselves out of nothing ...

HANNAH: Do you mean the world is saved after all?

VALENTINE: No, it's still doomed. But if this is how it started, perhaps it's how the next one will come.

Tom Stoppard has made it clear that his plays are events to be enjoyed, insisting that he does not write for scholars to pore over his work in search of dissertation topics. But the fact that the playwright has little use for the academic does not make the opposite proposition any less true. A journey through the rich ideas concealed in Stoppard's plays brings to life the notion that mathematics and art and science and poetry are related in ways we do not acknowledge and ultimately matter greatly in the way we choose to understand ourselves. *Arcadia* is a celebration of this point. Where Bernard is mostly bluster, Hannah and Valentine are redeemed by their own passionate reasons for what they do. Stoppard often hides behind his characters saying that "writing dialogue is the only respectable way of contradicting yourself," but it is impossible not to believe we are hearing the playwright's voice when Hannah confides to Valentine, "It's all trivial—your grouse, my hermit, Bernard's Byron. Comparing what we're looking for misses the point. It's wanting to know that makes us matter."

Acknowledgments: I would like to thank professor/director Cheryl Faraone and the 38 students whose contributions to the winter term course, "Stoppard, Science and Spirituality," (Middlebury College, January 1998) greatly shaped the content of this essay.

All quotations from Tom Stoppard in this article are taken from *Conversations with Stoppard* by Mel Gussow, Grove Press, New York, 1995. ■

Cycloidal Areas
without
Calculus

TOM M. APOSTOL and MAMIKON A. MNATSAKANIAN
California Institute of Technology

1. Introduction

For centuries mathematicians have been interested in curves that can be constructed by simple mechanical instruments. Among these curves are various cycloids used by Apollonius around 200 B.C. and by Ptolemy around 200 A.D. to describe the apparent motions of planets. The simplest cycloid is the curve traced out by a point on the circumference of a circular disk that rolls without slipping along a horizontal line; it forms a sequence of arches resting on the line, as shown in Figure 1.

Let S denote the area of the region above the line and below one of these arches (shown shaded in Figure 1). A routine use of integral calculus reveals that S is three times the area of the rolling circular disk, which we express symbolically as follows:

$$S = 3 \times \odot. \tag{1}$$

The derivation of this formula using integral calculus requires parametric or Cartesian equations for the cycloid.

This paper solves the more general problem in which the rolling circle is replaced by any regular polygon. The result is obtained by a geometrical method, and the area formula for the cycloid is obtained as a limiting case. We use the formula for the area of a circular sector, but there is no need to know the equations representing the cycloid.

2. Cyclogons

When a regular polygon rolls without slipping along a straight line, a given vertex on its circumference traces out a curve we call a *cyclogon*. Like the cycloid, a cyclogon consists of a sequence of arches resting on the line, as shown by the example of a rolling pentagon in Figure 2. Each arch, in turn, is composed of circular arcs, equal in number to one fewer than the number of vertices of the polygon. The arcs need not have the same radius.

If S denotes the area of the region above the line and below one of these arches, we will show that, in place of (1), we have the elegant and surprising result

$$S = \otimes + 2 \times \odot, \tag{2}$$

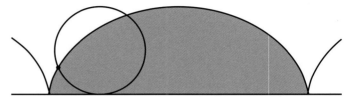

Figure 1. Cycloid traced out by a point on the circumference of a rolling circle.

where $\otimes$ denotes the area of the rolling polygon and $\odot$ is the area of the disk that circumscribes the polygon. The circle can be regarded as the limiting case obtained by letting the number of edges increase without bound in a regular polygon. Similarly, the cycloid is the limiting case of a cyclogon. Equation (1) for the area of the region under one arch of a cycloid is now revealed as a limiting case of Equation (2).

We begin with two simple examples, a rolling triangle, and a rolling square.

3. Rolling Equilateral Triangle

Figure 3 shows one arch of a cyclogon traced out by a rolling equilateral triangle whose edges have length a. The region under this arch and above the line consists of two equal circular sectors of radius a and one equilateral triangle. Each circular sector has area $(\pi/3)a^2$ which is also the area of the circular disk that circumscribes an equilateral triangle of edge-length a. Therefore

$$S = \text{area of } \Delta + 2 \times \frac{\pi}{3} a^2 = \text{area of } \Delta + 2 \times \odot,$$

which proves (2) in this case.

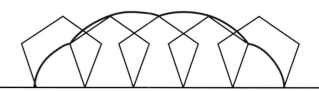

Figure 2. Cyclogon traced out by a vertex on the boundary of a rolling regular pentagon.

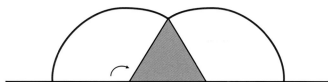

Figure 3. One arch of a cyclogon traced out by a rolling equilateral triangle.

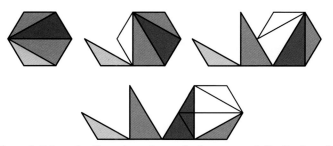

Figure 6. Triangular dissection of a regular hexagon and distribution of its footprints.

4. Rolling Square

Figure 4 shows a cyclogon traced out by a rolling square. The region under this arch consists of two right triangles plus three circular quadrants, two of radius a (the edge-length of the square), and one of radius $a\sqrt{2}$ (the diagonal of the square). The two right triangles have total area a^2, the area of the rolling square, and the total area of the three circular quadrants is

$$2 \times \frac{\pi}{4}a^2 + \frac{\pi}{4}\left(a\sqrt{2}\right)^2 = 2 \times \pi\left(a\frac{\sqrt{2}}{2}\right)^2 = 2 \times \odot.$$

Therefore we have $S = a^2 + 2 \times \odot$, which proves (2) in this case as well.

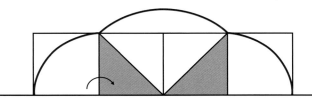

Figure 4. One arch of a cyclogon traced out by a rolling square.

5. Rolling *n*-gon

In the general case of a regular polygon with n vertices, the region under one arch of the cyclogon consists of $n-2$ triangles and $n-1$ circular sectors, each subtending an angle of $2\pi/n$ radians.

The $n-2$ triangles can be regarded as "footprints" left by the triangular pieces obtained by dissecting the original polygon with $n-3$ diagonals from a given vertex to each of the nonadjacent vertices, as illustrated in Figure 5 for $n=6$. The sum of the areas of these triangles is equal to the area of the region enclosed by the regular polygon. This is illustrated for the regular hexagon in Figure 6.

The radii of the circular sectors are the lengths of the segments from one vertex to each of the remaining $n-1$ vertices. A sector of radius r_k subtending an angle of $2\pi/n$ radians has area $\pi r_k^2/n$, so the sum of the areas of the $n-1$ sectors is equal to

$$\frac{\pi}{n}\sum_{k=1}^{n-1}r_k^2.$$

In the next section we will show that the sum of the squares of these radii is equal to $2nR^2$, where R is the radius of the circle that circumscribes the polygon. Therefore the sum of the areas of the sectors is equal to $2\pi R^2$, which is twice the area of the circumscribing disk. In other words, (2) is a consequence of the relation

$$\sum_{k=1}^{n-1}r_k^2 = 2nR^2. \tag{3}$$

6. An Extension of the Pythagorean Theorem for Regular Polygons

The result in (3), which is needed to calculate the sum of the areas of the circular sectors in the foregoing section, is of independent interest because it reduces to the Pythagorean theorem when the regular polygon is a square. The authors have not been able to locate this surprising theorem in any published work, so it may be new.

Theorem. *The sum of the squares of the $n-1$ segments drawn from one vertex of a regular n-gon to the remaining vertices is equal to $2nR^2$, where R is the radius of the circumscribing circle.*

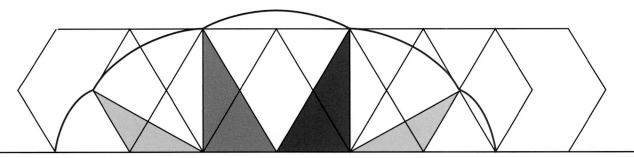

Figure 5. "Footprints" left by triangular pieces of a rolling hexagon.

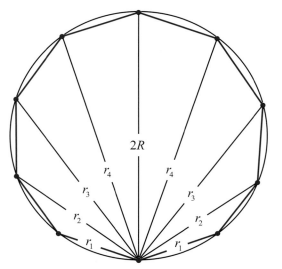

Figure 7. Nine segments drawn from one vertex of a regular decagon to the other nine.

7. Proof for Regular Polygons with an Even Number of Sides

The proof for even n makes repeated use of the Pythagorean Theorem. It is illustrated for the case $n = 10$ in Figure 7, which shows nine segments drawn from one vertex of a regular decagon to the other nine vertices.

In Figure 7 there are four segments, labeled as r_1, r_2, r_3, r_4, and four mirror images, plus the diameter of length $2R$, so the sum in question is

$$2\sum_{k=1}^{4} r_k^2 + 4R^2. \tag{4}$$

Figure 8 shows two segments from the opposite extremity of the diameter to consecutive vertices. By symmetry with respect to a horizontal diameter, these segments have lengths r_1 and r_2. The new segment r_1 meets the old segment r_4 on the circle and, together with the diameter, forms a right triangle with hypotenuse $2R$. (Here we use the fact that any triangle inscribed in a semicircle is a right triangle with the diameter as hypotenuse.) Applying the Pythagorean Theorem to this right triangle we find

$$r_1^2 + r_4^2 = 4R^2. \tag{5}$$

Similarly, the new segment r_2 intersects the old segment r_3 and forms another right triangle with hypotenuse $2R$. Applying the Pythagorean Theorem once more we find

$$r_2^2 + r_3^2 = 4R^2 \tag{6}$$

so the sum in (4) is equal to

$$2\sum_{k=1}^{4} r_k^2 + 4R^2 = 16R^2 + 4R^2 = 20R^2$$

which proves the Theorem for $n = 10$.

In the general case of even n, one of the $n - 1$ segments is the diameter $2R$ of the circumscribing circle, and the other $n - 2$ segments form $(n - 2)/2$ pairs symmetrically located with respect to the diameter. The same argument just given for the case $n = 10$ shows that

$$2\sum_{k=1}^{(n-2)/2} r_k^2 + 4R^2 = \frac{n-2}{2}\left(4R^2\right) + 4R^2 = 2nR^2$$

which proves the theorem for every even n. This proof does not work if n is odd.

8. Proof for Regular Polygons with an Odd Number of Sides

A different method that applies to all regular polygons with an odd number of sides is illustrated for a regular heptagon in Figure 9. The three segments and their mirror images in the diameter are the 6 segments drawn from one vertex of a regular heptagon to the other 6 vertices. We wish to prove that

$$2\left(r_1^2 + r_2^2 + r_3^2\right) = 14R^2,$$

or that

$$r_1^2 + r_2^2 + r_3^2 = 7R^2. \tag{7}$$

We apply the law of cosines to each of three isosceles triangles in Figure 9 having a vertex at the center of the heptagon, two edges of length R and base of length r_k, where $k = 1, 2, 3$. The corresponding vertex angles are θ_k, where $\theta_1 = 2\pi/7$, $\theta_2 = 4\pi/7$, $\theta_3 = 6\pi/7$. The law of cosines for the isosceles triangle with vertex angle θ_k states that

$$r_k^2 = 2R^2 - 2R^2 \cos\theta_k \tag{8}$$

so the sum of these equations gives us

$$\sum_{k=1}^{3} r_k^2 = 6R^2 - 2R^2 \sum_{k=1}^{3} \cos\theta_k. \tag{9}$$

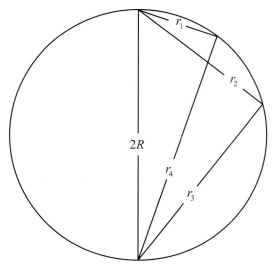

Figure 8. Rearrangement of segments r_1 and r_2 in Figure 7.

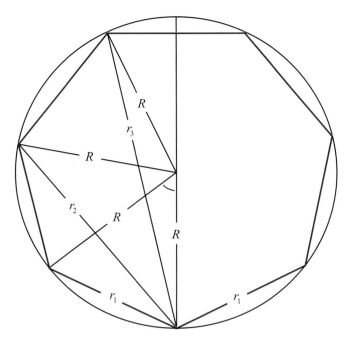

Figure 9. Regular heptagon with edge r_1 inscribed in a circle of radius R.

But, by a trigonometric identity described below in (12), the sum of cosines is equal to $-\frac{1}{2}$, so (9) implies (7).

In the general case of a regular polygon with $2n+1$ sides we wish to prove that

$$\sum_{k=1}^{n} r_k^2 = (2n+1)R^2. \qquad (10)$$

In this case we apply the law of cosines to n isosceles triangles, using (8) for each of these triangles, where now

$$\theta_k = 2\pi k/(2n+1).$$

Instead of (9) we have the equation

$$\sum_{k=1}^{n} r_k^2 = 2(n-1)R^2 - 2R^2 \sum_{k=1}^{n} \cos\theta_k. \qquad (11)$$

In this case we have the following identity (which we prove below in Section 9),

$$\sum_{k=1}^{n} \cos\theta_k = -\frac{1}{2}, \qquad (12)$$

so (11) reduces to (10).

9. Origin of the Trigonometric Identity (12)

The trigonometric identity in (12) can be written as

$$2\sum_{k=1}^{n} \cos\theta_k + 1 = 0 \qquad (13)$$

where $\theta_k = 2\pi k/(2n+1)$. This is a consequence of a more gen-

eral trigonometric identity

$$\sum_{k=1}^{m} \cos(2k\theta) = \frac{\sin m\theta \cos(m+1)\theta}{\sin\theta}, \qquad (14)$$

which holds for any positive integer m and any θ that is not an integer multiple of π. (See Exercise 32, p. 106, of Apostol's *Calculus*, Vol. I, 2nd ed., John Wiley & Sons, Inc, 1967.) If we take $\theta = \pi/m$ the right member vanishes and (14) becomes

$$\sum_{k=1}^{m} \cos\left(\frac{2\pi k}{m}\right) = 0. \qquad (15)$$

When m is odd, say $m = 2n + 1$ the last term in the sum is equal to 1. The remaining $2n$ terms can be arranged in n pairs, by coupling the terms with k and $m - k$, which have the same cosine. Consequently (15) can be written as

$$2\sum_{k=1}^{m} \cos\left(\frac{2\pi k}{2n+1}\right) + 1 = 0,$$

which is the same as (13).

Note. The foregoing method, using the law of cosines, also works if the polygon has an even number of sides, say $2n+2$ sides, but one minor change is needed. There are now $2n + 1$ segments from a given vertex to the remaining vertices. One of these is a diameter, and the other $2n$ can be arranged in pairs by coupling each segment with its mirror image in that diameter. However, we do not give further details because the proof presented in Section 7 is more elementary.

10. Alternate Proof for a Regular Pentagon

An alternate proof for a regular pentagon can be given by a method that is of interest because it makes use of Ptolemy's remarkable theorem on cyclic quadrilaterals (quadrilaterals inscribed in a circle). Ptolemy's theorem states that, for any cyclic quadrilateral, the product of the lengths of the diagonals is equal to the sum of the products of the lengths of opposite sides. (A simple proof of Ptolemy's theorem can be found in the Workbook that accompanies the videotape *Sines & Cosines, Part III*, produced by *Project MATHEMATICS!*, Caltech, 1994. The videotape also gives a computer animated version of this proof.)

Figure 10 shows a regular pentagon imbedded in a regular decagon with edges of length r_1. The segments r_2 and r_4 and their mirror images in a diameter $2R$ are the four segments drawn from one vertex of the regular pentagon to the other four vertices. We are to prove that $r_2^2 + r_4^2 + r_4^2 + r_2^2 = 10R^2$ or the equivalent statement

$$r_2^2 + r_4^2 = 5R^2. \qquad (16)$$

Apply Ptolemy's theorem to the cyclic quadrilateral in Figure 10 with two intersecting diagonals of length r_2 to obtain

$$r_2^2 = r_1 r_3 + r_1^2. \qquad (17)$$

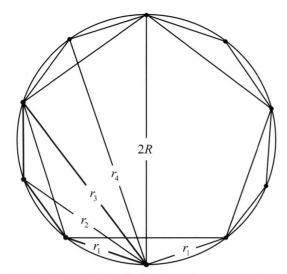

Figure 10. A cyclic quadrilateral with three edges r_1, one edge r_3, and two diagonals r_2.

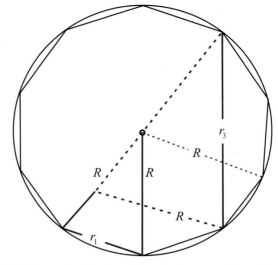

Figure 11. The isosceles triangle with equal edges r_3 and base R is similar to that with equal edges R and base r_1.

Next, refer to the two similar isosceles triangles shown in Figure 11, and equate ratios of corresponding sides to get $R/r_1 = r_3/R$, or

$$(18) \qquad r^1 r^3 = R^2.$$

Substitute (18) in (17) and then use the Pythagorean relation $r_1^2 + r_4^2 = (2R)^2$ to obtain

$$r_2^2 = R^2 + r_1^2 = R^2 + 4R^2 - r_4^2 = 5R^2 - r_4^2,$$

which implies (16). ∎

A Bicentennial for the Fundamental Theorem of Algebra

BARRY CIPRA
Northfield, Minnesota

Two hundred years ago, a promising young mathematician at the (later defunct) University of Helmstedt submitted a doctoral dissertation with the lengthy Latin title, *Demonstratio nova theorematis omnem functionem algebraicam rationalem integram unius variabilis in factores reales primi vel secondi gradus resolvi posse*: "A new proof of the theorem that every rational integral algebraic function in one variable can be resolved into real factors of first or second degree." Today, we might shorten things to "A new proof of the Fundamental Theorem of Algebra."

The student's name? Carl Friedrich Gauss.

Born in 1777, Gauss came to be regarded as one of the greatest mathematicians in history. His influence pervades nearly every nook and cranny of mathematics, from differential geometry (Gaussian curvature) to linear algebra (Gaussian elimination) to error analysis (Gaussian noise). It also extends to the study of magnetism (the *gauss*) and, most famously, astronomy: his predictive analysis of the orbit of the asteroid Ceres in 1801 is what first brought him popular acclaim [4].

Gauss is also credited with bringing a new level of rigor to mathematics. The salad days of analysis, when guys like Euler could get away with nonsense like $1 + 2 + 3 + \cdots = -1/12$, were coming to an end. Gauss saw that future progress would depend on careful attention to the underlying assumptions of algebra, analysis, and geometry.

Small wonder then that his doctoral dissertation, which appeared in August, 1799, was an attempt to bring rigor to the study of polynomial equations. First, though, a little background and some history.

From a modern viewpoint, the Fundamental Theorem of Algebra makes a simple claim: *Every polynomial has a root*. Actually we should be a little more precise: *Every nonconstant polynomial with complex coefficients has a complex root*. To flesh this out a bit, let $P(z) = a_0 + a_1 z + \cdots + a_n z^n$ be a polyno-

Gauss as a young man (circa 1803). Reprinted from Gauss: A Biographical Study *by Walter Kaufman-Bühler, used by permission.*

mial of degree n (so that $a_n \neq 0$), with $n > 0$. Then the claim is that the equation $P(z) = 0$ has a solution $z = x + iy$.

An equivalent, corollary form of the Fundamental Theorem says *Every polynomial of degree n has exactly n roots*. But again this is slightly imprecise, since, for example, $P(z) = z^2 - 2z + 1$ has only the one root $z = 1$. A more correct version says *Every polynomial of degree n has a unique factorization into n linear factors*, a linear factor being of the form $z - r$, where r is a root of the polynomial.

The equivalence of these assertions is relatively easy to prove by induction on n, using the fact that r is a root of $P(z)$ if and only if $z - r$ is a linear factor of $P(z)$ — i.e., if and only if $P(z) = (z - r) Q(z)$ for some polynomial Q of degree $n - 1$. This induction argument by itself suffices to show that a polynomial can't have *more* roots than its degree suggests. But it's not enough to show that any exist.

Why did Gauss refer to factors of the second degree? It's because of the key word "real." At the time Gauss wrote his dissertation, imaginary numbers were still somewhat suspect.

Mathematicians preferred to think in terms of polynomials with real coefficients—Gauss's "rational integral algebraic functions." If you want to avoid (or disguise) the use of complex numbers, you have to stop when you get to quadratic factors whose roots are complex conjugates (with nonzero imaginary part).

It was, in fact, Gauss who finally took the mystery out of complex numbers, by giving them their now-standard geometric interpretation. At least he gets much of the credit. Several people, including Caspar Wessel and Jean-Robert Argand, had also had the idea of interpreting complex numbers geometrically, but it was Gauss who had the clout to carry the day. (Wessel's paper, published in 1797, went unnoticed by mathematicians until 1895, at which point it was mainly of historical interest.) The turning point can be pegged to a paper he wrote in 1831, on biquadratic reciprocity, in which he coined the term "complex number."

The seemingly elusive nature of imaginary numbers plagued the early discussion of the Fundamental Theorem of Algebra. The most famous case of confusion occurred in 1702, when Leibniz looked at the factorization of $x^4 + a^4$ into

$$\left(x + a\sqrt{\sqrt{-1}}\right)\left(x - a\sqrt{\sqrt{-1}}\right)\left(x + a\sqrt{-\sqrt{-1}}\right)\left(x - a\sqrt{-\sqrt{-1}}\right).$$

So far so good. But, Leibniz went on to say, these imaginary roots cannot be recombined to give a pair of real quadratics. In short, the polynomial $x^4 + a^4$ *cannot be factored over the reals.*

That's just not so, as Johann Bernoulli pointed out in 1719: $x^4 + a^4 = \left(x^2 + \sqrt{2}ax + a^2\right)\left(x^2 - \sqrt{2}ax + a^2\right)$. How Leibniz could have missed this is a mystery (for one thing, it's implicit in the method for solving quartic polynomials that Descartes gave in his famous treatise *La Geometrie*), but the fact that someone smart enough to invent calculus stumbled on a "simple" point of algebra shows just how shaky was the early understanding of complex numbers.

Leibniz wasn't the last to goof. In 1742, Johann's nephew Nicolas Bernoulli came up with another "unfactorizable" quartic, $x^4 - 4x^3 + 2x^2 + 4x + 4$. He included this counter-example in a letter to Euler, who had conjectured the linear–quadratic factorization version of the Fundamental Theorem. The substitution $x = u + 1$ changes this into $u^4 - 4u^2 + 7$, with complex roots $\pm\sqrt{2 \pm \sqrt{-3}}$. But how do you factor it into a pair of quadratics with real coefficients? You can't, Bernoulli said.

"This example appeared at first to overturn my theorem," Euler wrote his friend Christian Goldbach a few weeks later, "until I thought about the matter more closely." He gave the quadratic factorization, which we'll leave to the reader to try his or her luck at. He also reasserted the linear–quadratic factorization conjecture, which he said he could "nearly, but not with complete rigor, prove."

Goldbach still had doubts. He wrote back with the "counterexample" $x^4 + 72x - 20$. Euler replied with a thorough analysis of quartic polynomials. He also repeated the claim that whenever you factor a polynomial completely, you can pair up

the (linear) factors with imaginary roots to get real quadratic factors. "To be sure, I can't prove this in general, but I can prove it for all equations of degree less than six, and also for equations of the form

$$\alpha x^{5n} + \beta x^{4n} + \gamma x^{3n} + \delta x^{2n} + \varepsilon x^n + \zeta = 0.\text{"}$$

By 1749, Euler had his general proof—or thought he did. He presented it in his paper *Recherches sur les racines imaginaires des équations*. His strategy was elegant: it suffices to show that every polynomial of degree 2^n (with real coefficients) can be factored into a pair of polynomials (with real coefficients) of degree 2^{n-1} (if $n > 1$). If you start with a polynomial of any other degree, just multiply it by an appropriate x^k to get up to a power of 2. The theorem, applied recursively, descends through the powers of 2, at the end of which you just remove the padding. (A more complete account of Euler's proof can be found in William Dunham's book *Euler: The Master of Us All.*)

However, to make the proof work, Euler assumes he has roots to start with. All he really establishes is that if you have roots (and if you can do ordinary arithmetic with them), then you can pair up the imaginary roots to produce real quadratic factors. In particular, imaginary numbers like $\sqrt{2 + \sqrt{-3}}$ can always be rewritten in the form $x + iy$, with *real* numbers x and y. But at the time no one really questioned the existence of roots; all that seemed doubtful was the form they took.

At roughly the same time that Euler came up with his proof, Jean le Rond d'Alembert found an alternative proof for the Fundamental Theorem. D'Alembert's approach was to show that if $P(z_0) \neq 0$, then there's a nearby point z_1 with $|P(z_1)| < |P(z_0)|$. His proof uses fractional power series—a device introduced by Newton—to solve the equation $P(z) = w$ for z in terms of w. The desired conclusion follows more or less immediately—provided, of course, you can trust the series to converge.

The first rigorous proof of convergence for fractional power series was given in 1850 by Victor Puiseux. His proof *assumes* the Fundamental Theorem of Algebra.

Various mathematicians during the latter half of the eighteenth century endeavored to improve on d'Alembert's and Euler's proofs, fixing up minor gaps. But it fell to young Gauss to really set things straight.

Only about a third of his dissertation is devoted to his "new" proof. "The rest," he wrote to his friend Wolfgang Bolyai, "is mainly a history and criticism of the work of other mathematicians (namely d'Alembert, Bougainville, Euler, de Foncenex, Lagrange, and the people who write compendia—of which these last will probably not be very happy) on the subject, along with various remarks on the insipid shallowness that's so prevalent in mathematics these days." (Gauss's word for "insipid shallowness" is *Seichtigkeit*.)

The strategy of Gauss's proof is to separate a given polynomial, say $a_0 + a_1 z + \cdots + a_n z^n$ (with $a_n \neq 0$), into its real and imaginary components, say $P(x, y)$ and $Q(x, y)$. The goal then

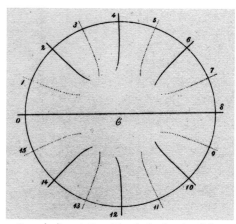

Figure 2 from Gauss's dissertation. The dashed and solid curves are portions of the zero sets of, respectively, the real and imaginary parts of the polynomial. These curves must alternate intersections with a circle of large radius.

is to show that the zero sets of these two functions of two variables intersect.

The trick is to look at "large" values of z written in polar coordinates $z = r(\cos\theta + i\sin\theta)$. The highest power of z dominates the behavior of the polynomial when r is large. Since $z^n = r^n(\cos n\theta + i\sin n\theta)$, this means that $P(x, y)$ and $Q(x, y)$ each have $2n$ zeroes on the circle of radius r. Moreover, the zeroes alternate: The ones for Q occur at approximately $\theta = \pi k/n$ for $k = 1, \ldots, 2n$, while those for P occur at approximately $\theta = \pi(k + 1/2)/n$.

The situation so far is illustrated by Figure 2 from Gauss's dissertation. The zero sets of P and Q form curves heading in at roughly right angles from the circle of radius r. At this point, the proof of the Fundamental Theorem hinges on the fact that if a branch of an algebraic curve *enters* a region, it has to *exit* the region. Consequently the $2n$ branches of P in Figure 2 have to

hook up, and likewise for the branches of Q. Finally, no matter how the hook-ups occur, there will be some intersections.

Gauss's dissertation is rightly heralded for bringing a new level of rigor to the attempts to prove the Fundamental Theorem of Algebra, especially because he pinpointed the *existence* of roots as the crucial element. But oddly enough, Gauss's proof has a gap of its own.

The gap lies in that "fact" about algebraic curves entering and exiting a region. To be sure, it really is a fact, but it needs proof—and the first completely rigorous proof wasn't given until 1929! If Euler and d'Alembert can be faulted for making unwarranted assumptions, then so can Gauss.

Actually Gauss was well aware the fact required proof. In a footnote he says "It seems sufficiently well established that an algebraic curve can neither break off suddenly somewhere (as, for example, happens for the transcendental curve with equation $y = 1/\log x$), nor lose itself, so to speak, in some point after an infinite number of spirals (as with the logarithmic spiral), and as far as I know, no one has ever doubted it. Nevertheless, if anyone insists, I will undertake on another occasion to give a proof which will leave no doubt."

He never did produce this indubitable proof. (Elsewhere in his dissertation, remarking on the then-open problem of solving quintics and higher-degree polynomials in terms of radicals, Gauss ventures the opinion "Perhaps it won't be very difficult to prove, with all rigor, the impossibility for the fifth degree. I'll lay out my investigations of this in greater detail in some other place." He never did this either.) But he did return to the Fundamental Theorem itself, giving three more proofs—the last in 1849, at the 50th-anniversary celebration for his own dissertation.

By the end of the nineteenth century, the foundations of real and complex analysis were firmly enough established that completely rigorous proofs of the Fundamental Theorem of Algebra were easy to come by. The development of abstract algebra trivialized the theorem in the sense that every (irreducible) polynomial $P(z)$ over a field K automatically has a root in the quotient field $K[z]/(P(z)K[z])$, but that doesn't explain why the roots of complex polynomials are necessarily complex. One remaining question was why was there no purely algebraic proof of the whole shebang? This got answered in 1927, by Emil Artin and Otto Schreier, who showed that the notion of continuity is essential to any proof of the fundamental theorem. Gauss, who would have been 150 years old by then, would no doubt have approved. ∎

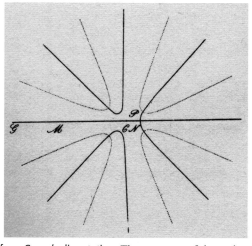

Figure 4 from Gauss's dissertation. The zero sets of the real and imaginary parts, again, respectively, dashed and solid, of the polynomial $z^4 - 2z^2 + 3z + 10$. Note that the horizontal axis increases from right to left, contrary to modern-day convention.

References

1. C. F. Gauss, *Werke* Vol. 3, Hildesheim, New York, 1973.
2. Walter Kaufman-Bühler, *Gauss: A Biographical Study,* Springer-Verlag, New York, 1981.
3. D. J. Struik, *A Source Book in Mathematics: 1200–1800,* Princeton University Press, 1986.
4. Donald Teets and Karen Whitehead, "The Discovery of Ceres: How Gauss Became Famous," *Mathematics Magazine,* vol. 72, 1999, no. 2, pp. 83–93.

Was Gauss Smart?

THOMAS E. MOORE
Bridgewater State College

Photo courtesy of Lester S. Levy Collection of Sheet Music, Special Collections, Milton S. Eisenhower Library, The Johns Hopkins University.

In my avocation as music director at St. Joseph the Worker Church in Hanson, Massachusetts, I often search for choral music both traditional and contemporary. The World Wide Web has turned out to be a terrific resource in this endeavor with many music publishers, composers, and music notation software companies providing links to sheet music or sound files.

Sitting at my desk at Bridgewater State College with a few hours to go before my first class of the new semester, I was surfing the web for material like this when I stumbled on the repository at Johns Hopkins University known as the Lester S. Levy Collection of Sheet Music.

Looking for Hymns or Sacred Music, I was disappointed not to find any such category. Instead I found links to boxes of sheet music whose contents were indicated by titles such as U.S. Presidents and Presidential Candidates; Patriotic Songs; Drinking, Temperance, and Smoking; Maritime; and so on. There was nothing among them that appealed to me. However, curious to see the contents of a typical box, I decided to choose a link from those visible on my screen. I arbitrarily chose the third-listed Maritime box.

The thumbnail-size JPEG images of the cover pages of 150 pieces of sheet music began to fill the screen. As I started scrolling through them I was astounded to see a familiar image. Enlarging this image (Item 38 of Box 183) confirmed that the cover art was a lithograph of Carl Friedrich Gauss! This is the image with which everyone in the scientific and mathematical community is familiar and virtually the only image of him to be found in texts on the history of mathematics.

The music itself was titled *Sailing Home* by Charles Osborne, composer and lyricist. It told a tale of Britons who had struggled to make their fortune in the gold fields of Australia and were bound for home aboard a ship that, unfortunately, foundered at sea.

Unfortunately, the music has no date of publication, but we may deduce from the lyrics that *Sailing Home* was penned in the late 1850s or in the 1860s. For, according to the first stanza, the ship embarks from Melbourne, Victoria, Australia, and the gold strike in Victoria began in the 1850s. (Osborne was definitely composing around this time as can be seen from some of his other work including *Bonnie Jean*, dated 1858, also in the Levy Collection.)

Wilton Smart was the performer of the ballad. The bibliographical data on *Sailing Home* tells us that the engraved image on the cover page is "an unattributed lithograph of Wilton Smart," but we know it is Gauss. What can this mean? We are told that Gauss died in 1855. Maybe not. Perhaps he survived to enjoy another career, as a balladeer! Gauss was smart, but was he actually Smart in the 1860s? Is Gauss actually a time-traveller visiting us at different times under different guises? (Come to think of it, doesn't the comedian Robin Williams look a little too much like our Gauss?)

Leaving aside such fancy, we wonder how Gauss came to be the cover art for a piece of popular music. Was Osborne or his publisher, White-Smith of Boston, putting one over on the American public? Did the engraver lose the image of Smart and substitute that of another smart guy?

I also wonder what serendipity led me, idly surfing for hymns, to choose to look only in the third box of Maritime-related music and to find Gauss, the Prince of Mathematics, looking back at me. ∎

Chorus

Over the sea like a bird she skimmed,
Back from the Land of Gold.
Pulses were beating, and eyes were dimmed,
Thinking of friends of old.
Soon they would stand in their Fatherland,
Never again to roam;
But little they thought that their ship so taut
Would founder while sailing home.

Reprinted from November 1999, p. 24

Adoption and Reform of the Gregorian Calendar

EDWARD L. COHEN
Ottawa, Ontario

Everyone knows that George Washington was born on February 22. Every mathematician knows that the year was 1732 (because 1.732 is the beginning of $\sqrt{3}$). As are so many facts known to everyone, neither of these is wholly correct. In 1752 we changed the way we keep track of dates because our calendar had drifted from its astronomical benchmarks. The problem is the earth does not go around the sun in a considerate number of days, but rather approximately 365.2422. In addition the moon goes around the earth in an inconsiderate 29.53059 days (approximately). This is 29 days and about 12 hours, 44 minutes, and 3 seconds. In other words a calendar based on celestial phenomena is necessarily imperfect.

Julius Caesar had reformed the somewhat incoherent Roman calendar in 45 BCE. By the time of Pope Gregory XIII (1502–1585) astronomical knowledge had improved greatly and the Julian calendar was noticeably incompatible with solar events. Gregory introduced a new calendar (the Gregorian), which, confusingly, was adopted at different times by different countries. We will use the notation OS (Old Style) and NS (New Style) to distinguish between Julian and Gregorian dates where confusion is possible: that is, for dates between October 15, 1582 (the introduction of the Gregorian calendar in Italy and surrounding countries) and September 14, 1752 (the changeover in England and its American colonies). For example, in Italy April 15, 1586 is the NS date of the day that in England was April 5, 1586 OS. To add to the confusion March 25 was the beginning of the new year in England until 1752. So, we write George Washington's birthday as February 11, 1731/2 OS, and February 22, 1732 NS. In his family Bible it is written that he was born on "ye 11th day of February." He apparently celebrated both after 1752.

Gregorian Reform of the Julian Calendar

Easter is defined by the Catholic Church as the Sunday following that fourteenth day of the calendar moon which happens upon or next after the twenty-first of March: so that if the said fourteenth day be a Sunday, Easter Day is not that Sunday, but the next. The twenty-first of March was set by the Council of Nicæa in 325 as the vernal equinox. The Julian calendar had a leap year every fourth year and thus worked perfectly for a year of length 365.25 days. The error in this approximation had pushed the date of the vernal equinox back by a full 10 days by the sixteenth century. The fixing of the date of Easter was a serious problem. Pope Gregory charged a group of astronomers/mathematicians with resolving the Easter/equinox problem. Aloisius Lilius (1510–1576) and Christoph Clavius (1537–1612) proposed a calendar with 303 365-day years and 97 366-day years. Leap years were to be those years divisible by 4 except those century years not divisible by 400; e.g., 1700, 1800, 1900. Century years divisible by 400, e.g., 1600 and 2000, would remain leap. The Gregorian calendar was launched with a Papal Bull, *Inter Gravissimas* [7], on February 24, 1582 and instituted on October 5 OS = October 15 NS of that year. The Easter problem seemed nicely solved.

We know today that this scheme overstates the length of the year by about 26 seconds, so that now, four hundred years later, we are almost eleven thousand seconds ahead of the sun (417 × 26 = 10842). This amounts to about one-eighth of a day. We'll return to this point.

Adoption of the Gregorian Calendar

The Protestant countries of Europe were reluctant to modify their calendars at the direction of the Pope. Thus only the Catholic countries adopted the new calendar in 1582. Most of the countries served by the Eastern Orthodox Church, in fact, still use the Julian calendar for ecclesiastical purposes though the Gregorian is used for secular purposes. For the record here's a list of some major adoption dates:

1582　Italy, Spain, Portugal, Luxembourg, France, Belgium, German Catholic States, Catholic Netherlands, Poland

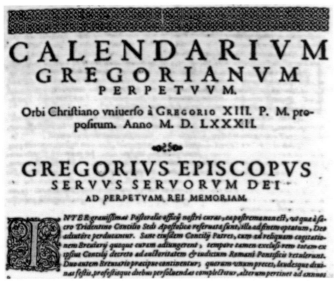

First page of the Papal Bull, Inter Gravissimas, *ordering Christians throughout Europe to adopt the Gregorian calendar on October 15, 1582.*

1583 Austria
1584 Catholic Switzerland
1587 Hungary
1600 Scotland
1700 Protestant Netherlands, Denmark, German Protestant States
1752 Britain and Empire (including American colonies), Quakers
1753 Sweden
1812 Rest of Switzerland
1867 Alaska
1873 Japan
1875 Egypt
1912 China, Albania
1917 Turkey
1918 Russia
1919 Yugoslavia, Romania
1923 Greece
1924 Eastern Orthodox Church in Romania, Yugoslavia, and Greece

Naturally adding ten or eleven days to the current date caused some confusion and upset: try to look up Isaac Newton's birthday in an encyclopedia published shortly after his death. (For the record, the correct date is December 25, 1642 OS; he died on March 20, 1726/7 OS.) There's even confusion about when some countries switched. Some sources say that Sweden skipped February 29 in 1700, reconsidered and went back to Julian dates by adding an extra leap day (February 30) in 1712, then re-reconsidered and converted to Gregorian in 1753. The *Journal of Calendar Reform* reported in 1938 that "Sweden (gradually) by the omission of 11 leap days, 1700 to 1740" switched. In this case a person born February 29, 1696 would not have had a birthday until 1744! This is probably a misinterpretation of an article by Ginzel [6], but that has not prevented the story from entering calendrical folklore.

Philip Dormer Stanhope, Earl of Chesterfield, in 1751 pushed through Parliament a bill to change September 3, 1752 to September 14, 1752 and to start the new year with January 1 instead of March 25. The archives of Swarthmore College contain a pamphlet printed in 1751 by Benjamin Franklin recording the acceptance of the Parliamentary action by the Quakers. The English painter and engraver William Hogarth (1697–1764) depicted some of the confusion in his *An Election Entertainment* [8], which contains a placard reading "Give us our eleven days!" Some people thought 11 days had been taken from their lives. There was widespread confusion about rents and mortgages with some landlords allegedly demanding a full month's rent for the 19 day September. Reportedly, riots left several dead in Bristol [4].

The *Gentleman's Magazine* for September 1752 contained the following communication from a perplexed reader.

I desire some way of setting my affairs to rights, or, I believe I shall run mad. I went to bed last night, it was Wednesday, September 2, and the first thing I cast my eye upon this morning at the top of my paper was Thurs-

A detail from William Hogarth's satiric An Election Entertainment. *Note the placard "Give us our eleven days" in the foreground. By courtesy of the Trustees of Sir John Soane's Museum.*

Not Silent, tells you something about her personality.

The early part of the twentieth century was the zenith of American optimism and can-do confidence. Americans felt that hard work, technological ingenuity, and devotion to a Christian God would truly lead us to a better, in fact perfect, world. Moreover, America was God's instrument to bring this new world about and reverse the decline of the Old World (Europe) and raise up the savages elsewhere. You can hear this spirit in this excerpt from Achelis's autobiography.

> To help America's onward march to better days, the proposed new calendar, The World Calendar, that begins every year with a Sunday, every quarter with a Sunday, and every week with a Sunday, will keep America on the straight and onward path. And what is equally important is the significance that the Christian Sunday, like the sun in the sky after which ancients named that day, will shine undimmed upon all peoples of the world. The World Calendar for all Nations, and Sunday, the religious day, will sow seeds for more harmonious living such as the world has never known before.

And again here, from her *Of Time and the Calendar*.

> This small book recounts the complexities and confusions in the present calendar, recalls its origin in remote history, traces its development in diverse civilizations, and points to its limitless potentialities for effective service through the simple revisions proposed in The World Calendar. The World Calendar develops the present calendar into a scientific, civil instrument for measuring and recording the flow of time in an ordered manner. It is a stabilizer of our days and years in perpetuity, a harbinger of peace and goodwill.

Breaking the Sabbath

The *World Calendar* is obviously more logical and neater than the Gregorian. It's perfect. No wonder Achelis was so zealous in promoting it. So, why don't we use it? Despite Achelis's firmly held conviction that her Christian god approved of her plan, calendar reform failed primarily because of opposition from religious groups disturbed by the blank day; every seventh day was no longer a Sabbath. For many people this was simply unacceptable. (We won't dwell on Achelis's conflation of the Protestant Christian Sunday with other religion's sabbatical ideas.)

A very potent foe of calendar reform was Dr. Joseph Herman Hertz, Chief Rabbi of England. He spoke out forcefully against reform, an idea whose prominence he blamed on American commercial and financial interests. In 1931 he (and many likeminded others) addressed the League of Nations Conference on Calendar Reform in Geneva. The League decided to postpone action. Afterwards, Hertz wrote,

> That agitation has shown how easily the League can be entangled in partisan, faddist propaganda; and it also shows that a considerable portion of its Secretariat has

Elisabeth Achelis, the Calendar Lady

forgotten that the League's most sacred charge is the protection of minorities. The agitators themselves have suffered a shattering, let us hope, annihilating, defeat. It is doubtful whether in the lifetime of any of them the question will ever be raised again. Three months before, nothing seemed more certain than that the Conference would pass an eight-day week Calendar.

Hertz underestimated Elisabeth Achelis—she was far from shattered and annihilated. She was just getting started. The matter continued to come before the League and later on, its successor, the United Nations. Many nations continued to support the *World Calendar* in these bodies. Achelis succeeded in getting bills brought before the U.S. House (1946 and 1947) and the U.S. Senate (1949). All, however, died. Finally in response to a 1955 question from the Secretary-General of the U.N. the U.S. Department of State said that it was no longer interested in considering changing the calendar mainly because of minority religious opposition. On April 2, 1956 the Economic and Social Council of the United Nations voted to postpone *sine die* world calendar reform from its agenda. No official action has been taken in the United Nations on the topic since that day.

The Catholic Church, having started the whole business of calendar reform back in 1582, weighed in on the occasion of the Second Vatican Council (1963) [9]:

> The sacred Council declares that it is not opposed to proposals intended to introduce a perpetual calendar into civil society. But among the various systems which have been

mooted for establishing a perpetual calendar and introducing it into civil society, the Church raises no objection provided that the week of seven days with Sunday is preserved and safeguarded, without intercalating any day outside of the week, so that the succession of weeks is preserved intact—unless very serious reasons are forthcoming, of which the Holy See will be the judge.

Keeping the Sabbath

There is another type of reform calendar out there that allows us to preserve the seven-day week. Notice that in 400 Gregorian years there are $400 \times 365 + 97$ days, but this obviously equals $400 \times 364 + 497$ days and, amazingly, $497 = 7 \times 71$. So, instead of having 303 365-day years and 97 366-day years in every 400, we could have 329 364-day years and 71 371-day years by adding "leap weeks." No blank days, and only two different kinds of calendar. Father Evarist Kleszcz (1902–1985) was a recent champion of this calendar and he worked out which of the years from now until 36,900 would need to have "Jubilee" weeks. (The year 36,899 would be leap, in case you were wondering.)

One big disadvantage of his scheme is that nobody but astronomers could figure out which years needed to be leap. The years in the second half of the twentieth century would have been 1953, 1959, 1965, 1970, 1976, 1982, 1987, 1993, 1998—notice the apparently pattern-less sequence of differences 6, 6, 5, 6, 6, 5, 6, 5. Remember that one important constraint is to keep the vernal equinox at or near March 21. Adding seven days to a year obviously has to push the equinox back and forth from a fixed date by at least 3 days. Father Kleszcz's calculation made it possible to have the equinox always fall between March 18 and March 24.

Professor Cecil L. Woods proposed that the Jubilee week years occur every five years with the exception of those years ending in '25 or '75 and then choose one of the century years ('00) also not to be leap. So that the four hundred years from 2001 to 2400 would be as follows:

18 leap weeks in century 1: '05, '10, …, except '25 and '75;
18 leap weeks in century 2: '05, '10, …, except '25 and '75;
17 leap weeks in century 3: '05, '10, …, except '25, '75, and '00;
18 leap weeks in century 4: '05, '10, …, except '25 and '75.

Actually the '00 non-leap year could be any of the four. Choosing 2300 best constrains the vernal equinox, however. In this scheme the equinox always lies between March 12 and March 29. These schemes have never attracted forceful supporters like Achelis and Cotsworth and seem somewhat more unwieldy than their elegant proposals.

A Technical Addendum

As we noted above the Gregorian calendar overstates the length of the year by approximately 26 seconds (more precisely, but still not exactly, 25.92 seconds). There are 86,400 seconds in a day. Thus $86400/25.92 \approx 3333.33$ years after the institution of the Gregorian calendar we will be a full day ahead of the sun and need to cancel a scheduled leap year. Figure this to happen around 4916 or so.

The earth orbits the sun in approximately 365.2422 days. The Julian calendar with its leap year every fourth year assigns a length of 365.25 days per year. Subtracting, there are $365.25 - 365.2422 = .0078$ days per year too many. The reciprocal of .0078 is 128.205. That is, to fix the Julian calendar we should skip a leap year about every 128 years. Such calendars have been proposed. The Gregorian calendar, of course, skips three leap years every 400 years, or one leap year every 133 1/3 years. Had Gregory chosen to skip four leap years every 500 years, i.e., skip one every 125 years, our calendar would be much more accurate requiring no extra leap years or leap year skipping until sometime around the year 14600.

It's unlikely that we'll see worldwide calendar reform anytime soon. Change of something so fundamental requires a strong leader with worldwide influence (like a first-century BCE Roman emperor or a Renaissance pope). Besides, no matter what we do there can never exist a perfect calendar. The length of the solar year is not a constant. This year (2000) is very nearly 365.24219265 days long; the year 4000 will be closer to 365.24206985 days long. (This variability in the length of the year effects the timing of the skipped leap year discussed above: we'll actually need to omit a leap year several centuries before 4916.)

So I'll never be able to remember which day of the week George Washington was born on, let alone his actual birthday: it looks like the universe gets the last laugh. ■

References

1. Elisabeth Achelis, *Be Not Silent*, Pageant Press, Inc., NYC, 1961.

2. Edward L. Cohen, A Short History of Gregorian Calendar Reform, *Canadian Society for the History and Philosophy of Mathematics*, **10**, 1997, 102–116.

3. Edward L. Cohen, Elisabeth Achelis: Calendar Reformer, *Canadian Society for the History and Philosophy of Mathematics*, **11**, 1998, 97–109.

4. David Ewing Duncan, Calendar, *Smithsonian Magazine*, February 1999, 48–58.

5. Joachim Ekrutt, *Der Kalender im Wandel der Zeiten*, Kosmos; W. Keller & Co., 1972.

6. F. Ginzel, *Handbuch der mathematischen und technischen Chronologie*, 1914.

7. Gordon Moyer, The Gregorian Calendar, *Scientific American*, **246**, May 1982, 144–152.

8. Robert Poole, *Time's Alteration: Calendar reform in early modern England*, UCL Press, London, 1998.

9. François Russo, SJ, *Gregorian Reform of the Calendar*, Pontificia Academia Scientiarum, Specola Vaticana, 1983.

Quadrilaterally Speaking

WILLIAM DUNHAM
Muhlenberg College

Among the figures of plane geometry, the triangle holds a special place. Triangles are simple, basic, and unembellished, yet their geometric importance cannot be overemphasized. All have three medians, three altitudes, and one centroid, and all possess both inscribed and circumscribed circles. When two or more triangles get together, they can be the spitting image of one another (i.e., be congruent) or at least bear a strong family resemblance (i.e., be similar). And, as everyone knows, congruent and similar triangles are critical to the logical development of geometry.

But triangles have one major shortcoming when compared to their polygonal cousins: they lack diagonals. After drawing a triangle's three sides, the mathematician finds that no unconnected vertices remain, and consequently the "diagonal" of a triangle is the geometric counterpart of one hand clapping.

In order to explore the interconnection among sides and diagonals, it is necessary to step up to quadrilaterals. For the purposes of this article, all quadrilaterals will be convex, with both diagonals falling neatly inside. As denoted in Figure 1, quadrilateral $ABCD$ will have sides of length $a, b, c,$ and d and diagonals of length $x = AC$ and $y = BD$. (We will use AC to represent both *the line segment* from A to C and *the length of this segment.* Which is meant should be clear from the context.)

We shall examine a trio of theorems relating the sides and diagonals of quadrilaterals. Although far from new, they seem to be relatively unknown, even among the mathematically preoccupied. Subscribing to the adage that good things bear repeating, we consider them in turn.

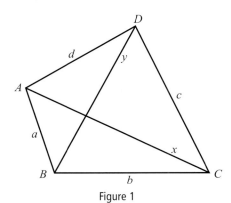

Figure 1

Ptolemy's Theorem

This result appears in Book I of the *Almagest,* the astronomical masterpiece of Claudius Ptolemy (ca. 85–165). A word of warning: the likeness we have provided of Ptolemy is not to be trusted. Indeed, it looks suspiciously like the images we have of Homer, Aeschylus, and Socrates—not to mention those of Noah, Nebuchadnezzar, and Neptune. Portraits of classical males tend to be interchangeable, with everybody more or less resembling Moses, or perhaps Charlton Heston.

But Claudius Ptolemy does hold one legitimate distinction: he is the earliest major mathematician who possessed both a first and a last name. His illustrious predecessors—from Euclid to Archimedes to Apollonius—got by quite well with one name. (Whether "Euclid" was his first name or his last name is a matter that concerns only chronic insomniacs.)

In any case, before stating Ptolemy's theorem, we need

Definition: A *cyclic* quadrilateral is one that can be inscribed in a circle.

Put another way, a quadrilateral is cyclic if a circle can be circumscribed about it. It should be clear that this is a very re-

Claudius Ptolemy

Reprinted from February 2000, pp. 12–16

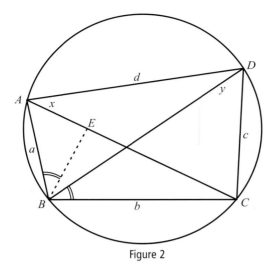

Figure 2

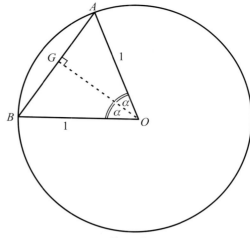

Figure 3

strictive condition. For, suppose we refer to the quadrilateral in Figure 1 and circumscribe a circle about $\triangle ABC$. Then quadrilateral $ABCD$ is cyclic if and only if the point D falls exactly upon the circle just drawn— an unlikely circumstance to be sure.

In spite of the fact that cyclic quadrilaterals are far from general, Ptolemy proved a beautiful result about them, namely:

Theorem 1. *If $ABCD$ is a cyclic quadrilateral, then the product of the diagonals is the sum of the products of the opposite sides. That is, in Figure 2, $ac + bd = xy$.*

Proof. To begin his elegant argument, Ptolemy constructed $\angle ABE$ congruent to $\angle DBC$, where E lies on diagonal AC. Because $\angle BAC$ and $\angle BDC$ intercept the same arc of the circle, they are congruent. Therefore $\triangle ABE$ is similar to $\triangle DBC$, from which it follows that $AB/AE = DB/DC$. That is, $a/AE = y/c$, or equivalently

$$ac = (AE)y. \qquad (1)$$

Likewise, $\triangle ABD$ is similar to $\triangle EBC$ because $\angle ADB$ is congruent to $\angle ECB$ (they intercept the same arc) and $\angle ABD = \angle ABE + \angle EBD = \angle DBC + \angle EBD = \angle EBC$. Thus $BD/AD = BC/EC$. That is, $y/d = b/EC$, and so

$$bd = (EC)y. \qquad (2)$$

Now simply add equations (1) and (2) to get

$$ac + bd = (AE + EC)y = (AC)y = xy,$$

and the proposition is proved. QED

Ptolemy used this theorem in the *Almagest* to generate his "Table of Chords," the precursor of our trigonometric tables. Moreover, a famous identity follows as an easy corollary. To see it, we first observe that if 2α is the measure of central angle AOB in the unit circle (see Figure 3) and if we draw OG perpendicular to chord AB, then $\sin\alpha = AG/OA = AG$, and so the length of chord AB is $2(AG) = 2\sin\alpha$.

Now for the identity. Suppose we begin with a pair of angles having measures 2α and 2β, as shown in Figure 4. Place these

as central angles within a unit circle, one on either side of diameter BD, and connect points A, B, C, and D to generate a cyclic quadrilateral and its diagonals. By our previous observation,

$$AB = 2\sin\alpha, \quad BC = 2\sin\beta, \quad \text{and} \quad AC = 2\sin(\alpha + \beta).$$

Furthermore, because DAB and DCB are inscribed in a semicircle, the corresponding triangles $\triangle DAB$ and $\triangle DCB$ are right, and so the Pythagorean theorem implies that

$$AD = \sqrt{BD^2 - AB^2} = \sqrt{4 - 4\sin^2\alpha} = 2\sqrt{1 - \sin^2\alpha} = 2\cos\alpha$$

and similarly that $CD = 2\cos\beta$.

Thus when we apply Ptolemy's theorem to $ABCD$, we get

$$(2\sin\alpha)(2\cos\beta) + (2\sin\beta)(2\cos\alpha) = 2[2\sin(\alpha + \beta)],$$

which reduces to

$$\sin(\alpha + \beta) = \sin\alpha\cos\beta + \cos\alpha\sin\beta,$$

the justly famous identity from trigonometry. Claudius Ptolemy was surely on to something.

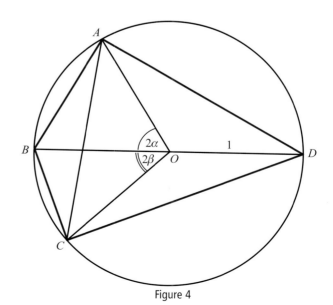

Figure 4

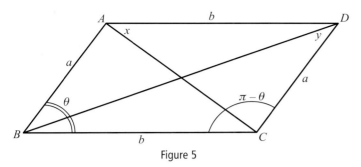

Figure 5

Whereas Ptolemy's theorem is restricted to cyclic quadrilaterals, our second is specific to parallelograms. Yet, as we shall see, it points the way towards the general case.

Theorem 2. *If ABCD is a parallelogram as in Figure 5, then*
$$2a^2 + 2b^2 = x^2 + y^2.$$

Recalling that the opposite sides of a parallelogram are congruent, we can restate this as: the sum of the squares of the four sides of a parallelogram (i.e., $a^2 + b^2 + a^2 + b^2$) equals the sum of the squares of the diagonals.

Proof. With θ as the measure of $\angle ABC$—and thus $\pi - \theta$ as the measure of $\angle BCD$—apply the law of cosines to $\triangle ABC$ and $\triangle DBC$ to get

$$x^2 = a^2 + b^2 - 2ab\cos\theta \quad \text{and} \quad y^2 = a^2 + b^2 - 2ab\cos(\pi - \theta).$$

Then, because $\cos(\pi - \theta) = -\cos\theta$, we need only add these equations to reach the desired end. QED

We now have two theorems at our disposal—one pertaining to cyclic quadrilaterals, the other to parallelograms—and it is natural to ask under what conditions they are simultaneously applicable. That is, when is a quadrilateral cyclic *and* parallelogramic (if there is such a word)?

A moment's thought suggests that this happens if and only if the quadrilateral is a rectangle, and the proof of this double implication is easy. In one direction, we note that a cyclic quadrilateral that is also a parallelogram must satisfy both theorems 1 and 2, and so (with our prior notation)

$$xy = aa + bb = \frac{2a^2 + 2b^2}{2} = \frac{x^2 + y^2}{2}.$$

Cross multiplication yields $0 = x^2 - 2xy + y^2 = (x - y)^2$, which implies that $x = y$. But a parallelogram with equal diagonals is a rectangle, so this direction is proved. Conversely, a rectangle is already a parallelogram and is obviously cyclic, for a circle with center at the intersection of the diagonals and with radius half the length of the diagonal passes through all four vertices.

In the case when $ABCD$ is a rectangle with $AC = BD = x$, then theorems 1 and 2 reduce (respectively) to

$$aa + bb = x^2 \quad \text{and} \quad 2a^2 + 2b^2 = x^2 + x^2 = 2x^2.$$

In short, when both apply, they jointly collapse into the Pythagorean theorem—a most significant crossroads.

Leonhard Euler

Alas, the aforementioned results hold only for special kinds of quadrilaterals. The big challenge is to prove a theorem relating the sides and diagonals of a *general* convex quadrilateral. What this relationship might be, and how to prove it, are far from obvious.

Fear not. In a 1748 paper, Leonhard Euler (1707–1783) rose to the challenge [1]. We shall present the theorem—as he did—in stages, building toward the general result after a pair of lemmas. (Note in passing that Euler looks nothing like Moses, although he does sport a natty turban.)

For the purposes of the remaining proofs, we shall explicitly draw our quadrilateral as though it is *not* a parallelogram. The reader can check that the results become trivialities in the case of parallelograms.

Lemma 1. *Given quadrilateral ABCD in Figure 6, complete parallelogram ABCE and draw DE. Then*

$$a^2 + b^2 + c^2 + d^2 = AC^2 + BD^2 + DE^2.$$

In words, this says that the sum of the squares of a quadrilateral's four sides equals the sum of the squares of its diagonals *plus* the square of the length of this newly created segment DE. Think of DE as measuring how far $ABCD$ deviates from a parallelogram.

Proof. In Figure 6, construct CF parallel to AD with $CF = AD = d$ and draw segments $AF, BE, BF, DF,$ and EF. It can be proved (exercise!) that quadrilaterals $ADCF$ and $BDEF$ are themselves parallelograms. Thus, by theorem 2,

$$2c^2 + 2d^2 = AC^2 + DF^2 \quad \text{and} \quad 2BD^2 + 2DE^2 = BE^2 + DF^2.$$

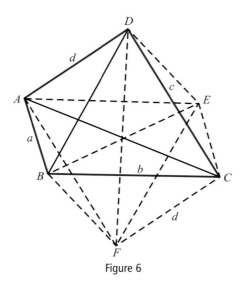

Figure 6

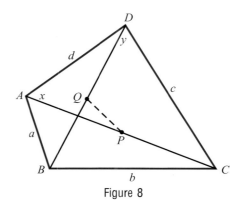

Figure 8

Consequently,

$$2c^2 + 2d^2 - AC^2 = DF^2 = 2BD^2 + 2DE^2 - BE^2,$$

and so

$$2c^2 + 2d^2 = 2BD^2 + 2DE^2 + AC^2 - BE^2.$$

But from parallelogram $ABCE$, we have $2a^2 + 2b^2 = AC^2 + BE^2$. Adding these last two equations yields

$$2a^2 + 2b^2 + 2c^2 + 2d^2 = 2AC^2 + 2BD^2 + 2DE^2,$$

and thus $a^2 + b^2 + c^2 + d^2 = AC^2 + BD^2 + DE^2$.   QED

As desired, this lemma relates the four sides and two diagonals of a general quadrilateral. But it is less than optimal, requiring as it does the "extraneous" segment DE. To tidy up this defect, Euler pushed on:

Lemma 2. *For quadrilateral ABCD with completed parallelogram ABCE as above, bisect diagonal AC at P and diagonal BD at Q and draw segment PQ connecting these midpoints (see Figure 7). Then $DE^2 = 4PQ^2$.*

Proof. We first assert that P lies on BE. This follows because the diagonals of parallelogram $ABCE$ bisect each other, and so P—the midpoint of AC—must also be the midpoint of BE. But then segment PQ connects the midpoints of sides BE and

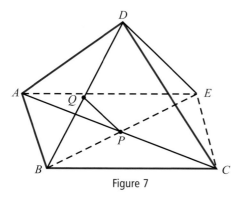

Figure 7

BD in $\triangle BED$, and by similar triangles we conclude that $2(PQ) = DE$. Squaring both sides proves the lemma.   QED

At last, we establish the general theorem (see Figure 8). Euler stated it as:

Theorem 3. *In any quadrilateral, the sum of the squares of the four sides equals the sum of the squares of the diagonals plus four times the square of the line connecting the midpoints of the diagonals.*

Proof. This is now trivial, for we need only combine lemma 1 and lemma 2:

$$a^2 + b^2 + c^2 + d^2 = AC^2 + BD^2 + DE^2$$
$$= AC^2 + BD^2 + 4PQ^2$$
$$= x^2 + y^2 + 4PQ^2.   \text{QED}$$

Notice how the point E, which Euler introduced somewhat artificially at the outset, has disappeared from the final theorem—truly, Euler giveth and Euler taketh away. With this argument, he had established a curious property of quadrilaterals that was, as he put it, "neither enunciated nor proved to this point."

Euler noted a pair of interesting consequences:

The sum of the squares of the four sides of a convex quadrilateral is always greater than or equal to the sum of the squares of its diagonals.

The sum of the squares of the four sides of a convex quadrilateral *equals* the sum of the squares of its diagonals if and only if the quadrilateral is a parallelogram.

The latter follows because, under the condition of equality, the length of the segment PQ is zero. Consequently, the diagonals bisect each other, and the quadrilateral is a parallelogram.

For the sake of variety, we conclude with an alternate—and shorter—demonstration of theorem 3. Although Euler needed the extraneous segment DE, this argument involves only the original quadrilateral. And, whereas Euler used synthetic geometry, this proof rests upon trigonometry, in particular upon a five-fold (!) application of the law of cosines.

Theorem 3 Revisited. [2] *In any quadrilateral, the sum of the squares of the four sides equals the sum of the squares on the*

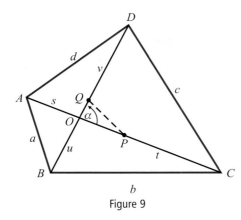

Figure 9

diagonals plus four times the square of the line connecting the midpoints of the diagonals.

Proof. As illustrated in Figure 9, we again consider quadrilateral $ABCD$. With O as the intersection of the diagonals, we let $OA = s$ and $OC = t$, and we stipulate without loss of generality that $s \leq t$. Likewise, let $OB = u$ and $OD = v$, with $u \leq v$. Finally, we let α be the measure of $\angle COD$. Of course it follows that α is also the measure of $\angle AOB$ and $\pi - \alpha$ is the measure of $\angle BOC$ and $\angle DOA$.

Apply the law of cosines individually to ΔAOB, ΔBOC, ΔCOD, and ΔDOA, while remembering that $\cos(\pi - \alpha) = -\cos\alpha$:

$$a^2 = s^2 + u^2 - 2su\cos\alpha$$
$$b^2 = u^2 + t^2 + 2ut\cos\alpha$$
$$c^2 = t^2 + v^2 - 2tv\cos\alpha$$
$$d^2 = v^2 + s^2 + 2vs\cos\alpha. \qquad (3)$$

Upon adding, we have

$$a^2 + b^2 + c^2 + d^2$$
$$= 2s^2 + 2t^2 + 2u^2 + 2v^2 - 2(su - ut + tv - vs)\cos\alpha$$
$$= (s+t)^2 + (u+v)^2 + (t-s)^2 + (v-u)^2 - 2(t-s)(v-u)\cos\alpha$$
$$= AC^2 + BD^2 + 4\left[\left(\frac{t-s}{2}\right)^2 + \left(\frac{v-u}{2}\right)^2 - 2\left(\frac{t-s}{2}\right)\left(\frac{v-u}{2}\right)\cos\alpha\right].$$

But recall that P is the midpoint of diagonal AC and Q the midpoint of diagonal BC, as shown in Figure 9. As a consequence,

$$OP = OC - PC = t - \frac{t+s}{2} = \frac{t-s}{2}$$

and

$$OQ = OD - QD = v - \frac{v+u}{2} = \frac{v-u}{2}.$$

Therefore, the law of cosines applied to ΔOPQ yields

$$PQ^2 = OP^2 + OQ^2 - 2(OP)(OQ)\cos\alpha$$
$$= \left(\frac{t-s}{2}\right)^2 + \left(\frac{v-u}{2}\right)^2 - 2\left(\frac{t-s}{2}\right)\left(\frac{v-u}{2}\right)\cos\alpha,$$

and this last is *exactly* the expression within the square brackets of equation (3) above. A final substitution gives us, as before,

$$a^2 + b^2 + c^2 + d^2 = AC^2 + BD^2 + 4PQ^2. \qquad \text{QED}$$

Of course much more could be said about quadrilaterals. There is, for instance, the fact that if the sums of the squares of opposite sides of a quadrilateral are equal (i.e., if $a^2 + c^2 = b^2 + d^2$), then its diagonals must be perpendicular. Or there is Brahmagupta's wonderful formula giving the area of a cyclic quadrilateral in terms of the lengths of its four sides.

For now, however, we must leave this topic, reminded once again of the unexpected patterns lurking beneath the surface of Euclidean geometry. It was Howard Eves [3] who perceptively observed that this subject "though elementary, is often far from easy." As the preceding theorems reveal, quadrilaterals can hold their share of surprises.

Note. I wish to thank Drs. Penny Dunham and Elyn Rykken of Muhlenberg College for their helpful suggestions in the preparation of this article. ∎

References

1. Leonhard Euler, *Opera Omnia*, Ser. 1, Vol. 26, pp. 29–32.

2. Thanks to Elyn Rykken for showing me this proof.

3. Howard Eves, *A Survey of Geometry*, Allyn and Bacon, Boston, 1963, p. 64.

Stopwatch Date

PHIL GRIZZARD
Illinois State University

Sophie Claire Meuth of Normal, Illinois was born on the last day of November 1999. What her parents, Jeremy and Alison, didn't realize at the time was that their daughter was born on a special date—the date that was overlooked throughout all the millennium hubbub—November 30, 1999: The Stopwatch Date.

Anyone who got married or (more likely, since it was a Tuesday) had a child as the Meuths did on November 30, 1999 needs to take note. Why? Because every date you see from now on will tell you exactly how long it has been since November 30, 1999.

When a stopwatch is cleared, the screen has all zeros. The convenience of this is that after it is started, the digits themselves show precisely how long it has been since the timing began. With a calendar, we don't use zero for months or days, so it doesn't appear that a stopwatch date would be possible. However, November 30, 1999 does the trick.

How does Sophie's birthday do this? Let's look at some examples. To start off, how old was Sophie on her first New Year's Day? Well, she had lived all the month of December 1999, plus one day. Hence, she was one month and one day old. How many years old? Zero years old, just one month and one day. Hey, look at the date—1/1/00! The date is Sophie's age exactly. Now, how old will Sophie be 5 months and 16 days after her first New Year's Day? Her folks would then be happy to tell you that she's 6 months and 17 days old. Take a glance at the date, and it's 6/17/00. Let's go from the other perspective. If Sophie gets married on

July 10, 2027, how old will she be then? We know that she has not yet turned 28, so she is 27 years old. Also, she turned 27 the past November, and has lived 7 full months (December through June) since her last birthday, plus an additional 10 days in the current month. Thus her age is 7 months, 10 days, and 27 years. Her wedding date? 7/10/27. In fact, as you read this, today's date is exactly how old Sophie is. This is true because Sophie was born on The Stopwatch Date.

Another convenient feature of Sophie's birthday is that she automatically knows how much older she is than anyone who is younger than she is. For example, suppose that in a couple of years, her folks have a baby boy. As siblings often do, Sophie and her brother will want to figure out exactly the difference between their ages. (Age difference is an important issue to siblings!) For Sophie, the answer will always be apparent. Her brother was born February 9, 2003, so his birthdate is 2/9/03. And Sophie will know exactly how old she was that day—2 months, 9 days, and 3 years old. Her brother's birthday is the difference between their ages. They will always know exactly how much older Sophie is, because it is her brother's birthday. This is true for anyone Sophie meets who is younger than she is. The other person's date of birth is exactly how much older Sophie is than the other person.

"So," you ask, "How do you figure Sophie's age every December?" Well, just like three strikes make an out and ten dimes make a dollar, twelve months make a year. Hence whenever the month value is 12, we "make change" by clearing out

the months and adding one to the years. The months are counted modulo 12, but we must keep track of time by adding one to the year whenever the months convert to zero. For example, let's say Sophie has a daughter born on 12/15/30. How can we get Sophie's age out of 12/15/30? We make change by turning the 12 into a zero and adding one to the years. Thus our "date" now reads 0/15/31. A quick check will confirm that Sophie is 31 years, zero months, and 15 days older than her daughter.

Another detail needs to be worked out for us precise mathematicians. Every November 30 it will be necessary to make change twice: once from the months to years, but first from the *days* to months. Since November has 30 days, the modulus for the days in November is 30. Thus every November 30, we clear the days but keep track of time by adding one to the months. However, we then have twelve months, so we make change again and get zero for the months and add one to the years. Hence later in her life, 11/30/74 becomes 12/0/74, which becomes 0/0/75, Sophie's seventy-fifth birthday.

So for anyone who regards this past November 30 as a special date, it is automatic to remember how much time has passed since then. For parents like the Meuths who gave birth to a Stopwatch Baby, you and your child are in for eternal calendar convenience, and here's how to figure it easily. Counting the years modulo 100, do *three* making change steps on 11/30/99 and you'll get 0/0/00. That's why it's The Stopwatch Date! ■

Reprinted from February 2000, p. 22

A Very Simple, Very Paradoxical, Old Space-Filling Curve
(or "Wrong-Way" Corrigan Rides Again!)

IRA ROSENHOLTZ
Eastern Illinois University

In July 1938, Douglas "Wrong-Way" Corrigan flew nonstop from California to New York in 28 hours (a record!?!). On his return flight to California he apparently became disoriented in the impenetrable fog and ended up, 24 hours later, landing in Ireland [1]! (His plane had a faulty compass and no radio.) He was given a hero's welcome when he returned to the States.

Mathematicians have long known of a space-filling curve which has certain similarities to Corrigan's flight. The example is due to Lebesgue, although, as you'll see, it might just as easily have come from Cantor. And while it might not be as pretty as the examples of Peano and Hilbert, its simplicity more than compensates for this. Plus its paradoxical nature (above and beyond the fact that it is a space-filling curve) should alone make it worth the price of admission—it provides a "flight plan" for going northeast, while spending 100% of the time going south and west.

Shortly after Georg Cantor shocked the world with an example of a one-to-one correspondence between the unit interval [0, 1] and the unit square $[0, 1]^2 = [0, 1] \times [0, 1]$, Giusseppi Peano exhibited in 1890 the first "space-filling curve," a continuous function from the unit interval onto the unit square. (Think of this as a way for a point-sized airplane to fly around a square in one hour touching each and every point of the square, and doing so in a continuous fashion.) Just as Cantor made people rethink the notion of infinity, Peano made people rethink the notion of dimension. After all, shouldn't a curve or path be one-dimensional? The list of contributors to the area of space-filling curves reads like a Who's Who of famous mathematicians: in addition to Peano, Hilbert, and Lebesgue, there are Sierpinski, E.H. Moore, Cesaro, Wunderlich, Knopp, Weierstrass, Pólya, Osgood, Steinhaus, Schoenberg, Mandelbrot, Hahn and Mazurkiewicz [2]. More re-

cently, Bartholdi and Platzman have applied space-filling curves to the famous Traveling Salesman problem.

Without further ado, let us proceed with the construction. Let C denote the usual "middle-thirds" Cantor set. Then C consists of those numbers in [0, 1] which can be written in base 3 using only 0's and 2's. Thus, for example, 1/3 belongs to C because, while 1/3 can be written in base 3 as 0.1000..., it can also be written as 0.0222..., and 1/4 belongs to C because 1/4 = 0.0202.... We define a function g from the Cantor set C to the unit square $[0, 1]^2$ by

$$g((0.x_1 x_2 x_3 x_4 x_5 x_6 \ldots)_3) =$$
$$((0.(x_1/2)(x_3/2)(x_5/2)\ldots)_2, \ (0.(x_2/2)(x_4/2)(x_6/2)\ldots)_2).$$

That is, we take our string of 0's and 2's, make two strings of 0's and 1's by separating the odd and even "decimal" places and dividing by 2, and then interpret the resulting strings of 0's and 1's as an ordered pair of numbers in base 2. For example, $g(1/3) = g((0.0222...)_3) = ((0.0111...)_2, (0.1111...)_2) = (1/2, 1)$. (To really understand what's going on, the reader should calculate $g(0)$, $g(1/9)$, $g(2/9)$, $g(2/3)$, $g(7/9)$, $g(8/9)$, and $g(1)$ before proceeding.)

To see that g is onto, let (y, z) be an element of $[0, 1]^2$, write y and z in base 2 (writing 1 as 0.1111... as need be), double all of the "decimal" places, and "shuffle" them together appropriately to get an element of the Cantor set which hits (y, z) under the function g.

Also, g is continuous since if s and t belong to C and $d(s, t) < 1/9^n$, then $d(g(s), g(t)) < \sqrt{2}/2^n$, which is easy to check. To see it, first note that the ninths of the Cantor set go to the quarters of the square. For example, if t belongs to C and $2/9 < t < 1/3$, then t must begin $(0.02...)_3$, so $g(t) = ((0.0...)_2, (0.1...)_2)$,

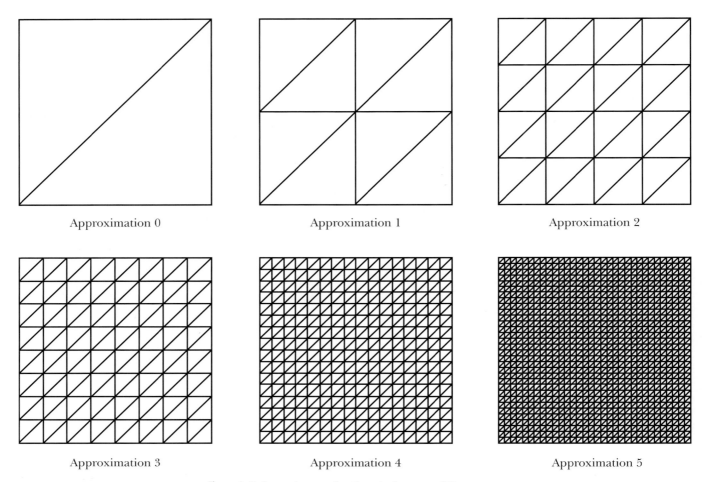

Approximation 0   Approximation 1   Approximation 2

Approximation 3   Approximation 4   Approximation 5

Figure 1. Polygonal approximations to the space-filling curve.

an element of $[0, 1/2] \times [1/2, 1]$. More generally, if $d(s, t) < 1/9^n$, then the first $2n$ ternary (base 3) places of s and t are the same, so the first n binary places of the x and y coordinates of $g(s)$ and $g(t)$ are the same, and $d\left(g(s), g(t)\right) < \sqrt{2}/2^n$.

Now to obtain our space-filling curve, we simply extend the function g defined on C to a function G defined on $[0, 1]$ by linearly "filling-in-the-gaps." For example, since $g(1/9) = (1/2, 1/2)$ and $g(2/9) = (0, 1/2)$, and g is not defined at any point of the open interval $(1/9, 2/9)$, we let G take the interval $[1/9, 2/9]$ linearly onto the segment from $(1/2, 1/2)$ to $(0, 1/2)$ in the square —we tell the plane to fly straight from $(1/2, 1/2)$ to $(0, 1/2)$ from time $1/9$ to time $2/9$ at constant speed.

Polygonal approximations to this space-filling curve G are not difficult to visualize. The nth polygonal approximation simply connects the dots

$G(0)$, $G(1/9^n)$, $G(2/9^n)$, $G(3/9^n)$,…, $G((9^n-1)/9^n)$, $G(1)$

in order. Approximation 0 simply goes NE. Approximation 1 goes NE, W, NE, S, S, S, NE, W, NE on the ninths of $[0,1]$. Approximation $n + 1$ replaces each NE segment (Approximation 0) in Approximation n with NE, W, NE, S, S, S, NE, W, NE (Approximation 1). The function G is certainly onto, since g is. And G

is also continuous because if s and t belong to $[0, 1]$ and $d(s, t) < 1/9^n$, then $d\left(G(s), G(t)\right) < \sqrt{2}/2^n$. So G is a space-filling curve.

Finally, notice that $G(0) = (0,0)$ and $G(1) = (1,1)$, so G goes northeast. But G goes south on the intervals $[1/3, 2/3]$, $[1/27, 2/27]$, $[7/27, 8/27]$, $[19/27, 20/27]$, $[25/27, 26/27]$, ... or

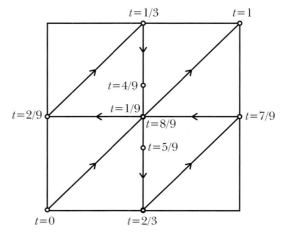

Figure 2. A look at approximation 1, up close and personal.

Cartoon by Loel Barr

$1/3 + 4/27 + \cdots = 3/5$ or 60% of the time. And G goes west on the intervals [1/9, 2/9], [7/9, 8/9], [1/81, 2/81], [7/81, 8/81], [19/81, 20/81], [25/81, 26/81], [55/81, 56/81], [61/81, 62/81], [73/81, 74/81], [79/81, 80/81], ... or $2/9 + 8/81 + \cdots = 2/5$ or 40% of the time. (Both of these are geometric series.) While the polygonal approximations do go northeast part of the time, all of these northeasterly segments get replaced as we go along— Approximation n goes northeast $(4/9)^n$ of the time, to be precise —so the final curve is traveling west and south 100% of the time! Indeed Wrong-Way Corrigan rides again! As do Cantor and Lebesgue and…. The author hopes that this clears away some of the fog surrounding space-filling curves. ∎

Notes and Further Reading

1. There's no record of what happened to his luggage.
2. Hans Sagan, *Space-Filling Curves*, Springer-Verlag, New York, 1994.

Coal Miner's Daughter

DEANNA HAUNSPERGER and STEPHEN KENNEDY
Carleton College

The first thing that strikes you is that she's charming. Only later do you realize that she is probably the smartest person you've ever met. Somehow, you don't expect mathematical genius and a charming, genuinely warm personality to coexist. Yet, here sitting down to lunch with us is an existence proof: Professor Ingrid Daubechies of Princeton University, MacArthur Fellow, member of the National Academy of Sciences, the mother of wavelets. Fuzzy-haired, bespectacled, quick to smile and a world-class storyteller, she's telling us about growing up in a coal-mining town in Belgium in the sixties.

"I grew up in a very small town. My father worked in a coal mine. He was an engineer in a coal mine, and of course while I was growing up I didn't notice, but in retrospect coal-mining towns are very special kinds of small towns. I mean, there is just one big employer, and he controls the whole life, even the social life, of the town. My mother, well, for her generation in Belgium it was not common to have university education, but she did. She had expected to have a career, but after marrying my father she didn't. Partly because there was no opportunity, but also because in this very small paternalistic town, two generations behind the wide world, it just was not done for wives of engineers to work. There was one wife who worked, she was a nurse, and everybody knew that that was why her husband never got a promotion. It wasn't said that way, of course, and even then they couldn't write things down that way. So my mother didn't work.

"I remember when we were little she did a lot with us kids. As we grew up we became more independent because she wanted to give us our independence, but she was also very bitter at not having any bigger framework for her own life. She went back to college when I went to university. We had moved by then. She went

for a different degree because her first degree, in economics, had become obsolete — she hadn't worked for twenty years. She got a second degree in criminology and worked for about twenty years as a social worker. She worked with troubled youths, trying to monitor them, and to help them, and to give them decent lives.

"She had met my father while they were students at different universities at a meeting that brought together students from universities in Belgium. She met him, and later they decided to get married. They married in 1952, and he started work at the coal mine. At this coal mine they would only hire engineers who were either married or engaged to be married very soon. They did not want trouble with single men around.

"My father really would have liked, himself, to become a scientist, to become a physicist. He was really mostly interested in physics, but he became a mining engineer. His parents were very poor. They came from a coal-mining region, and for them an educated person was a coal-mining engineer. They had never seen any

other profession. Also, there was a very good engineering school in that area, so they made a deal with him when he was growing up that they would not, like their friends, save for their retirement. My grandfather actually didn't work at the coal mine, he worked in a glass factory, and they lived in a house that was owned by his employer. In these company towns people didn't live in houses that belonged to them, so everybody would save so that they could buy a small house or a small apartment to live in after their retirement. But my grandparents made a deal with my father that they wouldn't do that; instead they would pay for his education and then he would take care of them when they retired. So, that's how it happened that he became a coal-mining engineer —

because it had all been planned that way, and he only discovered while at the university that there were other choices. He was at a school which was an engineering school. He wouldn't have been able to explain a change to his parents; they would have been so worried. As a coal-mining engineer they knew he would be able to support them.

"The region where my father's parents were born and lived was a very poor region. The reason my grandfather didn't work in the coal mines was that his mother really wanted one of her children not to go down the coal mine because that was the time when you would die young if you went into the coal mines. They didn't know how to prevent black lung disease; everybody died young. The coal mines were just disputing the fact that it was anything to do with work in the mines, so you didn't get any compensation either. My grandfather had been sickly when he was little and he was the first-born, so his mother wanted him to work elsewhere. Generations in my family are very long, so I'm talking the end of the 19th century when my grandfather was born. He left school when he was nine. I said this was a poor region, and this was before child-labor laws. His mother, my great-grand-mother, had arranged a job for him in a big glass factory; he was in packing. But then, these things are incredible, by the time he was 14, he had in various accidents lost a finger and an eye in the glass factory. I think of my children, and I think — how is this possible? My father always says of his father that he was really very

smart. He went to evening school at some point. He was in packing all of his working life, and he became a foreman. At some point, for a very complicated delivery, they had to make a case the inside of which was the intersection of two cylinders, but he had to make it in wood, of course, and fold it out of plywood. He had tried with ellipses and somehow it never fit, so he actually went back to evening school to study mathematics in order to learn how to do that. Another foreman had explained how to do it, but he wanted to know what it looked like. If you took a cylinder and then unfolded it, what would it look like? How do you actually compute that? So, he went back to school."

An Education in Physics

Daubechies attended the Dutch Free University in Brussels where she studied physics. She held a research position in physics at that same university until 1987 when she came to the United States. Today, though, she asserts, "I'm a mathematician."

"My father was always interested in mathematics, and he was always interested in explaining things to me, and I liked it. I would ask questions. I would usually get answers which were much longer than I hoped for, so I am trying

with my children to yes, give answers, but maybe not go beyond so far. I remember liking mathematics when I was little, but I actually did major in physics. I think I majored in physics because it was my father's dream to become a physicist; he explained to me things about physics. He went to extra Open University courses whenever he could. Sometimes they would organize a series of physics lectures, and he would go.

"Physics just seemed to be a very noble choice because of my father's influence. It was something intermediate between what I really wanted—mathematics—and what my mother really wanted—which was that I would have become an engineer. She was a bit worried about all this science. She thought scientists were like artists, they really cannot make a good living. An engineer can always find a good job.

"With a free choice, I think I probably would have chosen mathematics. I don't know. I liked physics very much. I especially liked some physics classes that we had. At some point I was considering switching between math and physics, and I decided to stay in physics because of one particular course which I thought was wonderful, in which we were going beyond geometric optics. If you go to the Kirchoff-Fresnel theory of optics, you actually see, and can compute, that a lens, in fact, computes a Fourier transform, which I think is wonderful. I think this is mind-boggling, that a lens would, in fact, compute a Fourier transform and this is used in some optical computing. This was marvelous. In fact, this course was really a course in applied mathematics. It was labeled as a physics course, and it was wonderful, and so I stayed with physics. I don't regret it. As a result I have learned a whole lot of things that I wouldn't have learned in a standard math curriculum. And the math that I wanted and needed I have learned by myself anyway.

"I think I think like a mathematician; I switched from theoretical particle physics to more mathematical physics because I felt that people who were really good at particle physics had an intuition about which I

ED ABOUFADEL

A Short Primer on Wavelets

Wavelets have many applications, including the processing of finger-print images by the FBI, the analysis of earthquake data, and the development of a text-to-speech system. As a window into understanding wavelets, consider the problem of approximating complicated functions with simpler functions. In calculus, we learn how to approximate functions with Maclaurin and Taylor polynomials. For example, if f is a function defined near $t = 0$ then we can approximate it near $t = 0$ with a Maclaurin polynomial such as:

$$f(t) \approx f(0) + f'(0)t + \frac{f''(0)}{2!}t^2 + \frac{f'''(0)}{3!}t^3.$$

In Figure 1, we see how $\ln(t + 1)$ can be approximated by $0 + 1t - t^2/2 + 2t^3/6$.

When using Maclaurin polynomials, we call the functions $\{1, t, t^2, t^3, \ldots\}$ *basis functions*, and our approximation is created by adding multiples of the basis functions together.

A more powerful version of function approximation is *Fourier analysis*. In Fourier analysis, sine and cosine functions are used for the basis functions, instead of polynomials, and we attempt to have a good approximation on a fixed interval such as $[0, 2\pi]$. The goal is to decompose a function by thinking of it as a combination of trigonometric functions with different *frequencies*. For example, we can approximate $\ln(t+1)$ with the following Fourier series:

$$\ln(t+1) \approx 1.301 - .155\cos(t) - .053\cos(2t) - .026\cos(3t)$$
$$- .528\sin(t) - .294\sin(2t) - .202\sin(3t).$$

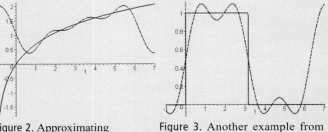

Figure 1. Approximating $\ln(t + 1)$ with a Maclaurin polynomial

For a graph of $\ln(t+1)$ and the Fourier approximation, see Figure 2.

This approach gives a good approximation in general, although the error is worse near $t = 0$ and $t = 2\pi$. The results are even worse for jagged functions, such as the characteristic function on the time interval $[0, \pi]$. (This is the function that is equal to 1 on the interval $[0, \pi]$ and 0 everywhere else.) In Figure 3, this function is approximated by a sum of sines and cosines.

Wavelet analysis is designed to better handle this type of function, because of the focus on the *time intervals* where functions are defined. In the example above, $[0, \pi]$ is an important time interval, so wavelet

Figure 2. Approximating $\ln(t + 1)$ with a Fourier series

Figure 3. Another example from Fourier analysis

had no clue. I felt like I could learn how to read those papers, but it was like learning a language without understanding the meaning of the words, which I didn't like at all. It's hard to describe how I think. Even in analysis, I don't think in formulas. Although when I work something out, I *do* compute a lot. I have some kind of mechanical or geometrical way of thinking, I don't really know where that comes from.

"Anyway, in Belgium, undergraduate education is really different from here in that you track very, very early on. When you register for the university, you have to say what you're going to major in. So you get very few courses outside your major or outside things related to your major. For physics, you get a lot of math, you get some chemistry, but you don't get any liberal arts courses. I think you could go sit as an auditor in some of these courses, but really there's no time; you don't choose your own courses. You say

'I will major in physics,' and then the courses are specified except that in later years, you have some choices, you get to choose one of four. In the first two years everything is completely chosen for you, and it's quite a heavy schedule. It's a heavier schedule than I see here, but the result is that you can do much less independent work. I think a schedule where you put together a combination yourself and where you're encouraged to do a lot of independent work is actually better.

"I was tracked with physics, so I had a lot of math courses, especially the first two years. And when I had majored, I had seen a lot of physics courses that would be at graduate level in the States because you cannot cram four years full of physics courses and not get to that level. Things are not organized so much by semester as they are for a whole year, so many courses were a full year. In the third year, that was really the heaviest year, we

had 13 different physics courses, and we had 5 weeks of labs. Lots of that physics I have forgotten.

"At the graduate school level, however, in most universities in Belgium, you don't get courses any more. There is a movement there now to get some what-they-call third-cycle courses which are graduate courses, but mostly you're left to learn extra things on your own or with your advisor. You're assigned or find a research topic; right away when you arrive you start working on papers. You also have a teaching schedule, and you're expected to have your PhD in 5 or 6 years. I never had a teaching schedule because I had a special fellowship, but on a teaching assistantship, you would teach (be in the classroom) 30 hours a week. I would typically have 8–10 hours a week. The things you would be teaching would be problem sessions. Of course, you were there for the students, and typically the

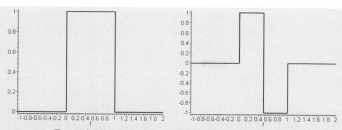

Figure 4. The "father" and "mother" Haar wavelets

basis functions are needed that will emphasize this interval in the decomposition. An easy-to-understand set of basis functions that have this property are the *Haar wavelets*. The "father" and "mother" Haar wavelets can be found in Figure 4.

The Haar wavelets have properties which are different than sine and cosine. For instance, these wavelets are not periodic. They are zero for most of the real line $(-\infty, \infty)$.

There is also a way of creating these functions that is different. The basis functions for Fourier analysis come from *scaling* the sine and cosine function (in other words, we start with $\sin x$, and $\cos x$ and then scale these functions by creating $\sin 2x$, $\cos 2x$, $\sin 3x$, $\cos 3x$, ...).

The basis functions for wavelet analysis come from *dyadic scaling*, which means that the scaling coefficients are only powers of 2. If Fourier analysis was done with dyadic scaling, then we would only use functions like $\sin 2x$, $\cos 2x$, $\sin 4x$, $\cos 4x$, $\sin 8x$, $\cos 8x$, etc. We also use *translating*, so that we can slide the scaling function to any important time interval. The function that is scaled and translated is called, not surprisingly, the *scaling function*, or "father" wavelet. We use ϕ to stand for the scaling function, and some of the "children" of $\phi(t)$ are $\phi(2t)$, $\phi(2t-1)$, $\phi(4t)$, $\phi(4t-3)$ and $\phi(8t)$. Other wavelets are created by combining these wavelets. For example, the "mother" wavelet $\varphi(t) = \phi(2t) - \phi(2t-1)$.

In Figure 5, we see how $\ln(t+1)$ can be approximated on the interval $[0,1]$ by a series of Haar wavelets.

During the 1980s, Ingrid Daubechies developed a special type of scaling function ϕ that had three properties. First, the function is equal to zero outside of the interval $[0, 3]$. Second, for any two different integers k and l

$$\int \phi(t-k)\phi(t-l)\,dt = 0.$$

This condition is called the *orthogonality condition*. Third, you can approximate constant and linear functions with no error, which is actually quite remarkable.

Combining these requirements, Daubechies deduced the following identity:

$$\phi(t) = \frac{1+\sqrt{3}}{4}\phi(2t) + \frac{3+\sqrt{3}}{4}\phi(2t-1)$$
$$+ \frac{3-\sqrt{3}}{4}\phi(2t-2) + \frac{1-\sqrt{3}}{4}\phi(2t-3).$$

From this identity, you can generate Daubechies's scaling function, which is pictured in Figure 6.

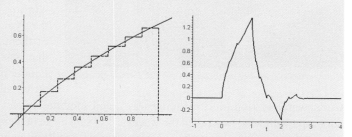

Figure 5. Approximating $\ln(t+1)$ with a wavelet series.

Figure 6. Daubechies's scaling function.

only person there for the students to ask questions from because they would not dare to approach a professor. You don't have control over what you're doing; you'd be given a list of problems, and you'd answer questions about them. It was very frustrating because it meant you had zero input into what you did. That persists even after the PhD, for a while."

A New Paradigm

Wavelets are everywhere these days. Wavelets are a new method for encoding and compressing information (see box). They are being used in image compression (the FBI's files of approximately 200 million fingerprints are being converted to wavelet-compressed electronic images), also sound and video compression, medical imaging, and geological exploration. Ronald Coifman of Yale University used wavelet techniques to remove the noise

from a century-old recording of Brahms playing one of his own compositions.

The most exciting thing about wavelets might just be the way that they are drawing together people and ideas from so many different fields of science: mathematicians, physicists, geologists, statisticians, computer scientists, engineers of all kinds. In fact the history of the idea has roots in all of these fields and more. Yves Meyer has identified precursors to the idea in mathematics, computer science, image processing, numerical analysis, signal processing, studies of human and computer vision, and quantum field theory. The short version of the history has geophysicist Jean Morlet and mathematical physicist Alexander Grossmann introducing the idea in the early 1980s. Meyer and Stephane Mallat pieced together a mathematical framework for wavelets in the mid-eighties. In 1987 Daubechies made her famous contribution of a family of

wavelets that are smooth, orthogonal, and equal to zero outside a finite interval. Thus, in a stroke, accomplishing what everyone supposed impossible and making wavelets very much more applicable.

"How did I get started on wavelets? For my PhD work, I had worked on something in quantum mechanics which are called coherent states. This is a tool to understand the correspondence between quantum mechanics and classical mechanics. So you try to build functions that are well-localized, that live in Hilbert space, but that correspond as closely as you can with being in one position, in one momentum in classical mechanics. I had worked in Marseilles with Alex Grossmann, and Alex was one of the people who really started the whole wavelet synthesis. There are roots in pure mathematics, in many different fields, but the synthesis really, I feel, was created by Alex

Daubechies discussing mathematics with R. Gundy (Rutgers)

Grossmann and Jean Morlet. Morlet was not happy with what was called the wavelet transform, and he wanted something else, so he invented an algorithm. But he had no mathematical theory behind it. Alex Grossmann realized there was an analogy with the coherent state formalism in quantum mechanics. He recast it in terms of a group representation and then people started saying 'But we've been doing that all along in a different context.' And they were right; they had been doing it all along. So that's what made the synthesis happen that made the jump from mathematicians to engineers.

"I knew Alex Grossmann very well from my thesis work, and I was looking for something else to start working on. This was a time when I had many changes in my life. Before my husband, I had a long relationship, and I had just left him, and I was looking for something else, for changes. So I changed research topics. I started working on wavelets in '85. I started working on wavelets and I met Robert [Calderbank, the mathematician to whom she is married] within a period of six weeks in '85.

"At this time geologists were doing windowed Fourier transforms, so they would window [look at a short time segment of a signal], then Fourier transform [approximate that bit of signal with combinations of sines and cosines]. This means that for very high frequency things that live in very short time intervals — these are called transients — if you've determined your window to be this wide, then you need a whole lot of high frequency functions to capture that behavior. So, you could of course make your window very narrow, but then you don't capture a lot of them. Morlet didn't like that aspect of the windowed Fourier transforms, so he said 'After all, I'm using location and modulation, let's do it differently. Let's take one of these functions that have some oscillations and let's put that in different places and squish it so that I have a different thing,' really wavelets. Now he didn't really formulate this precisely. Actually the name comes from there because in geology when you have different windows, which determine the shape of these multi-layered functions, they call them wavelets. So he called his transform 'a transform using wavelets of constant shape' because the other ones didn't have constant shape. If you adjust the window, or you modulate this way, then of course they look different. In his case, the wavelets look the same, they were just dilated versions. He called them wavelets of constant shape, but then once they left that field, there was no other thing around called wavelets, so he just dropped the 'of constant shape' to the great annoyance of geophysicists because in their field it has another meaning.

"Wavelets were not really something that were a trend in geophysics, Morlet just came up with it. In harmonic analysis, people had been looking at, not exactly the same way, but something similar for ages. It goes back to Littlewood-Paley theory [circa 1930], and even the integral transform formula that Grossmann and Morlet wrote, because Morlet had no real formulas, is a transform that you find in Calderón's work in the sixties. So in some sense they had reinvented the wheel. In another very real sense they had looked at it completely differently. For Calderón it was a tool to carve up space into different pieces on which he would then use different techniques for estimates. Grossmann and Morlet gave these wavelets some kind of physical meaning, in a certain sense, viewing them as elementary building blocks which was a different way of looking at it. And Yves Meyer later told me that when he read those first papers by Grossmann, it was very hard for him because it was a different style. It took a while before he realized in what sense it was really different, because at first you see the formulas there, and you say, 'well, yeah, we've been doing that for 20 years,' but then you realize that here was a different way of looking at it: a new paradigm shift."

A Link in a Chain

To most mathematicians it appears that wavelets sprang full-grown from the foreheads of Grossmann, Morlet, and Daubechies and then were immediately grabbed by engineers and scientists. It is unusual for a piece of mathematics to find so many applications so quickly. Daubechies with her ability to talk the languages of physics and engineering and mathematics is, to a large degree, responsible for the building of so many bridges between the groups.

"In pure mathematics the idea was developed starting in the thirties, then in greater detail in the sixties. It was a very powerful tool which lived in a relatively small community in mathematics, and outside the small community, I felt it spread rather slowly. For example, in quantum mechanics I think some of these techniques would have been useful to mathematical physicists earlier than they penetrated. I think it's because through Grossmann and Morlet there were intermediate people. I treasure every single electrical engineer I meet and with whom I can talk. I'm interested in talking with them, I think many mathematicians aren't, but I am. Even so, I find it hard to talk with many of them because we've been trained in completely different ways and the words mean different things. But I have found some I can talk to and I think it's very valuable when they are also interested in talking with me. I think it's easier for me because of this physics background I have and because I have learned at least some of their language. I think Alex Grossmann played a very important role that way. I have met Jean Morlet several times, I think he's a very interesting man, but I find it very hard to talk with him. I mean, of course, not talking socially, but really understand his ideas. Because it's not even that I can see that here's an idea and I know that I don't understand the formal mathematics, it's that I don't even understand the idea: that I don't even notice or can't tell if there is something there or not.

"The problem is you don't know what are ideas and what aren't: he probably knows the different layers of what he's saying, but for me it's impossible. Alex Grossmann can talk to him. It's good to have a chain of people. I think so. And I think that's the role that I played. In some sense you could say that I didn't discover anything that anybody didn't know because there's this one aspect of the mathematical roots, but then there's the other one which has to do with an algorithm for implementing the whole thing. If there wasn't an algorithm, then none of this would be happening anyway. But that algorithm existed in electrical engineering: it's called sub-band filtering. There was no connection with any of the pure mathematics, and I don't know that that connection ever would have been made if it weren't for this chain of people. I mean, once this connection was made, then you had mathematicians interested in hearing about the algorithms and electrical engineers interested in hearing about the mathematics.

"I'm a mathematician. I feel like one. I feel like a mathematician, but I am very much motivated by applications. I like to go off on a mathematical tangent, but I like to get back to applications. So in that sense, I'm an applied mathematician. At one time, and still today for some people, applied mathematics meant only certain types of results obtained by solving certain types of partial differential equations. I'm not that type of mathematician at all. So I very much feel I'm an applied mathematician, but what I apply is functional analysis, rather than PDE theory. Actually, there's no such field as applied mathematics: I think there are subfields within mathematics and that, as a mathematician, you always really like it when different subfields get into contact with each other. I think that virtually all of these mathematical subfields can have contacts with applications, so in some sense I'm an applied harmonic analyst, not an applied PDE person, but one can just as easily be, say, an applied number theorist.

"Ideally, I think there's an important place for pure mathematics and an important place for pure mathematicians; I see my role as identifying and bringing to more pure mathematicians than myself very interesting problems coming from applications. I think that's an important role to play and that it is good for pure mathematics. There was a while when pure mathematics wasn't open to this, but I think that mathematicians are starting to open up more. It's very important to remember that whole fields in pure mathematics have come from applications. That doesn't mean that all the pure math that was done in that area can therefore be described as applied. No, it's just mathematics. But it also suggests that it's very well possible for other fields of mathematics to start to be fostered by applications. I mean applied mathematics is not just learning some nice mathematics and not being upset by getting your hands dirty on some problems where things are not as neat but they will have an application—it's also identifying opportunities for mathematical thinking, which can lead to other fields of mathematics. I don't

know which other fields. I can't predict. I mean it's the ones that you cannot predict that are the most interesting."

A Life in Mathematics

Problems are the lifeblood of mathematics and Daubechies, like most mathematicians, has several going at once. All the ones she tells us about come from real world applications. It's quite clear listening to her explain her problems that she is a phenomenal teacher — the explanations are so clear and the problems sound so exciting that we're itching to get out of the interview and get to work on them.

"Before this meeting I was at a molecular biology meeting. I got really interested in people who can add proteins; for some proteins they know in which order all the atoms go, but then they don't know what the thing will look like, and they have to 'solve it' to know what it looks like, and these are really important to understand their functions. There are groups that try to predict the form of proteins from energetic computations from just the formula, and they have it organized so that they have a way to objectively compare how good their predictions are with what the grand truth is by finding out about proteins that will be solved. I mean you can reasonably predict when things will get solved, but are not solved yet. So they use all those formulas, and they all work on it: they have a deadline. At some point it's clear when the thing will be solved, and they say 'now, you have to submit.' And you submit at that time, and it gets compared with the grand truth, and in that way you can score different prediction programs.

Okay, so one guy was explaining about his prediction program: because it's just too big a space to exhaustively search, he searched in a multi-resolution way. He first tried to build a coarse model, then find the best coarse model given, then build it up from there. Now his first-level coarse model, was indeed very coarse, but he was putting it on a regular lattice, and his method was doing very well. But it struck me that if we have a good idea of how to compress, how to find subdivision schemes for curves, we could look at all the proteins that they know, and try to find that subdivision scheme that will be adapted to the protein world. I've been always looking at smoothness; they don't care about smoothness. These proteins actually do all kinds of strange things. But one could try to find a subdivision scheme that would adapt to their goal, and that could give you a good idea of what kind of coarse things to start from. So that's something that hasn't started, but something I'm very excited about and I hope to work on this spring.

"Another thing that I'm very involved in is understanding the mathematical properties of coarsely quantized but very oversampled audio signals, modeled by so-called band-limited functions. There are really neat links to dynamical systems. I'd like to do that for other wavelet transforms. I think we can do it. If we can do it, I think it's going to have very useful applications, plus I think mathematically it's going to be very interesting. Already for the band-limited functions it's much more interesting that I had expected a year ago.

"I have a graduate student with whom I work on applications of wavelets to the generation and compression of surfaces. People represent surfaces with triangulations with tens and hundreds of thousands of triangles, so you'd like to compress that information. Well, you can do that via multi-resolution, and then you can wonder what kind of wavelets are associated with that. Then you can think about smoothness. In some applications, smoothness again is very important. So that's another project.

"What else? I look at my students and collaborators because everything I work on I work on with a collaborator. I'm still working with one of my former students on a way of using frames for transmitting information over multiple channels, but that's very theoretical work, and I'm not sure how close it's going to get to applications. I have the impression I'm forgetting something.

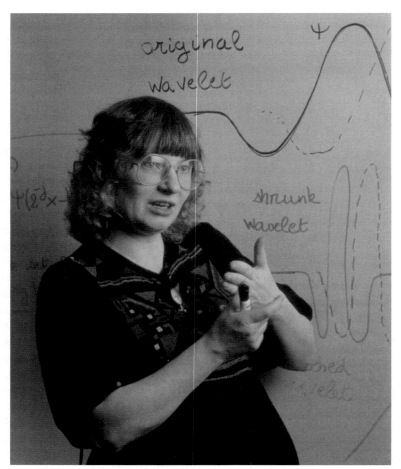

Daubechies, pregnant with daughter Carolyn, lecturing on wavelets.

"I love to talk about mathematics and so I enjoy all the courses I teach. I teach regular undergraduate math courses, I developed a course of mathematics for non-math majors. For many students in our calculus classes this will be their last contact with mathematics. I don't think this is a very good idea. Many of them are really not turned on by calculus, and it's hard to get a really meaningful application in a course if you also want to teach calculus tools. You can try to fit in some applications, but they really feel very contrived. The real applications of calculus are all the physics and other math courses, but they won't ever see that, and they're not interested in seeing that. I wanted to get mathematical ideas across without teaching technique.

"I call the course Math Alive; in two-week units we visit different concepts. I do one on voting and fair share, and one on error correction and compression, and I have one on probability and statistics,

and one on cryptography, and there's one (actually not taught by me, it's a course we co-teach) on Why Newton Had to Invent Calculus, that's a unit students have more trouble with. Then there's one on dynamical systems and population explosion, it gets into population models. I've enjoyed this course a lot. I'm teaching it this spring for the fifth time, and I'd like to document it so that it can be taught by somebody else the next time, because that's the best way to prove a concept, if it can be done by somebody else.

"I want these students to go away knowing that mathematics is really important, that it turns up in lots of things where they may be impressed by the technology, but they don't realize there's very deep mathematics in there. I want them to see that mathematics is neat in that you really solve a problem. You think your way out of it. And basically, if they remember that, that's fine. I'm sure that when people meet historians, they don't say, "you must know all the dates." They know it's something else. Well, they don't know that it's something other than balancing a checkbook in mathematics. I'd like them to know that.

"I also teach graduate courses — often a starting graduate course in wavelets, sometimes a more advanced course. I enjoy teaching undergraduate courses more than graduate courses. Not because I don't like teaching graduate courses, but because at graduate study level, the starting graduate courses I can see work well, but I think an advanced graduate course works better as a reading course than as a lecture course."

An American Life

Daubechies is married to Robert Calderbank, a distinguished British mathematician. They have two children, Carolyn and Michael. If you're wondering what it must be like to have a genius for a mom, well, it sounds a lot like having a mom.

"I go to my children's school and the tables are in groups and the classroom is full of wonderful things. In my school days classrooms might have had some stuff, but we had little desks which were all lined up in rows, maybe that's the way things were here as well in the sixties. I think the new way is much more fun. It may well be different in Belgium now; I haven't visited the elementary schools. All my elementary and secondary education was in single-sex schools. That was the way it was in Belgium at that time. I went to public schools, but all public schools at that time were gender-separated.

"I have mixed feelings about that. At the time I thought it was a bad thing; it's better if you don't see another gender as a different species because at some point you will start looking for a companion, and if you haven't really met any people of the other gender until you're 18, it's very artificial. So I didn't like it at the time, but then after I got to university I realized that people in classrooms were less likely to ask me to give an answer

than some boys that were there. At least in the beginning. After a while, when they knew I was interested, then it was different. But you always felt that there was a bigger hurdle to get over as a girl to get noticed than as a boy. I hadn't thought about that in great detail, but then coming to this country and hearing all the debate here about it — I can see how it might be good for some girls to have separate gender schools. In an ideal world there would not be this effect, but given it exists, I can see how it might help in building self-assurance in girls. My daughter goes to public school and it's mixed gender, and she's happy, but I'm wondering if at some point there might be a problem. I think she's a smart little girl, and if at some point I feel that because she is in a mixed-gender school she is not getting as much of an opportunity, I might consider a single-gender school. I haven't, she's only seven, and it's not an issue at this time. It's something that ten years ago I would not have thought I would ever consider, but now I would.

"I don't know if going to all-girl schools had an effect on me. My parents always made it clear I could do anything. It didn't occur to me until I went to university that people could think I was less good at something because I was a girl. I think I was very fortunate, because at that age, you're too old to take that prejudice seriously, and when you encounter someone with that attitude, you think, 'You're a jerk.'

"As I said, my parents, especially my father, really influenced my education. So did popular psychology actually. When I was little the prevailing theory was that it wasn't good to mix languages too early. You might really confuse children and then they wouldn't really be able to use any of the languages in great depth and it would leave marks on them the rest of their lives. My parents were in an ideal situation to bring my brother and I up bilingually, because they spoke French to each other and we lived in the Flemish part of Belgium, where people speak Dutch. But since there was this myth that mixing languages early was not good, they decided to bring us up in Dutch, which is my mother's tongue;

my father's fluent in it since he went to school in Dutch.

"Theories having changed now; I bring my children up bilingually in Dutch and English. I thought it would be too hard doing it in a language that is not my mother tongue. My French is fluent, but in the beginning especially it took quite an effort to have a bilingual household because my husband is British and he didn't speak Dutch. He has learned together with the children. So I speak English with him, but Dutch with the children.

"My husband since he has learned Dutch would like to have practice speaking, but my children won't allow it. They roll on the floor when he tries it. They think it's quite incredibly funny. My son went through a stage where when my husband would say something in English that he knew was of interest to me too, he would turn to me and translate for me.

"English is their first language because they go to school in English. So in Dutch they sometimes have more difficulty finding words, but I really try to encourage them to find the words rather than switch to English, and so sometimes before my daughter tries to tell me something, she'll say 'how do you say that word in Dutch?' They had been talking in school about different languages, and she really wanted to tell me, so she said, in Dutch 'How do you say 'Dutch' in Dutch?' But at the end of the sentence the name was already there.

"Our son is talented in math, and I think our daughter might be too. We, of course, like to stimulate that when we ask questions, but we are not pushing them hard. What we are pushing is that they have to cooperate and work at school and do the best they can, not just in math, but in everything. In fact, I'm more concerned about writing and things like that. Trying to get ideas in an organized way on paper I think is important in mathematics as well as elsewhere.

"I like to go to my children's school and help out, especially on science day. When I'm there I'm not that woman professor mathematician, I'm Michael's mom and Carolyn's mom, and I like that. I like that." ■

Beware of Geeks Bearing Grifts

ALLEN J. SCHWENK
Western Michigan University

What is a grift? It is circus slang for a swindle, a game rigged so the customer is at a disadvantage. The guys operating the games in the midway are known as grifters. As for geeks, I don't know how to define them, but I recognize them when I see them. If you are wondering what this has to do with mathematics, read on.

Three Nontransitive Dice

Let's play a game. We are given three nonstandard dice. One die is amber in color, the second is blue, and the third is crimson. Each die has six sides with six numbers, but these numbers are not the traditional 1, 2, 3, 4, 5, 6. Instead we see:

amber = A = {2, 2, 2, 11, 11, 14}
blue = B = {0, 3, 3, 12, 12, 12}
crimson = C = {1, 1, 1, 13, 13, 13}.

Now we are each to select a die and roll it once. Whoever rolls the higher number will win our simple game. Always the gentleman, I offer you first choice, so if you believe that one of the dice is superior to the other two, you certainly can select it and I will have to accept one of the two remaining. So which die do you select? You may think, "I'll take the one with the highest average number. That ought to be best in the long run." Plausible reasoning, but unfortunately you quickly confirm that each die will produce an average roll of 7. So perhaps they are equally good. Pick any one and let's see what happens.

Perhaps you decide to take A. If so, I will select B. Now there are 36 pairs that we can roll. Brute force listing shows that I will win 21 times and you will win 15. That is a 58.33% probability in my favor. Not overwhelming, but Las Vegas prospers on a smaller edge than this. We say die B dominates die A, or B → A.

OK, so now you realize A was not the best choice. You ask for B. If so, I now choose C. Again listing all 36 pairs, we find that I win 21 times and you win 15. Die C dominates B, or C → B.

Now you see your best strategy, you ask for C. Fine, I will take A. The listing shows … 21 wins for me and 15 for you! Die A dominates C.

What we have just discovered is that A → C → B → A. Domination is a nontransitive relation on this set of dice. And if a geek bearing dice offers you first choice, beware!

The concept of nontransitive dice is not new, although the triple shown here is. Martin Gardner [1] presented several sets of four nontransitive dice credited to Bradley Efron. One of these sets gives the second player a two-to-one advantage. Gardner [2] continued the discussion of nontransitivity in games and bets where we might expect transitivity. Tenney and Foster [3] generalized to construct sets of d dice with s sides for numerous pairs (d, s) with d and s at least 3.

Let's play again. Being a good sport, I will select first, and take A. I suppose you would like to have B, correct? And just for fun, let's roll twice each and take our total, the higher total wins. Recall that you have a 58.33% chance of beating me on the first roll, and also a 58.33% chance of beating me on the second roll. When we find our totals, is your probability of winning greater than 58.33%, equal to 58.33%, or less than 58.33%? Pause a moment and try to predict the answer. Does extending the game to two rolls enhance your expectation, leave it unchanged, or reduce it? Computing these probabilities is a lot more tedious. There are now 36 ordered pairs for each of us, so together we have $6^4 = 1296$ possible outcomes. Curiously, none are ties. We find 675 wins for A and 621 wins for B! That's right, not only has your winning edge been reduced, but it has been reversed to a winning margin of 52.08% for me! I call this a perverse reversal. In a single roll, B → A, but for a pair of rolls, A → B! To keep the conditions clear, we'll write the latter as 2A → 2B.

What happens when B opposes C? We have an even stronger reversal, 2B → 2C by a margin of 53.47%! And A → C also reverses to 2C → 2A by the same margin of 53.47%. For two rolls, our set of three dice is still nontransitive, but the dominating cycle has been reversed to

$$2C → 2A → 2B → 2C.$$

Perverse reversal, indeed.

Reprinted from April 2000, pp. 10–13

Illustration by Greg Nemec.

After that surprise, are you ready to predict what will happen with three rolls? Here ties are possible, so we shall take the following point of view. If both players produce the same total for three rolls, we simply start over from scratch. They roll three more times, and either produce a winner or have another equal total leading to yet another repetition. In this way the game never ends in a tie. There are $6^6 = 46656$ possible outcomes. Let's say that t produce ties, a are wins for A and b are wins for B. Evidently $46656 = a + b + t$. Our tie-breaking rule effectively removes the t tying cases, giving A the probability $a/(a + b)$. In the case of three rolls, we find $a = 19818$, $b = 20358$, $t = 6480$. These perverse dice have tipped the scales back in the original direction. We find 3B → 3A by 50.67%. And also 3A → 3C → 3B, both dominating by the margin of 50.28%. Another nontransitive perverse reversal.

Several questions come to mind. For r rolls, will we always have a nontransitive triple? Which values of r will give the original order rA → rC → rB → rA? And which will perversely reverse to rA → rB → rC → rA? So far we have the original order for $r = 1$ and 3 and the reversed order for $r = 2$. It may seem that the margins are quickly dying out, but the first three cases may be misleading.

Finding the winning probabilities for r rolls is not as hard as it may first appear. For the amber die, we have sides of 2, 2, 2, 11, 11, 14. We represent this by a polynomial

$$a(x) = 3x^2 + 2x^{11} + x^{14}.$$

Similarly, the blue die gives

$$b(x) = 1 + 2x^3 + 3x^{12},$$

and the crimson one has

$$c(x) = 3x + 3x^{13}.$$

When A faces B in a single roll, B's winning margin is seen by observing that the term of x^{14} beats all six terms in $b(x)$. The two terms x^{11} each win three times, and the three terms x^2 each win one time. That's $1 \times 6 + 2 \times 3 + 3 \times 1$. Thus A has 15 winning combinations, similarly, B wins $3 \times 5 + 2 \times 3 = 21$ times. For two rolls we use

$$a^2(x) = 9x^4 + 12x^{13} + 6x^{16} + 4x^{22} + 4x^{25} + x^{28}$$

and

$$b^2(x) = 1 + 4x^3 + 4x^6 + 6x^{12} + 12x^{15} + 9x^{24}.$$

The winning pairs for A count up to be

$$1 \times 36 + 4 \times 36 + 4 \times 27 + 6 \times 27 + 12 \times 15 + 9 \times 5 = 675,$$

and the winners for B give

$$9 \times 31 + 12 \times 21 + 6 \times 9 + 4 \times 9 + 4 \times 0 + 1 \times 0 = 621.$$

By using a computer algebra system such as *Maple,* we can analyze r rolls quickly by computing the polynomials $a^r(x)$, $b^r(x)$, and $c^r(x)$. Of course it is still a chore to carry out the term-by-term comparisons.

However, using *Matlab* permits a matrix-vector approach that is even more convenient. The three polynomials are represented by column vectors whose $(i + 1)$st coordinate stands for the coefficient of x^i. Thus we have three vectors

$$\mathbf{a} = [0\ 0\ 3\ 0\ 0\ 0\ 0\ 0\ 0\ 0\ 0\ 2\ 0\ 0\ 1]^T$$
$$\mathbf{b} = [1\ 0\ 0\ 2\ 0\ 0\ 0\ 0\ 0\ 0\ 0\ 0\ 3\ 0\ 0]^T$$
$$\mathbf{c} = [0\ 3\ 0\ 0\ 0\ 0\ 0\ 0\ 0\ 0\ 0\ 0\ 0\ 3\ 0]^T.$$

Now the polynomial $a^2(x)$ is represented by the convolution vector

$$\mathbf{a}^2 = \text{conv}(\mathbf{a},\mathbf{a}) = [0\ 0\ 0\ 0\ 9\ 0\ 0\ 0\ 0\ 0\ 0\ 0\ 0\ 12\ 0\ 0\ 6\ 0\ 0$$
$$0\ 0\ 0\ 4\ 0\ 0\ 4\ 0\ 0\ 1]^T,$$

and in general the power $a^r(x)$ is represented recursively by the convolution

$$\mathbf{a}^r = \text{conv}(\mathbf{a}^{r-1}, \mathbf{a}).$$

To analyze totals for r rolls, we need the strictly lower triangular matrix of order $n = 14r + 1$ given by $L_n = (l_{i,j})$ where $l_{i,j} = 1$ if $i > j$ and 0 otherwise. The number of times A beats B with r rolls is given by $\mathbf{a}^{rT} L_n \mathbf{b}^r$. Similarly, the number of times B beats A is given by $\mathbf{b}^{rT} L_n \mathbf{a}^r$. We do not need it, but the number of ties is $\mathbf{a}^{rT} \mathbf{b}^r$. For verification, we can sum these three numbers to get the total 6^{2r}. The table below summarizes the results of summing r rolls for each $r \le 15$. Notice that ties occur if and only if 3 divides r. This is not surprising once we note that each face of A is congruent to 2 mod 3, on B it is 0 mod 3, and on C we have 1 mod 3. Thus ties can happen only when 3 divides r.

Table 1 shows the probability that the left competitor of each pair wins. For $r \le 14$, the three possible pairs always form a nontransitive triple. Moreover, it is the original triple $rA \rightarrow rC \rightarrow rB \rightarrow rA$ for $r \equiv 0$ *or* 1 mod 3 and it is perversely reversed to $rA \rightarrow rB \rightarrow rC \rightarrow rA$ whenever $r \equiv 2$ mod 3. We certainly might wonder whether this pattern continues for all positive r, but the pattern fails for $r = 15$. Here B dominates both A and C, with C

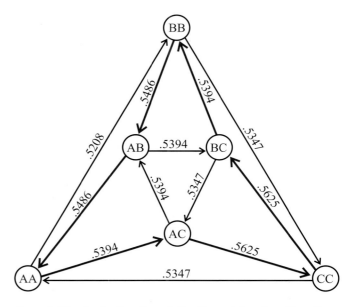

Figure 1. The domination digraph for all possible pairs of two dice.

$\rightarrow$ A. Thus, the first player could select B and, for the first time, transitivity guarantees a winning advantage. Of course the second player can hold his disadvantage to a minimum by selecting C.

Incidentally, the set presented here is about the fifth I investigated. I tried various ways to try to extend the number of nontransitive triples through the largest possible value for r. Earlier sets had their first failures at 3, 9, and 12. Upon finding the present set, I chose to go no further. I do not know if it is possible to find a set that is nontransitive for every possible number of rolls r. Perhaps you can find a better set.

A Nontransitive Variation

Suppose we have a large supply of amber, blue, and crimson dice with the faces given above. A variation of the game would be to let you select any two dice, of like or different colors, then I select my pair. We roll our chosen pairs and compare totals. If a tie occurs we roll over. This means you have six selections available: AA, BB, CC, AB, AC, and BC. We have already seen that B $\rightarrow$ A and AA $\rightarrow$ BB. But, when I have second choice, is AA the pair I should choose to gain the greatest advantage over BB? The results can be presented by the directed graph shown in Figure 1. Here we place an arc joining WZ to YZ whenever WZ $\rightarrow$ YZ. If a pair happens to be perfectly fair, we have no arc joining them. For example, there is no arc between AA and BC because these pairs happen to be totally fair. Moreover, at each vertex we have selected the incoming edge with the largest margin to appear as a blue arc. The blue arcs identify the best selection for the second player for each possible first choice. Observe that the strongest dominations give rise to a nontransitive cycle among the six possible pairs, namely

$$AA \rightarrow AC \rightarrow CC \rightarrow BC \rightarrow BB \rightarrow AB \rightarrow AA.$$

r	A versus B	B versus C	C versus A
1	$\frac{15}{36} \approx 0.41667$	$\frac{15}{36} \approx 0.41667$	$\frac{15}{36} \approx 0.41667$
2	$\frac{675}{1296} \approx 0.52083$	$\frac{693}{1296} \approx 0.53472$	$\frac{693}{1296} \approx 0.53472$
3	0.49328	0.49725	0.49725
4	0.47468	0.47555	0.47555
5	0.50962	0.51533	0.51533
6	0.49658	0.49858	0.49858
7	0.48545	0.48577	0.48577
8	0.50744	0.51004	0.51004
9	0.49773	0.49949	0.49949
10	0.48903	0.49021	0.49021
11	0.50646	0.50808	0.50808
12	0.49821	0.49993	0.49993
13	0.49069	0.49246	0.49246
14	0.50579	0.50711	0.50711
15	0.49846	0.50025	0.50025

Table 1. Probability that Left dominates Right for the total of r rolls.

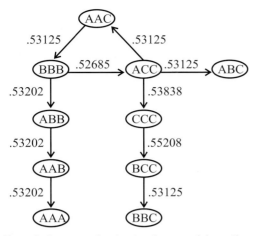

Figure 2. Strongest domination for sets of three dice.

The first player, seeking to minimize his opponent's advantage, needs to select either BB or AC, both of which give up only a 53.94% edge.

If we alter the rules to select three dice each, repetitions allowed, Figure 2 shows the strongest dominations for each choice. While each selection wins against certain other selections, and each loses, some selections, AAA for example, never win by enough to be the dominant choice. Here the first player can choose ACC to minimize his opponent's advantage to 52.685% via BBB. Similarly, the strongest dominations are also shown for a selection of 4 dice in Figure 3. Here the first player minimizes his disadvantage by selecting AACC to hold his opponent to 52.127% via AAAC.

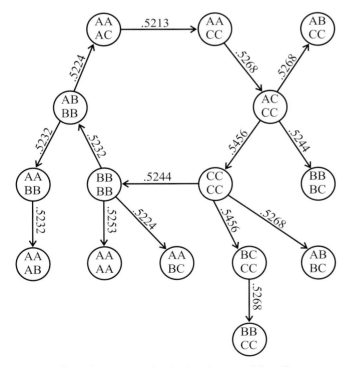

Figure 3. Strongest domination for sets of four dice.

We have presented these possible variations of the game to illustrate how difficult it is to predict the best choices for each player.

Five Nontransitive Dice

Here is a set of five 5-sided dice, colored amber, blue, crimson, dayglo orange, and evergreen, that have been designed to demonstrate how much optimal choices can vary. Lest you worry about how to construct 5-sided dice, it is very simple. We use cubes, with one side labeled $*$. Whenever the star side lands facing up, the player rerolls.

$$A = \{4, 4, 4, 4, 4\}$$
$$B = \{3, 3, 3, 3, 8\}$$
$$C = \{2, 2, 2, 7, 7\}$$
$$D = \{1, 1, 6, 6, 6\}$$
$$E = \{0, 5, 5, 5, 5\}$$

By now it should not surprise you that these dice beat one another in a nontransitive fashion. For a single roll the strongest dominations occur in a cycle

$$A \to B \to C \to D \to E \to A.$$

Within this cycle, each domination is by at least 64%. But how does this cycle vary as the number of rolls r increases? When $r = 2$, the dominant cycle has changed to

$$2A \to 2D \to 2B \to 2E \to 2C \to 2A.$$

For $r = 3$, we reverse this to get

$$3A \to 3C \to 3E \to 3B \to 3D \to 3A$$

Next $r = 4$ reverses the original cycle to yield

$$4A \to 4E \to 4D \to 4C \to 4B \to 4A.$$

This is perverse reversal with a vengeance. After the first player selects his die, the second may optimally select any one of the remaining four dice, depending upon how many rolls are intended. And what happens for $r = 5$? Surprisingly we now have a transitive domination sequence

$$5E \to 5D \to 5A \to 5C \to 5B .$$

The first person can finally grab the advantage by selecting E. All the second player can do is minimize his losses by selecting D. Curiously, the domination pattern for each value of $r = 6$ to 10 repeats the pattern for $r - 5$.

Many mysteries remain in the realm of nontransitive dice. Why not invent your own and see what surprises you can discover? ■

References

1. M. Gardner, The paradox of nontransitive dice, *Scientific American*, 223, (1970), 110–111.
2. M. Gardner, On paradoxical situations that arise from nontransitive relations, *Scientific American*, 231, (1974), 120–125.
3. R. C. Tenney and C. C. Foster, Nontransitive dominance, *Math. Magazine*, 49, (1976), 115–120.

The Traveling Baseball Fan

RICK CLEARY, DAN FAGA,
ALEX LUI, and JASON TOPEL
Cornell University

In the summer of 1980, Rick Cleary and Mike Chuba had just finished Master's degrees in mathematics and they wanted to do a little traveling. Both were avid baseball fans so they decided to see a game in every major league ballpark. With a limited budget and no reliable car, Rick and Mike chose to get around the country by bus. Gathering some helpful friends, road maps, baseball schedules and bus schedules, they spent an afternoon coming up with a plan that would get them to each of the ball parks over a period of about seven weeks.

Their solution was feasible since they had enough time and money to make it work. However, it probably wasn't optimal since there was almost certainly a faster way that they didn't find. Having studied mostly classical pure mathematics, traveling baseball fans Rick and Mike applied only rudimentary problem solving skills to get an ad hoc solution. Had they known a little bit about the field of operations research, Rick and Mike might have recognized that they were working on a special version of the classic traveling salesman problem (TSP for short).

Twenty years later, Rick teaches statistics and operations research at Cornell University. To celebrate the anniversary of his trip across the country with Mike, he enlisted the help of three Cornell juniors majoring in Operations Research & Engineering (Dan Faga, Alex Lui and Jason Topel) who brought some specialized knowledge to the problem. While the problem's complexity had grown a little bit (there are now thirty major league teams, four more than in 1980), the issues remained the same. In this article we present some background about the TSP and its applications. We also discuss how to look for solutions to the extension we call the traveling baseball fan problem (TBFP).

The Traveling Salesman Problem

Here is a classic description of the TSP: A salesman, starting from his home city, must visit each of n cities exactly once, and then return home. Being an efficient individual, the salesman wants to visit the cities in an order which minimizes the total length of his tour (as measured by distance or time). Sounds pretty simple, right? Think again. The TSP has challenged mathematicians and operations researchers for decades. In fact, there is currently no algorithm in existence that can efficiently find the solution to the problem for a large number of cities. There are many algorithms and programs that find good solutions but the goal of having an *optimal* solution, the absolute best answer, is still elusive. One naive way our traveler could be sure to find the optimal solution would be to list every possible tour…but that might be a lot of work!

Each tour is one of the n! permutations of the cities the salesman needs to visit. Thus, as we increase the number of cities, the size—and therefore the complexity—of the problem increases at an exponential rate. It's possible to solve the TSP when we have only ten cities: in this case, we only have to examine 10! = 3,628,800 tours, which can be done by a computer in a few seconds. But suppose our traveler is a lobbyist who wants to talk to legislators in each of the 48 capital cities in the continental United States. Listing every tour would require us to look at a whopping 48! $\cong 2.59 \times 10^{59}$ tours! We'd better hope our salesman doesn't plan on leaving anytime soon. With all these computations, it's no wonder that listing every possible solution is not an efficient way to solve the TSP.

A good mathematical representation of the TSP can be expressed using the language of graph theory. A *graph* is a set of points (called *nodes* or *vertices*) and a collection of *edges* (or *arcs*) that connect pairs of points. The collection of cities and the paths connecting them can be thought of as a *weighted graph*, where the cities are the nodes of the graph and the weights on each edge correspond to the distance (or travel time) between the two cities. For the case of the TBFP, we will think of our graph as being *complete*, that is we can proceed directly from any node to any other and the distance between the cities is the weight of the edge between them.

The TSP is an example of a *combinatorial optimization* problem. Like all optimization problems, we have an objective in mind such as minimizing time, distance, cost or some other variable. Despite the computational difficulties associated with solving a TSP, there are a wide variety of real world problems

Reprinted from September 2000, pp. 18–22

that can be solved using this model. The most obvious use is in transportation and distribution, but the problem can also be applied to many other, seemingly unrelated, areas. Some applications of the TSP include:

Vehicle routing: Vehicles based at a central location are to serve a given number of customers dispersed throughout the region. The object is to determine which vehicles should visit which customers, and the order of the visits.

Job sequencing: A given number of jobs must be completed on a single machine. The jobs can be done in any order, but the machine must be set up for each new job after the previous one is completed. The goal is to complete all of the jobs in the shortest possible time.

Cutting wallpaper: Suppose we need to cut a given number of sheets of different lengths from a roll of wallpaper. The objective is to cut the pieces in an order which minimizes the total amount of wallpaper that is wasted.

Formulation of the TBFP

The TBFP is a slightly harder version of the TSP. The main difference between them is that while the salesman can usually visit a customer during a wide range of business hours, the fan can only go to a stadium when the team is home. There is no reason for our traveling baseball fan (TBF) to visit St. Louis while the Cardinals are playing in Atlanta! Many people who have written about the TSP (see [1]) call this a "time windows" problem, although we like to think of this as a graph with "blinking nodes."

Another important difference arises when considering the objective. In the classic TSP, we seek to minimize the distance or time spent traveling. Here, however, we will try to minimize the number of days it take us to complete the tour. If a TBF had enough money to fly (if expense was not a constraint) then it is likely that the fan could see a game in all 30 ball parks in 30 days or fewer. With clever scheduling, a wealthy TBF might find an afternoon game in Baltimore and a night game in Philadelphia on the same day, hiring a limousine to get between these two ball parks in a couple of hours. We did not consider this case but chose to consider the more cost-conscious fan who has decided to drive or take the bus between cities. In summary, we define the TBFP as follows:

Objective: Minimize the number of days it would take a TBF to see each major league team at home during the 2000 major league baseball season, and then return to his/her home city. We chose not to specify a particular starting point or time.

Constraint: We are assuming that the fan can travel up to 750 miles per day. (This allows roughly 12–15 hours for travel…enough to get from a night game in one city to a day game in another by car or bus.)

Data: The 2000 major league schedule was available from many sources, including team by team schedules at `majorleaguebaseball.com`. Distances between cities were found by using a United States road map with a matrix showing the mileage between major cities.

As in any attempt to use a mathematical model to study a real world problem we have to make a few simplifying assumptions. Some potential problems a TBF could face on the road: a rainstorm postponing a game, a lack of available tickets, car trouble, or bus scheduling. We will assume that all games will be played on their scheduled date, that tickets are available, and that our transportation is reliable.

After formulating the problem, we faced a decision that confronts researchers at many levels, and which undergraduate students might be familiar with from doing calculus homework. Should we try to find a solution by hand, or would it be more efficient to use a computer? In either case further questions arise. If we use a computer, can we find a software package that solves the type of problem we are interested in? Or do we need to write our own program? Deciding how to proceed on an optimization problem is itself an optimization problem! What is the least difficult method for finding a good (if not optimal) solution?

Since we had an entire semester in which to work, and a team of four people, we tried both methods to solve the problem. A description of how we proceeded and what we found follows.

Our Hand Solution and a Note on Optimality

When faced with a complex problem to solve, what's the mathematician's first reaction? Try an easier problem! (Students sometimes think this applies only to doing homework, but professors are masters at it.) Sitting down to find a solution for all 30 ballparks seemed a bit overwhelming, so we began by looking for efficient tours of the ballparks in the Northeastern United States and Canada, where the teams are fairly densely packed. Once we had a good solution there, we looked at other portions of the country and tried to paste the solutions together in reasonable ways.

Table 1 shows our best solution using the 2000 major league schedule. With a solution in hand, we begin to think about optimality. Table 1 shows a 44-day tour covering over 17,000 miles. Since there are 30 stops to make, you might think that 30 days is the "gold standard" we should seek to achieve. However, Seattle is a little over 800 miles from either San Francisco or Oakland, the nearest neighboring ballparks. With our assumption of 750 miles per day, this will add a day each way and make our best possible solution 32 days. In a slightly more subtle problem, the Tampa Bay Devil Rays are within 750 miles of only two other teams, the Atlanta Braves and the Florida Marlins, who play in the Miami area. The Marlins, in turn, are within 750 miles of the Devil Rays and Braves only. This adds another day (if you don't see why, try thinking about what order you'll visit these teams.) Our best possible solution is now 33 days and growing!

How large will this target value get? Our optimal solution to the TBFP can certainly not be shorter than the best possible solution we could find if each team was home every day. This, of course, is an ordinary TSP since the blinking nodes are all available. In this formulation, the weights between teams would be

Table 1. Hand Solution

Date		Home City	Estimated Distance from Last City
May	3	Baltimore	
	4	Philadelphia	98
	5	Boston	296
	6	New York (Yankees)	197
	7	Toronto	568
	8	Detroit	383
	9	Montreal	654
	10	—	
	11	Cleveland	900
	12	New York (Mets)	481
	13	Pittsburgh	422
	14	Chicago (White Sox)	486
	15	—	
	16	Milwaukee	91
	17	Chicago (Cubs)	91
	18	Cincinnati	285
	19	Atlanta	484
	20	—	
	21	Florida (Miami)	653
	22	—	
	23	Tampa Bay	266
	24	—	
	25	Houston	1021
	26	—	
	27	St. Louis	792
	28	Kansas City	250
	29	Colorado (Denver)	601
	30	—	
	31	Arizona (Phoenix)	926
June	1	—	
	2	Texas (Dallas)	1014
	3	—	
	4	Anaheim	1407
	5	Oakland	384
	6	—	
	7	Seattle	808
	8	—	
	9	San Francisco	808
	10	San Diego	502
	11	Los Angeles	118
	12	—	
	13	—	
	14	Minnesota (Minneapolis)	1993
	15	—	
	16	Baltimore (HOME)	1269
		TOTAL: 17440 miles	

the number of days it would take to get between the two ballparks with our constraint on mileage. Any edge connecting two cities up to 750 miles apart would have weight 1; 751 to 1500 miles would have weight 2; and so on. Since we could not be sure that the solution we find from an approximate algorithm would actually be optimal, we have not pursued this calculation.

What we can be sure of is that the optimal solution is at least 33 days, but less than or equal to 44 days since we have a solution in 44 days. If we are going to make progress using a computer solution, we at least have a number to beat.

Our Computer Solution

While a variety of commercial software is available to give good solutions to TSPs and scheduling problems, we never found one that we could apply precisely to our TBFP. Thus we wrote our own, using a "greedy" algorithm. After choosing a starting city and a starting date, the program looks for the closest city in which there will be a game the next day. It is often the case that greedy algorithms are not very good at producing optimal solutions since they may do a good job early in the tour but leave very poor choices late in the process. With the ability to check multiple starting dates and locations, we were still hopeful that our algorithm might beat our hand calculation.

Here are some details of our programming process:

1. Create a schedule matrix with teams and dates. This matrix has 30 rows (one for each team) and a column for each date from May 1 through the end of the season. (We skipped April as the schedule includes a lot of dates with no game early in the season.) The i,j entry in this matrix is 1 if team i is home on the date corresponding to j, and 0 otherwise.

2. Create a distance matrix (30×30 and symmetric) recording the distance between the different ball-parks. We enter a 0 along the main diagonal, the distance from a team to itself.

3. Create a progress vector z, (30×1) which keeps track of the teams that have been visited. Initialize to a vector with each entry equal to 1, change a team's entry to 0 after we have seen them play at their home park.

4. At a given day d, we form a vector in which each entry is the product of the corresponding entries in column $d+1$ of the schedule matrix, the column of the distance matrix corresponding to our current city, and the progress vector z. The resulting vector will be non-zero only in those cities that we have not yet visited and which have a game the next day. We proceed to the closest of these cities within 750 miles.

5. If no non-zero value is found within 750 miles, the program will move on to use the schedule column for the next day, and pick the closest city within 1500 miles.

The program runs through a variety of starting dates from each city and keeps track of the best tour. Table 2 shows our optimal solution using this method, a 41-day tour covering 15,732 miles, three days and about 1700 miles shorter than our hand solution. Given that greedy algorithms run into the most trouble when they are forced into bad choices late in the process, it is no surprise that our best computer solution began on the west coast where teams are sparsely distributed and there are usually few choices.

Some Generalizations and a Challenge

Our best solution stands at 41 days for the 2000 season. Is there room for improvement? Perhaps a less greedy and more clever programming algorithm would allow for a quicker tour. Or a hybrid solution, trying by hand to improve the best answer we get from our program, might allow us to save a few days. When solving optimization problems of this type, it is a common practice to mix programming and intuition to get as close as possible to the best answer. There may be a trip a little better than the one in Table 2, but not too much better, since it must be at least 33 days!

We chose a natural and simple objective in minimizing the number of days spent on the road, but there are many other

Table 2. Computer Solution

Date		Home City	Estimated Distance from Last City
May	29	San Francisco	—
	30	Los Angeles	384
	31	San Diego	118
June	1	Arizona (Phoenix)	368
	2	Anaheim	393
	3	Oakland	384
	4	—	
	5	Seattle	808
	6	—	
	7	—	
	8	—	
	9	Colorado (Denver)	1314
	10	Kansas City	601
	11	*Minnesota (Minneapolis)	462
	12	Milwaukee	348
	13	Chicago (Cubs)	91
	14	Detroit	288
	15	Pittsburgh	314
	16	Baltimore	290
	17	Philadelphia	98
	18	New York (Yankees)	99
	19	Boston	197
	20	New York (Mets)	197
	21	Montreal	347
	22	Toronto	333
	23	Cleveland	351
	24	Cincinnati	254
	25	Chicago (White Sox)	285
	26	—	
	27	Tampa Bay	1183
	28	—	
	29	—	
	30	St. Louis	931
July	1	Texas (Dallas)	629
	2	—	
	3	Atlanta	788
	4	Florida	653
	5	—	
	6	Houston	1287
	7	—	
	8	—	
	9	San Francisco	1927
		TOTAL:	**15732 miles**

* By coincidence, author Rick Cleary was at this game! Brewers 5, Twins 3.

Table 3. Rick and Mike's 1980 trip (only 26 ballparks)

Date		Home City	Estimated Distance from Last City
May	27	Boston	—
	28	—	
	29	Philadelphia	296
	30	—	
	31	New York Yankees	99
June	1	—	
	2	Montreal	347
	3	Toronto	333
	4	Pittsburgh	323
	5	—	
	6	—	
	7	Cincinnati	293
	8	—	
	9	—*	
	10	Detroit	260
	11	Chicago White Sox	288
	12	Milwaukee	91
	13	Cleveland	436
	14	Chicago Cubs	345
	15	—	
	16	St. Louis	302
	17	Kansas City	250
	18	—	
	19	Minnesota	462
	20	—	
	21	Houston	1311
	22	Texas	237
	23	—	
	24	San Diego	1289
	25	California (Anaheim)	118
	26	—	
	27	Oakland	384
	28	San Francisco	15
	29	—	
	30	Seattle	808
July	1,2,3,4,5	—	
	6	Los Angeles	1143
	7,8,9	— (All-Star Break)	
	10	Atlanta	2362
	11	—	
	12	Baltimore	727
	13	New York Mets	197
	14	Return to Boston	197
		TOTAL:	**12913 miles**

*A rain-out in Cleveland forced a change in plan that added several hundred miles, but no days. This schedule was constructed with the objective of maximizing time in cities where we knew people who would let us stay with them.

good choices. For travelers who like to linger in a city and want to save gas, minimizing the total distance might take precedence. This could be done by solving the usual TSP for the 30 ball parks, and simply waiting for a team to get home if they are out of town on the date of arrival. Those with a love of symmetry might want to see each team exactly twice (once on the road and once at home); this would add another set of constraints to the problem. Serious fans of competitive baseball might set the goal to maximize the winning percentage of the visiting teams. When Rick and Mike were TBFs in 1980, their objective was to lower costs by staying overnight in cities where they had friends or relatives! (See Table 3.)

Sticking to the original objective of fewest days, perhaps you can find an answer that is even better than ours. If so, send it to us and we will send a small prize (perhaps a cap from your favorite team) to the best solution we get before December 31, 2000. Then let's turn our attention to the 2001 season! ■

For Further Reading

1. E. L. Lawler, J.K. Lenstra, A. H. G. Rinnooy Kan and D. B. Shmoys (eds.), *The Traveling Salesman Problem*, John Wiley and Sons, 1985.

2. R. J. Wilson and J. J. Watkins, *Graphs An Introductory Approach*, John Wiley and Sons, 1990.

A Dozen Areal Maneuvers

JAMES TANTON
St Mark's Institute of Mathematics

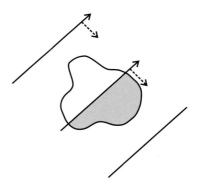

Your dozenal correspondent is at it again! This time I have put together a collection of twelve curiosities all to do with areas, and, in some cases, the perimeters that contain them. The questions about slicing pie and cake are technically ones about volumes, but we'll assume here all desserts are of uniform thickness so they may be reduced solely to analysis of area! Many of these results are classic (one even known by Archimedes) but hopefully the few extra twists I've put in shine these gems in a new and interesting light. I hope you have as much fun thinking about these as I did.

1. Plucky Perimeters

What curious property do the following figures share?

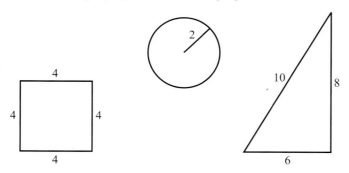

2. Irregular Pizza

Much to their dismay, Sam and Maggie receive from their local pizza parlor an irregularly shaped pizza. Both being mathematicians, they realize the Intermediate Value Theorem assures them the existence of a straight line cut that divides the pizza exactly in half: By sliding the knife across the pizza, first from a position with all the area of the pizza sitting to the right of the knife, to one with all the area sitting to the left, there must be some intermediate position where the area is split precisely in two.

Is there necessarily a straight line cut that not only divides the area of the pizza precisely in half but also the pizza crust (that is, the perimeter of the figure) in half as well?

3. Square Pie

Cutting a wedge emanating from the center, Beverly wants to take precisely one-seventh of a square pie. She doesn't like crust. Where should she position her cut so as to receive the minimal length of perimeter?

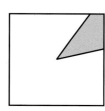

4. Hexagonal Pie

Beverly cut eighteen slices into a hexagonal pie. She missed the center of the pie but managed to ensure that each wedge-shaped piece possessed the same length of perimeter.

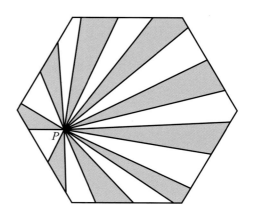

Reprinted from September 2000, pp. 26–30, 34

Prove that the total area of the shaded regions (every second piece) equals the total area of the unshaded pieces, and that this is always the case no matter where the "center" point P is placed.

5. Cake Sharing

Is it possible to share a cake among three people so that each person honestly believes she is receiving *more* than one-third of the cake?

6. Creating Area

Who said area is always preserved? Take an 8×8 inch square piece of paper and subdivide it as shown on the left.

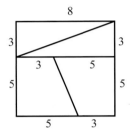 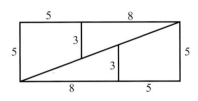

Now rearrange the pieces to form a 5×13 rectangle as shown on the right. (Try it!) This transforms 64 square inches of paper into 65 square inches.

What's going on?

7. Capturing Area

It is always possible to capture any given shape within a rectangular box. Simply slide in four straight lines, one from each direction of a compass, north, south, east and west, until they each just touch the given region.

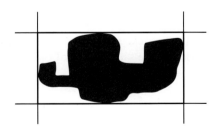

Is it always possible to capture a region within a perfectly square box?

8. Rational Replication

Four squares stack together to form a larger copy of themselves, as do four equilateral triangles, and four bent trominoes. The larger figures are scaled versions of the original tiles, each with rational scaling factor 2.

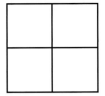 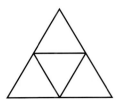 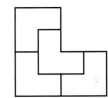

Is there any figure in the plane that replicates itself with fewer than four copies to produce a larger copy still with rational scale factor?

9. Bicycle Tracks

A bicycle of length r (measured as the distance between the points of contact of the two wheels with the ground) moves along a closed convex loop.

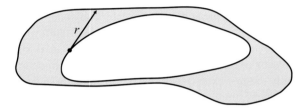

What is the area between the two tracks it leaves?

10. Spherical Bread

A spherical loaf of bread, n units in diameter, is sliced into n pieces of equal thickness.

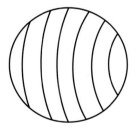

Which piece has the most crust?

11. Circles on Spheres

Which is greater: The area of a circle of radius r drawn on a plane, or the surface area of a circle of "radius" r drawn on a sphere? (Here "radius" is the length of the straight line segment

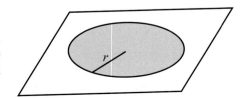

passing through the interior of the sphere connecting the center of the circle to its perimeter.)

12. Soap Film on Cubical Frames

A classic problem asks what system of roads connects four houses situated on the vertices of a square (one mile wide) using minimal total road length. The answer, surprisingly, is the wing-shaped design below. It uses $1 + \sqrt{3} \approx 2.732$ miles of road.

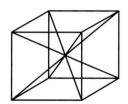

My aim here is to take this problem up a dimension. What design of surfaces, meeting somewhere in the center, connects the skeleton of a cube (namely its 12 edges and 8 vertices) with minimal total surface area?

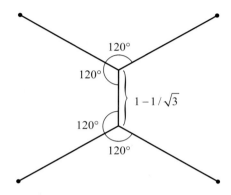

This problem is very difficult to analyze mathematically, but with the aid of soap solution the answer can be determined experimentally. Using pliable wire make a cubical frame and dip it into soap solution. The surface tension of the film acts to minimize surface area and so careful dipping (making sure the film is attached to every edge of the cube and meets in the center) will result in the desired solution. Try it! What's the answer?

Answers, Comments, and Further Questions

1. Their perimeters equal their areas! Of course this is just an artifact of scale. With the appropriate enlargement or reduction it is theoretically possible to scale *any* figure so that its perimeter equals its area. (Challenge: Use a photocopier to produce a reduced copy of this very page with perimeter equal to area, measured in inches and square inches.)

Taking it Further   Find a non-square rectangle with integer side lengths whose perimeter equals its area. Are there any more

such rectangles? There is only one other right triangle with integer side lengths having this property. What is it?

Taking it Even Further   Is there a rectangular box whose volume equals both its surface area and the total sum of its edge lengths?

2. Sam and Maggie's argument using the Intermediate Value Theorem shows for *any* given angle θ there is a unique directed line tilted at that angle dividing the area of the pizza precisely in half. (We regard a line at angle θ and a line at angle $\theta + 180°$ as distinct lines pointing in opposite directions.) Let's measure how successful these lines are in cutting the perimeter in half as well. For each angle θ set $f(\theta)$ to be the total length of the crust to the right of the line minus the length of the crust to its left. Our goal is to find a line with $f(\theta) = 0$.

Notice that $f(0°)$, whatever its value, equals $-f(180°)$. These angles represent the same line but pointing in opposite directions. Since f varies continuously with the angle θ, the Intermediate Value Theorem tells us there must indeed be an angle θ with $f(\theta) = 0$. (The continuity of f is subtle.) This does the trick.

Taking it Further   Prove there is always a (very long) single straight line cut that simultaneously slices any *two* irregularly shaped pizzas in half, no matter where they are placed on the table top. Can one always simultaneously divide *three* pizzas in half in a single straight cut?

3. It does not matter where she places her cut: All wedges from a square pie possess the same portion of perimeter! Pieces of pie are either triangular or a union of two triangles. As all these triangles have the same height, the area of any wedge is directly proportional to the length of perimeter it contains. Thus one-seventh of the area always means one-seventh of the perimeter. Note there is nothing special about the fraction "one-seventh" nor the square shape. The same phenomenon occurs when taking slices from the center of any regular polygon.

4. Assume the hexagon has unit side length. The height of the hexagon (shortest diameter) is thus $\sqrt{3}$. By the regularity of the situation, if $\lambda\%$ of the top edge length belongs to shaded wedges,

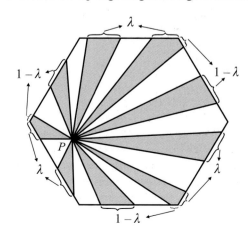

$(1-\lambda)\%$ of the opposite edge length belongs to them too. In fact, the portion of edge belonging to shaded regions alternates between $\lambda\%$ and $(1-\lambda)\%$ around the figure.

Changing track for the moment: Let a, b, and c be the distances to every other side of the hexagon. See the figure below. View these as line segments in the interior of a large equilateral triangle. A result from geometry says that the sum of these lengths equals the height of the triangle. Thus $a + b + c = 3\sqrt{3}/2$, no matter where P happens to lie. (To establish this, first consider the case where P lies on the base edge of the triangle. The result is clear from drawing a reflected image of the triangle across this edge. To establish the general case, raise the base of the triangle so that P lies on the base edge of a sub-equilateral triangle.)

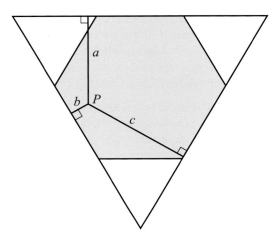

Regarding the shaded pieces encompassing a corner of the hexagon as a union of two triangles, we have that the total area of the shaded triangles touching the top and bottom edges is $\frac{1}{2}\lambda a + \frac{1}{2}(1-\lambda)(\sqrt{3} - a)$. Similarly for the remaining two pairs of edges. Summing and simplifying thus gives the total area of the shaded regions to be:

$$\frac{1}{2}\lambda(a+b+c)+\frac{1}{2}(1-\lambda)(3\sqrt{3}-a-b-c)$$
$$=\frac{1}{2}\lambda\frac{3\sqrt{3}}{2}+\frac{1}{2}(1-\lambda)\frac{3\sqrt{3}}{2}$$
$$=\frac{1}{2}\cdot\frac{3\sqrt{3}}{2}$$

which is precisely half the area of the hexagon.

Taking it Further   Every third region of the eighteen slices is selected. Prove these sum to one-third of the area of the hexagon. Suppose instead Beverly makes just 12 cuts. Prove every second piece, and then every fourth piece, account for precisely one-half and one-quarter respectively of the pie. Can you extend these results to other numbers of cuts? To other regular polygons?

5. If everyone possesses a different estimation of "one-third," this seemingly impossible task is then indeed possible! Here's one scheme: First have each person score a straight line across the cake, in parallel, at a position she honestly believes cuts off one-third of the cake from the left. Then make a cut anywhere between the two leftmost lines and hand that piece to the person who marked the line closest to the end. This person is receiving more than one-third of the cake in their estimation, and the two remaining folks believe more than two-thirds remains.

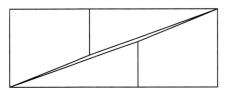

Have these two folks then each mark a line dividing the remaining portion precisely in half in their estimation. Cutting the cake between these two lines lands each person with more than half of more than two-thirds of the cake! This does the trick.

Taking it Further   Devise a cake cutting scheme between three people that not only assures everyone at least one-third of the cake in his estimation, but also the biggest (or tied for biggest) piece ever cut!

6. If you look carefully at the rectangular arrangement of the 8×8 square you will notice that the pieces don't quite line up correctly. We usually deem such discrepancies as due to imprecise cutting, but in this case the errors are inherent to the problem. There is a gap in the middle of the rectangle that accounts for the missing unit area.

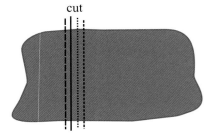

Taking it Further   Take any three consecutive integers F_{n-1}, F_n, F_{n+1} from the Fibonacci sequence 1, 1, 2, 3, 5, 8, 13, 21, 34, 55,... (defined recursively by $F_1 = F_2 = 1$, $F_{n+1} = F_n + F_{n-1}$ for $n \geq 2$). Show how to transform an $F_n \times F_n$ unit square into an $F_{n+1} \times F_{n-1}$ rectangle. Have you again lost track of a square inch of paper?

7. Every closed and bounded planar region can indeed be captured within a square box! For each angle θ we can certainly find a rectangular box tilted at that angle that captures the region.

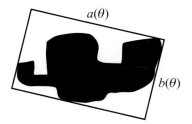

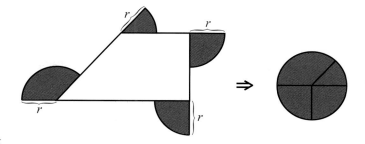

Let $a(\theta)$ and $b(\theta)$ be the side lengths of that box and set $f(\theta) = a(\theta) - b(\theta)$. As $f(\theta + 90°) = -f(\theta)$, the Intermediate Value Theorem guarantees an intermediate value θ^* with $f(\theta^*) = 0$. The rectangle at this angle is a square.

Taking it Further Can every closed, bounded planar region be captured by an equilateral triangle? By a regular pentagon?

8. Suppose lengths scale by a factor k. Then area scales as k^2. Suppose there is a figure in the plane that replicates itself with just two copies. Then

$$2 \times \text{Area (small figure)} = \text{Area (large figure)}$$
$$= k^2 \times \text{Area (small figure)}.$$

Necessarily $k = \sqrt{2}$, an irrational number. Similarly, we must have $k = \sqrt{3}$ for any three-self-replicating tile. We need four (or nine, sixteen, …) tiles to produce a rational scale factor.

Taking it Further Find examples of self-replicating figures that replicate with just two and three tiles (necessarily with irrational scale).

Note If we extend our notion of "planar figure" to include fractal figures (that is, objects whose "area" scales by a factor k^d with $d \neq 2$) then it is possible to find self-replicating objects that produce rational scaled copies of themselves with fewer than four copies. The Sierpinski triangle is an example of such an object, replicating itself with just three copies with scale factor $k = 2$. (Here $d = \ln 3 / \ln 2 \approx 1.585$.)

9. Note that because the back wheel is fixed in its frame, the tangent line to the inner curve (the back wheel track) always intercepts the front track a fixed distance r along the direction of motion. (See [5].)

First consider the case where the back wheel travels along the edges of a convex polygon, turning sharply at each corner

(in fact, pivoting about the point of contact.) The front wheel travels in straight lines as the back wheel follows the edges, and sweeps out sectors of a circle of radius r at each corner. These sectors fit together to form a complete circle of radius r, and hence the area between the two tracks in this polygonal case is πr^2.

Any curve can be approximated by a polygonal curve. By a limit argument we thus deduce the area between bicycle tracks is always πr^2.

Taking it Further What can you say about the area between two bicycle tracks along non-convex curves? (See [4].) If a bicycle follows only a portion of a curve, turning a total angle θ in the process, what can you say about the area between the two curve segments?

10. A sphere (of radius R) is obtained by revolving the graph

$$f(x) = \sqrt{R^2 - x^2}$$

about the x-axis. From calculus, the surface area of a segment between positions $x = a$ and $x = a + h$ is given by:

$$\int_a^{a+h} 2\pi f(x)\sqrt{1 + f'(x)^2}\, dx = \int_a^{a+h} 2\pi R\, dx$$
$$= 2\pi R h.$$

Thus all slices of thickness h have the same surface area. In terms of our spherical loaf of bread this means all slices have precisely the same area of crust! (This result was known to Archimedes.)

11. Suppose the sphere has radius R. Let h be the distance indicated. By the Pythagorean theorem (twice), $r^2 - h^2 = R^2 - (R - h)^2$. Consequently, $2Rh = r^2$. Now by question 10 the surface area of this slice of sphere of thickness h is $2\pi R h = \pi r^2$. This is the same area as the planar circle!

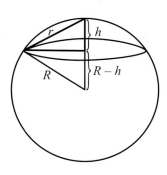

Taking it Further   What is the area between two bicycle tracks on a sphere?

12. In analogy to the two-dimensional problem, the soap solution forms a small square of film hovering in the center of the cube.

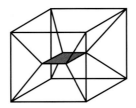

     What happens if you gently tap this structure?

Taking it Further   Notice that four edges of film meet at every interior vertex and that any two films meeting at an edge do so at an angle of 120°. In 1976 F. J. Almgren and J. E. Taylor proved that all such soap film structures behave this way. With this in mind would you care to predict what results when the frame of a tetrahedron or a triangular prism is dipped in soap solution?

Acknowledgments and Further Reading

Many of these puzzles appear in my forthcoming mathematical activities book (see [5]) along with further analysis connecting them to other branches of mathematics. Problem 4, in some sense, is a discrete version of a problem that appears in J. Konhauser, D. Velleman and S. Wagon's truly wonderful text [2] (problem 63). (How does the novel solution presented there apply to our situation?) Problem 11 is also from this text. Areas swept out by tangent line segments are examined in M. Mnatsakanian's delightful piece [4] (though bicycles are never mentioned). Combining problems 9 and 11 leads to interesting thoughts about bicycles on spheres. The sharing of cake is a classic topic in mathematics. For a very accessible account of this see [1], chapter 13. Other intriguing soap film questions and experiments can be found in F. Morgan, E. Melnick and R. Nicholson's fabulous article [3]. ■

1. COMAP Inc., *For All Practical Purposes: Introduction to Contemporary Mathematics*, 4th ed., W. H. Freeman & Co., New York, 1997.
2. J. Konhauser, D. Velleman, S. Wagon, *Which Way did the Bicycle Go? and other Intriguing Mathematical Mysteries*, Dolciani Mathematical Expositions No. 18, The Mathematical Association of America, Washington, DC, 1996.
3. F. Morgan, E. R. Melnick, R. Nicholson, "The soap-bubble-geometry contest," *The Mathematics Teacher*, **90**, No. 9 (1997), pp. 746–749.
4. M. Mnatsakanian, "Annular rings of equal areas," *Math Horizons*, November 1997, pp. 5–8.
5. J. S. Tanton, "A half-dozen mathematical activities to try with friends," *Math Horizons*, September 1999, pp. 26–31.

Update

James Tanton's prize-winning *SolveThis! Math Activities for Students and Clubs* was published by the MAA in 2001.

Suppose You Want to Vote Strategically

DONALD SAARI
University of California, Irvine

Be honest. There have been times when you voted strategically to try to force a personally better election result; I have. The role of manipulative behavior received brief attention during the 2000 US Presidential Primary Season when the Governor of Michigan failed on his promise to deliver his state's Republican primary vote for George Bush. His excuse was that the winner, John McCain, strategically attracted cross-over votes of independents and Democrats.

McCain's strategy was just the accepted behavior of encouraging supporters who can vote, to vote. But let's pursue this issue further; let's question whether the power of mathematics can help identify when and how you can strategically alter the election outcome of your fraternity, sorority, social group, or department to force a personally better conclusion.

First, a disclaimer: I am not interested in training a generation of manipulative strategists—the economists do a much better job of this. Instead, my goal is to demonstrate unusual applications of elementary mathematics while alerting the reader to subtle political actions which affect all of us. As we will see, each component of an election—the procedure, voting, and process—offers manipulative opportunities. So, I describe the mathematics behind the strategy of selecting an election procedure, casting a ballot, and debating or introducing amendments.

Choice of a Procedure

Remember those boring debates about the choice of a voting procedure? Someone wants to use the standard "vote for one" approach; others prefer to vote for two, while still others wish to distinguish between a voter's top- and second-ranked candidate by assigning, respectively, five and four points. Who cares? Let's just vote. Won't the answers be essentially the same?

To check, suppose five voters prefer the candidates Anita, Bonnie, and Candy in that order, denoted by *ABC*, six others prefer *CBA*, and the last four prefer *ACB*. While it doesn't seem like anything can go wrong, let's check.

- By voting for one candidate, the commonly used plurality system, Anita wins with a 60% landslide; the *ACB* outcome has the $9:6:0$ tally.

- Bonnie failed to receive a single plurality vote, yet she wins when each voter votes for her top two candidates where the *BCA* outcome has the $11:10:9$ tally.

- Candy? She wins with the procedure offering 5 and 4 points, respectively, to a voter's first and second choices; the *CAB* outcome has a $46:45:44$ tally.

This example illustrates the worrisome reality that *rather than representing the views of the voters, an election outcome may more accurately reflect the choice of the election procedure.* On the other hand, this setting presents strategic opportunities to a young aspiring Machiavelli to select a procedure which yields his preferred winner. While it may seem difficult to discover which procedure Machiavelli should select, the analysis is mathematically simple when there are only three or four candidates.

To introduce terminology, a *positional voting method* assigns weights $(w_1, w_2, 0)$ to candidates according to how a voter positions them on a ballot. As w_1 and w_2 points are assigned to a voter's top- and second-ranked candidates, the weights must satisfy $w_1 \geq w_2 \geq 0$. Thus, the standard *plurality vote*, where each voter votes for one candidate, corresponds to $(1, 0, 0)$. The Borda Count (BC) named after the eighteenth-century mathematician J.C. Borda, is given by $(2, 1, 0)$. A polite way to vote against one candidate is to vote for the other two; this *antiplurality method* is defined by $(1, 1, 0)$.

Each w_1 and w_2 choice defines a procedure, so there are an uncountable number of them. But notice, when voting for one candidate, the election ranking is the same if a candidate is assigned one point, or 500 points, for each vote. This suggests normalizing the tallying procedures so that $w_1 = 1$. By doing so, the points become

$$\mathbf{w}_s = (1, s, 0), \quad 0 \leq s \leq 1$$

where $s = w_2/w_1$. So, $s = 0$, ½, 1 represent, respectively, the plurality vote, the BC, and the antiplurality method. The earlier system giving five and four points, respectively, to a voter's first- and second-ranked candidates becomes $\mathbf{w}_{4/5} = (1, ^4/_5, 0)$.

To find personally beneficial procedures, tally the ballots to learn which s values deliver an outcome to our liking. By doing so, the above example yields

Number	Preferences	A	B	C
6	CBA	0	$6s$	6
5	ABC	5	$5s$	0
4	ACB	4	0	$4s$
Total		9	$11s$	$6+4s$

Which procedure, that is, which s should be used? For an election ranking to change, the tallies must pass through a tie. So, set pairs of tallies equal and solve for those s values which cause $\{A, B\}$, $\{A, C\}$, $\{B, C\}$ election ties. For instance, an A–C tie requires $9 = 6 + 4s$, so it occurs with $s = \frac{3}{4}$. The results, in Figure 1, prove that these seemingly innocuous preferences generate *seven* different election outcomes where four of them have no ties. Moreover, as each candidate wins with some procedure, this setting provides fruitful opportunities for our young Machiavelli! He just needs to justify using an appropriate choice of weights which ensures his preferred outcome. Since most people don't worry about which procedure is used, this offers no serious challenge.

The same approach identifies all outcomes for all specified preferences for any number of candidates. As a challenge, use this approach to find all $(1, s, t, 0)$, $1 \geq s \geq t \geq 0$, election outcomes for the following ten-voter example. (In graphing the outcomes, replace the "marks" with lines; 18 different rankings without ties emerge.)

Number	Preference	Number	Preference
2	ABCD	2	CBDA
1	ACDB	3	DBCA
2	ADCB		

A strategist might worry whether only rare, unlikely instances of voter preferences allow this phenomenon where the election ranking can switch with the weights. No; as Maria Tataru and I showed, even with conservative assumptions about the distribution of voters' preferences, about 69% of the time a three-candidate election ranking changes with the weights. Then, this likelihood and the number of possible election outcomes escalate with the number of candidates. For instance, examples can be constructed where, by choosing different weights for n alternatives,

Figure 1

as many as $n! - (n-1)!$ different election rankings emerge without ties. This suggests that the ranking of the top twenty collegiate football teams can differ drastically—$n = 20$ allows up to 2×10^{18} different rankings—depending upon how the ballots are tallied. Yet, not everything can happen; e.g., try as hard as I may, I cannot find weights which would rank the Northwestern University football team at the top.

By the way, this strategic behavior is not restricted to voting. For instance, suppose the head of a company producing a particular product wants his product to be the "best" in a statistical comparison. Can an appropriate choice of a statistical method skew the answer? Maybe; Deanna Haunsperger showed how a single data set can define a surprisingly wide array of different rankings by varying the choice of seemingly excellent statistical procedures.

Strategic Voting

As it is difficult to continually change procedures with each election, let's examine ways to be strategic in the privacy of the voting booth. This is commonly done. During the March 2000 primaries, for instance, a Keyes supporter from the State of Michigan confided that since Keyes had no realistic chance of winning, he voted strategically for Bush in an unsuccessful attempt to prevent McCain from winning. This behavior, which illustrates the commonly used "Don't waste your vote!" strategy, alters the legitimacy of any message based on election outcomes. As such, it is worth wondering whether a procedure can be invented where it never is in a voter's best interest to be strategic.

No such method exists. In the early 1970s, Alain Gibbard and Mark Satterthwaite independently proved that with three or more alternatives, all reasonable election procedures (e.g., not a dictatorship) provide opportunities for someone to strategically obtain a personally more favorable outcome. But while this result ensures that opportunities exist for our Machiavelli, it does not explain how he can recognize when they arise or how to take advantage of them. Not much help comes from the extensive literature on this topic as it favors existence theorems—proving that strategic opportunities exist—over the pragmatic issue of describing how to find them. So, let me indicate how to solve the more general problem for all procedures.

Notation Designate each voter type with a number as follows.

Type	Ranking	Type	Ranking
1	ABC	4	CBA
2	ACB	5	BCA
3	CAB	6	BAC

If p_j represents the number of voters of the jth type, then our introductory example has $p_1 = 5$, $p_2 = 4$, $p_4 = 6$ and $p_j = 0$ for all other voter types. These voter preferences—a *profile*—create the six-dimensional vector $\mathbf{p} = (5, 4, 0, 6, 0, 0)$.

Illustration by Loel Barr

Changing the outcome When Bob, a type-four voter, votes as though he is type six, there is one fewer type-four voter and one more type-six voter. This change is represented by $\mathbf{v} = (0, 0, 0, -1, 0, 1)$. If the original profile is $\mathbf{p}$, the profile adjusted by Bob's action is $\mathbf{p} + \mathbf{v}$.

Let $f(\mathbf{p})$ be the election procedure which determines the winner (or election ranking) for profile $\mathbf{p}$. The electoral difference Bob causes by adjusting the profile is

$$f(\mathbf{p} + \mathbf{v}) - f(\mathbf{p}). \tag{1}$$

Unfortunately, (1) does not make sense. For instance, if $f(\mathbf{p} + \mathbf{v}) = \text{Gore}$ and $f(\mathbf{p}) = \text{Bush}$, what does $f(\mathbf{p} + \mathbf{v}) - f(\mathbf{p}) = \text{Gore} - \text{Bush}$ mean? While it is tempting to explore the differences between these candidates, return to mathematics by noticing how, at least formally, (1) resembles the start of the definition for a directional derivative

$$f(\mathbf{p} + \mathbf{v}) - f(\mathbf{p}) \approx \nabla f(\mathbf{p}) \cdot \mathbf{v}. \tag{2}$$

To convert this nonsensical expression into an useful tool, note that $f^{-1}(\text{Gore})$ and $f^{-1}(\text{Bush})$ are defined; they represent all profiles which elect, respectively, Gore and Bush. So, if Bob's actions, $\mathbf{v}$, changes the election outcome, then $\mathbf{v}$ crosses a boundary separating these two profile sets. If ∇f did exist, it would be a vector orthogonal to an f level set pointing in the direction of greatest change. This suggests replacing ∇f in (2) with a normal vector $\mathbf{N}$ to this separating boundary. If $\mathbf{N}$ points into the set of profiles which elect Gore, then, if

$$\mathbf{N} \cdot \mathbf{v} \text{ is } \begin{cases} > 0, & \text{the change helps Gore,} \\ < 0, & \text{the change helps Bush,} \\ = 0, & \text{the change is neutral.} \end{cases}$$

In other words, a loose but useful interpretation of (2) is that the sign of $\mathbf{N} \cdot \mathbf{v}$ indicates the potential change in the election outcome.

To review—a profile $\mathbf{p}$ can be strategically manipulated if $\mathbf{p}$ is sufficiently close to a tie vote so that a voter can change the outcome. A useful strategic action is a $\mathbf{v}$ which changes the outcome in a desired manner. To illustrate the use of this tool, I calculate the strategies which change the A-B ranking of a $\mathbf{w}_s$ election.

First, the boundary separating where A and B wins are the profiles which define an A–B tie. A's tally for profile $\mathbf{p} = (p_1, ..., p_6)$ is $p_1 + p_2 + sp_3 + sp_6$, B's tally is $p_5 + p_6 + sp_1 + sp_4$, so an A–B tie occurs when the two tallies agree, or when

$$[p_1 + p_2 + sp_3 + sp_6] - [p_5 + p_6 + sp_1 + sp_4]$$
$$= (1 - s)p_1 + p_2 + sp_3 - sp_4 - p_5 + (s - 1)p_6 = 0.$$

To find a normal vector, treat p_j as a real variable rather than a nonnegative integer. The gradient of the above expression defines the normal vector

$$\mathbf{N}_s = ((1 - s), 1, s, -s, -1, -(1 - s))$$

which points toward the profiles which help candidate A.

This N_s makes it easy to determine who can be Machiavellian. Clearly, the type-one, two and three voters, who prefer A to B, have no interest in assisting B. If a type-four voter wants to change this A-B outcome, the change vector $\mathbf{v}$ has -1 in the fourth component reflecting that he no longer votes sincerely. The voter's strategy—the way he actually votes—determines which $\mathbf{v}$ coordinate has $+1$. Remember, to help B, this voter's strategy must ensure that $N_s \cdot \mathbf{v} < 0$.

As $\mathbf{v}$ has only two non-zero terms, a "$+1$" and a "-1," the dot product changes the sign of N_s's fourth coordinate to s and adds it to the N_s coordinate which represents the voter's strategy. So, finding a successful strategy reduces to finding N_s components which are less than $-s$. If our Machiavelli votes as though type six, so $\mathbf{v} = (0, 0, 0, -1, 0, 1)$, the dot product of $s - (1 - s) = 2s - 1$ certifies this strategy as useful (i.e., $N_s \cdot \mathbf{v} < 0$) if and only if the procedure is defined by $s < 1/2$. Voting as though type five, however, leads to $s - 1$ which is successful for all $s < 1$. Carrying out all computations, we obtain the following list of useful strategies.

Type	Strategy	methods	Strategy	methods	
4	5	$s < 1$	6	$s < \frac{1}{2}$	
5	None				(3)
6	5	$s > 0$	4	$s > \frac{1}{2}$	

For instance, if Machiavelli has sincere preferences CBA (type 4) and a burning desire for B to beat A with the $(6, 5, 0)$ voting procedure (so, $s = 5/6$), then, according to (3), his only manipulative strategy is to vote as though his preferences are BCA (type 5). On the other hand, if Machiavelli's sincere preferences were BCA (type 5), then he is powerless; according to (3) there is no way he can strategically vote to alter the conclusion. Of course, for a strategy to be effective, either the sincere election outcome must be close to an A–B tie, or enough voters must join forces. To determine how a mixed coalition should vote strategically, use the same analysis where the "change vector" $\mathbf{v}$ is a sum of the vectors describing the changes for each voter.

Other procedures

Somewhat surprisingly, learning how to politically connive reduces to a mathematics problem of finding certain normal vectors and taking appropriate scalar products. This description suggests that, in general, the more "tied" surfaces a procedure admits (so there are more normal vectors), the more varied the admissible strategies and opportunities. A procedure with several stages, for instance, admits "tied surfaces" at each stage.

Consider, for instance, a runoff election where after dropping bottom-ranked candidate(s) from a first election, a second (or third, or fourth) election is held. The "tied election" surfaces for the first election determine who will be advanced to the second stage, so the strategies represent trying to advance weaker candidates to help your preferred choice win in the final election. This is a well-used approach. After all, our election for the US President has this staged property with the primaries and a general election. The constant fear is that voters

from one party will manipulatively vote in the other party's primary to ensure a weaker, more vulnerable opponent.

Agendas

An example of this staged behavior is a meeting agenda $<A, B, C>$. Here, the first two specified candidates, A and B, are compared in a majority vote, and the winner is advanced to a majority vote with C to determine the overall winner. To illustrate by using an example, suppose the preferences for nine voters are

Number	Ranking	Number	Ranking
4	ABC	4	BCA
1	CAB		

where, to add intrigue, suppose your preferences are ABC. A quick computation shows that your favored candidate A beats B by 5:4 and advances to the final stage against C. Unfortunately, the candidate you dread, C, wins with a 5:4 vote.

What are your strategies? There are no options at the final stage because your only choices are to vote for A or for despised C. But at the first stage, by voting for B instead of your favored A, the final election is between B and C where B beats C. Although your favorite A does not win, neither does your least favored C.

As an aside, the $\mathbf{v}$ from (2) models much more than strategic action. It can be used to determine what happens with *any* vote change where say, voters refuse or forget to vote, new voters join, subcommittees join as a full committee, and so forth. All of these questions can be similarly analyzed, i.e., find the appropriate normals $\mathbf{N}$, change vectors $\mathbf{v}$, and compute the scalar product.

To illustrate, suppose you and a friend with the same ABC preferences either forgot to vote, or headed for the beach. Could your negligence be rewarded with a personally better election outcome? It can. In the agenda example, for example, by not voting, it is B who wins; had both of you voted sincerely, you would have suffered by having C as the winner. Ah, the perversity of voting procedures! I leave it to the reader to characterize the kinds of procedures which allow such a pathological behavior. (For this inverse problem, note that if type j voter doesn't vote, $\mathbf{v} = -\mathbf{e}_j$; this is the unit six-dimensional vector with unity in the j component. Next, characterize the appropriate kinds of normal vectors.) Another challenge is to determine which kinds of procedures can punish a candidate should she receive more support (they exist!); namely, the winning candidate loses only because more supporters voted.

The MAA election procedure

If all procedures can be manipulated, what about the one adopted by our organization the Mathematical Association of America? The MAA procedure, Approval Voting (AV), allows a voter to vote for one, or for two, or … candidates. As such, AV is ripe with strategic opportunities because each way to vote defines a "tied vote" surface offering new strategic options. Stated in another manner, a voter can use the strategies available to a "vote for one" procedure, a "vote for two" procedure, …. In addition, the voter has the strategy of selecting which procedure to use!

AV even encourages strategic voting. For instance, the 1999 Social Choice and Welfare Society AV election for their president involved three candidates so highly regarded that, most surely, most (if not all) voters sincerely approved of more than one of them. But as an election with popular, well-qualified candidates would be closely contested, the obvious AV strategy was to vote for one candidate. Thus, a reasonable measure of the strategic intent is the percentage of voters doing so. I wildly predicted that over half, maybe even 65%, would vote for one candidate. I was wrong; it was over 92%.

As another example, before I studied voting systems and discovered many worrisome AV properties, I was sufficiently intrigued by AV to propose it as the procedure to elect the committee to select the new President of Northwestern University. Later, the Chair of another department asked, "Did you see the silly election procedure? Is it easy to manipulate!" His strategy was for his department to propose some candidates, for the Math Department to propose others; with both departments voting for these candidates, we could determine the committee. I leave it to you to wonder whether we did this.

Debate and Selecting Amendments

Often I joke during lectures that

> "For a price, I will serve as a consultant for your group for your next election. Tell me who you want to win. After talking to the members, I will design a 'democratic procedure' which ensures the election of your candidate."[2]

To show that my joke goes beyond bad humor to be a realistic warning, suppose a 30-member department is evenly split over the eight candidates for a job; 10 each prefer

$$A \ B \ \mathbf{C} \ D \ E \ F \ G \ \mathbf{H},$$
$$B \ \mathbf{C} \ D \ E \ F \ G \ H \ A, \qquad (4)$$
$$\mathbf{C} \ D \ E \ F \ G \ H \ A \ B.$$

C appears to be the departmental favorite; only she is high ranked by all voters. H, on the other hand, is so poorly appreciated that *all* voters prefer $C, D, E, F,$ and G to H. The challenge is to elect H with a procedure which involves all candidates and where all voters are satisfied that H truly reflects the departmental wishes. Let me encourage you to try to design such a method before reading any further.

The approach is simple; it is an agenda where the winning candidate in each pairing is advanced to be compared with the next candidate. As true with basketball tournaments, the strategic action involves the seeding. The seeded agenda of Figure 2 not only elects H, but as each election outcome is decided either unanimously or by a landslide two-thirds vote, nobody should object.

The smug reader may feel confident that his or her organization is immune to this behavior if only because the decisions are made by consensus. But, this bothersome phenomena can fully

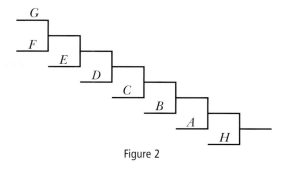

Figure 2

arise even during a friendly discussion or debate. As an illustration, think back to a recent meeting where someone suggested calculus book G only to have it compared with F. As everyone prefers the graphics in F, G is dismissed without a vote. But F is heavy while book E will not induce hernias by lugging it to class, so the discussion shifts from F to E. The debate continues by covering one detail after another; D handles the limit process better than E; C describes derivatives more intuitively than D, and so forth. At each stage, the strong support for the new alternative makes votes unnecessary. The result? Inferior H wins with a strong consensus. The problem is inherent in the dynamics of the discussion; the sad fact is that the organization may never recognize why they made such an inferior choice.

Where are the opportunities for our young Machiavelli? They come from governing the dynamics of the discussion; by finding ways to compare stronger competitors at an early stage to eliminate them. But, always, always, try to introduce your personally desired choice at the end of the discussion.

To indicate the associated mathematics, Example (4) is part of an orbit of the symmetry group Z_8. To explain, I will use what I call a "ranking wheel." As illustrated in Figure 3, this freely rotating disk is attached at its center to the wall; ranking integers are evenly distributed along its edge. On the wall, write the names of the alternatives; this determines the first ranking of $A \ B \ C \ D \ E \ F \ G \ H$. Next, rotate the disk so that "1" now is under the previously second-ranked candidate, and read off the ranking $B \ C \ D \ E \ F \ G \ H \ A$. The full "orbit" results when each alternative has been top-ranked once; this defines a profile with eight different rankings.

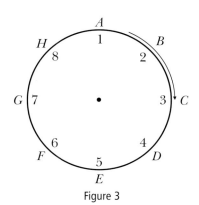

Figure 3

By construction, each alternative is in first, second, third, …, last place once. As such, no candidate should have an advantage; the societal outcome should be a complete tie. But pairwise comparisons cannot recognize nor capture this full circular symmetry, so it "breaks the symmetry" by creating cycles. These eight rankings generate the pairwise cycle $A\,B,\,B\,C,\,C\,D,$ $D\,E,\,E\,F,\,F\,G,\,G\,H,\,H\,A$ where each winning candidate receives seven-eighths of the total vote. Example (4) was created by noticing that cycles already emerge by using only three of the eight rankings. The manipulative agenda, or the strategic discussion, just exploits the cycle; place H last and arrange to have A, the only alternative H can beat, placed next to last, …

Perhaps it is time to tackle an "impossible problem." If a group is unanimous in their belief, then surely it is impossible to strategically elect the bottom-ranked candidate while keeping all voters content. Or, is it? Suppose the above example started with everyone preferring $C\,D\,E\,F\,G\,H$. To elect H, in addition to finding the correct seeding, just introduce two amendments A and B which divide the voters as above. Yes, this is more involved, but to solve the "impossible," we should try harder.

For a final seeding setting, recall the common basketball refrain, "On any given night, …." If true, if even the best team can be beaten by someone, then during tournament time we should expect that a strategic seeding will spawn upsets. To design an eight-candidate example, the three stages require the "Champ" to beat at least three other candidates. To create an example where the "weakest" candidate wins, select three of the above Z_8 eight rankings so that a weak contestant beats three candidates. (Other conditions are needed, but I will let you find them.) For instance, if the thirty members of a department are evenly split among the three rankings

$$A\quad B\quad C\quad D\quad E\quad F\quad G\quad H,$$
$$D\quad E\quad F\quad G\quad H\quad A\quad B\quad C,$$
$$F\quad G\quad H\quad A\quad B\quad C\quad D\quad E,$$

there are ways to seed a tournament so that H wins. I leave the design to the reader.

Summary and Further Reading

As probably true of many mathematicians, I have been accused of trying to detect mathematics in just about anything and everything. Maybe; but I do know that mathematics can shed light on a surprising number of events which play a crucial, central role in our lives. As for "voting" and the other social sciences, the above only hints at what is available; it does not even suggest the tip of the huge, fascinating, essentially unexplored iceberg of other issues. What follows are suggested references for the reader interested in learning more about difficulties in voting.

The mathematical approach used here was introduced in my book Saari, D. G., *Basic Geometry of Voting*, Springer-Verlag, 1995. Also, this book has the references for all of the different topics described in this article; this includes the Gibbard-Satterthwaite Theorem, Haunsperger's statistical analysis, the public debate Jill van Newenhizen and I had with Steve Brams and Peter Fishburn over the dangers of using Approval Voting, the problems of assigning seats to states based on the census, and so forth.

To learn about other kinds of problems suffered by voting procedures, let me suggest Hannu Nurmi's recent book *Voting Paradoxes and How to Deal with Them*, Springer-Verlag, NY, 1999, or my short article, "Are individual rights possible?" *Mathematics Magazine* 70 (April 1997), 83–92. A fairly complete list of references for this area can be found on Jerry Kelly's web page www.maxwell.syr.edu/maxpages/faculty/jskelly/biblioho.htm. ∎

Endnotes

1. This research was supported by NSF grant DMI-9971794 and my Northwestern University Arthur and Gladys Pancoe Professorship. The paper is based on an invited talk to the Illinois section of the MAA, March 2000.

2. Please, no calls, no letters, no offers. This is only a joke!

3. Saari, Donald, *Chaotic Elections! A Mathematician Looks at Voting*, AMS, Providence, RI, 2001.

TopSpin
on the
Symmetric Group

CURTIS D. BENNETT
Bowling Green State University

When I was about 10 I remember getting a puzzle in my stocking which consisted of a 4 × 4 grid with 15 square pieces in it. Of course there was one space in the grid that held no piece, and you could slide the pieces around so that a piece next to the "hole" could be slid into that space. This particular puzzle had the pictures of four comic book figures when solved. However, you could move the pieces around to give some of the figures different heads, which added a great deal of fun for me. The box the puzzle came in gave some "impossible" positions, and I recall that at the time I wondered how they knew this. Today I still look for puzzles like these whenever I visit a toy store. Now, though, I find that the mathematics behind the puzzles intrigues me as much as the challenge of solving them.

These puzzles on 4 × 4 grids first appeared in the middle of the 19th century. They were sold as "fifteen" puzzles, and the companies that manufactured them offered monetary prizes for anyone who could show how to rearrange the pieces so that a particular position could be reached. Of course, the advertised position was impossible to reach. Around 1980, the Rubik's cube was brought to this country and set off a puzzle craze, and this craze continues with many more recent puzzles, including the TopSpin puzzle which first came to my attention around 1993.

The TopSpin puzzle consists of twenty circular pieces, numbered 1–20, in an oval track. The puzzle has two basic moves. The pieces can be moved around the track maintaining their order, or four consecutive pieces originally numbered, say 1, 2, 3, 4, can be moved to the order 4, 3, 2, 1. Curiously, this puzzle has no impossible positions, although if there were either 19 or 21 pieces instead of 20 this would no longer be true.

The fifteen puzzle can be viewed similarly. That is, if we give the blank space the number 16, then we have that the 16 pieces are numbered by:

In contrast to the TopSpin puzzle, the fifteen puzzle has positions that cannot be obtained.

An Abstract Algebra Excursion

To understand why impossible positions occur, it is necessary to embark on a brief tour of the symmetric group. The symmetric group S_n on the set $A = \{1, 2,\ldots, n\}$ is the set of one-to-one and onto (bijective) functions from A to itself. Such functions are called *permutations*. For our purposes, we shall represent the permutation $\sigma \in S_n$ by the sequence $\sigma[1], \sigma[2],\ldots,\sigma[n]$. Thus the permutation in S_4 represented by 2, 3, 1, 4 is the function σ with $\sigma[1] = 2$, $\sigma[2] = 3$, $\sigma[3] = 1$, and $\sigma[4] = 4$. We define an *inversion* of σ to be a pair $i < j$ such that $\sigma[i] > \sigma[j]$. Hence in the example above, 1, 3 is an inversion as $\sigma[1] = 2$ and $\sigma[3] = 1$. Given our representation of permutations, we can find inversions by simply looking for pairs of integers that occur in the wrong order. We say that a permutation is *even* if it has an even number of inversions, and we say that it is *odd* if it has an odd number of inversions.

A transposition is a permutation $\sigma \in S_n$ that fixes all but two elements of $\{1,\ldots, n\}$ and switches those two. To save space, such transpositions are usually written in *cycle notation* [1] as (i, j). Given any permutation $\sigma \in S_n$, the *support* of σ is the set of elements of $\{1,\ldots, n\}$ **not** fixed by σ. For example, the transposition $\tau = (3, 5)$ has support $\{3, 5\}$ as these are the only two elements τ moves. Suppose σ is a permutation, and m is the greatest element in the support of σ. Then the greatest element in the support of $(\sigma(m), m) \circ \sigma$ is less than m. For example, if $\sigma = 1, 3, 4, 5, 2$, then 5 is the greatest element of the support of σ. In this case $\sigma(m) = 2$, so we have $(2, 5) \circ \sigma = 1, 3, 4, 2, 5$, which has support $\{2, 3, 4\}$. Continuing inductively, for any σ we obtain transpositions $\tau_1,\ldots,\tau_k$ such that $\tau_k \circ \tau_{k-1} \circ \cdots \circ \tau_1 \circ \sigma$ fixes everything. Hence

$$\sigma = \tau_1 \circ \cdots \circ \tau_k.$$

In the above example this corresponds to

$$(2,3) \circ (2,4) \circ (2,5) \circ \sigma = 1, 2, 3, 4, 5,$$

and hence

$$\sigma = (2,5) \circ (2,4) \circ (2,3).$$

Thus we have shown:

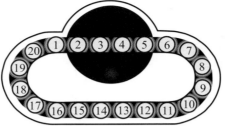
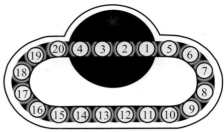

Figure 1. The TopSpin puzzle and the two basic moves.

Theorem 1. *Every permutation can be written as the product of transpositions.*

Theorem 1 is important because it enables us to show the alternating group, A_n, is a subgroup of S_n. The *alternating group* consists precisely of the even permutations σ, which we recall are the permutations having an even number of inversions. This makes it easy to tell if a permutation is in the alternating group, as we can just count inversions, but showing that the product of any two even permutations is even turns out to be difficult. For this we examine how composing a transposition (i, j) with σ affects the number of inversions. This is best seen with an example. Suppose $\sigma = a, 2, b, 4, c$, and consider the permutation $(2,4) \circ \sigma = a, 4, b, 2, c$. Then switching 2 and 4 has neither created nor deleted inversions involving a and c, since it didn't affect the order that 2 or 4 occurred in with respect to these two. Meanwhile, the switch will have changed the number of inversions involving b by an even number as our three possibilities for b are: (1) $2 < b < 4$ in which case the number of inversions increased by 2; (2) $2 < 4 < b$, in which case the number of inversions remains unaffected; or (3) $b < 2 < 4$, and again the number of inversions is unaffected. Of course, the pair $\{2, 4\}$ is now inverted where initially it wasn't. Thus the change in the total number of inversions is odd. The general argument looks much the same, except that there may be more than one element between the two transposed elements. However, this won't affect the parity. Thus we have:

Theorem 2. *A permutation σ is even if and only if σ can be written as the product of an even number of transpositions.*

Most texts use this product of transposition property as the definition of an even permutation; however, with this definition one must prove no transposition can be both even and odd. Our theorem implies the following important corollary:

Corollary 3. *Composing two even permutations yields an even permutation. Hence A_n is a subgroup of S_n (which is to say that it is closed under multiplication and inverses).*

Abstract Algebra Solves Puzzles

How do these theorems apply to our puzzles and the impossibility of certain positions? In the case of the fifteen puzzle, since the empty space changes places on every move, and every move is a transposition, Theorem 2 tells us certain positions cannot be obtained. In particular, recall that we number the squares of the fifteen puzzle as

and think of the empty square as being numbered 16. Then, whether the empty space is in a blue or white position determines whether an even or odd, respectively, number of moves has been made. Thus, if the desired position is:

so that the permutation has only interchanged 16 and 14, then the empty space is in the 14 position, implying an even number of moves must be made. As the per-

mutation of the squares is (14, 16), however, we need an even number of transpositions to make an odd permutation if we are to obtain this position. Our theorem tells us this is impossible. Conversely, showing positions not violating this even-odd condition are obtainable is a little harder.

What about TopSpin? In the case of a 19-piece TopSpin game, there are two basic moves: Cycling the 19 pieces around which is represented by 2, 3, 4,…, 19, 1 and interchanging four pieces which is represented as 4, 3, 2, 1, 5, 6, 7,…, 19. Counting, the first move has 18 inversions and the second has 6 inversions, both of which are even numbers. But by Corollary 3 this means that any permutation arising from compositions with only these two permutations as constituents must also be even. Hence, any position requiring an odd permutation must not be reachable. In particular, we can never simply interchange two adjacent pieces.

At this point we know certain positions are not reachable in the 19-piece TopSpin game, but we don't know that all positions are reachable in the 20-piece game. For this, we need to revisit the proof of Theorem 2. We showed that precomposing σ with a transposition τ changed the parity of the number of inversions. However, if τ exchanges consecutive numbers i and $i + 1$, the argument shows more. In particular in this case the number of inversions changes by exactly 1. It increases if i and $i + 1$ occur in the proper order in σ and the number of inversions decreases if i and $i + 1$ occur in the wrong order in σ. If σ has no consecutive numbers i and $i + 1$ occurring in the wrong order, then σ must be the identity and thus contains no in-

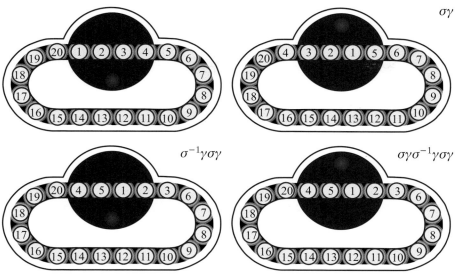

$\sigma^{-1}\gamma\sigma\gamma$

$\sigma\gamma\sigma^{-1}\gamma\sigma\gamma$

Figure 2. The permutation α slides 2, 3, 4, 5 past 1.

$\sigma^3\gamma\sigma\gamma\sigma^{-1}\gamma\sigma\gamma$

$\sigma\gamma$

achieved in the 20-piece TopSpin puzzle, then all permutations are reachable. At this point we have to get dirty and find a sequence of moves for the puzzle producing such a transposition. Let us name the two basic moves as follows:

$$\sigma = 2, 3, 4, \ldots, 20, 1$$
$$\gamma = 4, 3, 2, 1, 5, 6, 7, \ldots, 20.$$

A tedious check shows the permutation α given by

$$\alpha = \sigma^3\gamma\sigma\gamma\sigma^{-1}\gamma\sigma\gamma$$

has representation

$$1, 6, 7, \ldots, 20, 2, 3, 4, 5.$$

Of course the most convincing way to check this is to get a copy of the puzzle and try it out, remembering that σ is moving the circuit counterclockwise and σ^{-1} is moving the circuit clockwise. Figure 2 shows some of the steps along the way. In other terms, we would say this sequence of moves *cycles* the numbers 2 through 20 four to the left. Thus, repeating this sequence of moves again and again (Figure 3) yields:

$$\alpha^2 = 1, 10, 11, 12, \ldots, 20, 2, 3, \ldots, 9$$
$$\alpha^3 = 1, 14, 15, 16, \ldots, 20, 2, 3, \ldots, 13$$
$$\alpha^4 = 1, 18, 19, 20, 2, 3, 4, \ldots, 17$$

and

$$\alpha^5 = 1, 3, 4, \ldots, 20, 2.$$

The last of these is very nearly what we want. In particular, if we cycle all twenty

versions. On the other hand, if σ is not the identity permutation, then there must exist such an i and $i + 1$. Letting τ_1 be the transposition interchanging these, we would then have $\tau_1\sigma$ has fewer inversions than σ. Working inductively, we find transpositions $\tau_1, \ldots, \tau_k$, each interchanging consecutive numbers, such that $\tau_k \ldots \tau_1\sigma$ has no inversions and is hence the identity. This implies, in turn, that $\sigma = \tau_1 \ldots \tau_k$. Thus we have proven:

Theorem 4. *Given any permutation σ, then we can find transpositions $\tau_1, \ldots, \tau_k$ such that $\sigma = \tau_1 \ldots \tau_k$ where each τ_i interchanges consecutive numbers. Moreover, k can be chosen to be the number of inversions of σ.*

For our purposes the importance of this theorem is if we can show any transposition of consecutive numbers can be

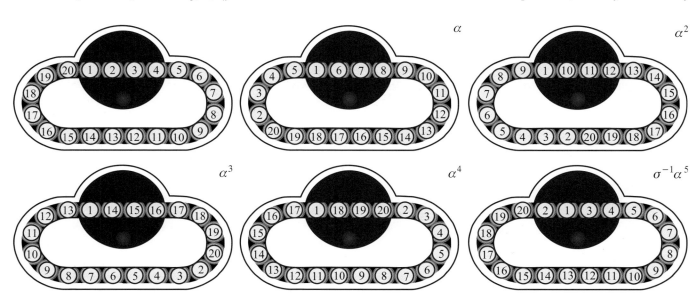

α

α^2

α^3

α^4

$\sigma^{-1}\alpha^5$

Figure 3. Iterating the permutation α slides 1 all the way around and interchanges 1 with 2.

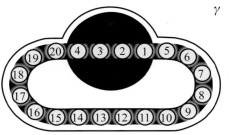  

Figure 4. The permutation β interchanges 2 with 20 and 3 with 4.

numbers one to the right, we obtain:

$$\sigma^{-1}\alpha^5 = 2, 1, 3, 4, 5, \ldots, 20,$$

the transposition interchanging 1 and 2. Using powers of σ, we can now put any number j in the first position to obtain the transposition interchanging j and $j + 1$. (The actual permutation switching j and $j + 1$ is given by $\sigma^{-j}\alpha^5\sigma^{j-1}$.) Thus any permutation can be reached in the 20-piece game.

Abstract Algebra Answers All Questions

Examining the preceding argument more closely, we can generalize it to tell us that for any **even** $n > 5$, every position in the n-piece TopSpin game is reachable. The key here is to note that when we change to the n-piece game, the permutations σ and γ change in only minor ways. In particular, here

$$\sigma = 2, 3, 4, \ldots, n, 1$$
$$\gamma = 4, 3, 2, 1, 5, 6, 7, \ldots, n.$$

As we require $n \geq 6$, the computations for α now yield that

$$\alpha = 1, 6, 7, \ldots, n, 2, 3, 4, 5.$$

If you have the puzzle in front of you, the easy way to see this is by performing the moves for α. Only pieces $1, \ldots, 5$ are moved out of the cyclic order.

The above representation for α tells us that α is cycling the numbers 2 through n around in steps of 4. Hence if $n - 1$ is relatively prime to 4, i.e., odd, performing α enough times will give

$$\alpha^m = 1, 3, 4, \ldots, n, 2$$

yielding that $\sigma\alpha^m$ is again the permutation interchanging 1 and 2.

What about the n-piece puzzle when n is odd? When there are an odd number of pieces, both γ and σ are even permutations. Corollary 3 then implies we cannot hope to get any position corresponding to an odd permutation. Can we even get all the positions corresponding to even permutations? The answer is yes, although to justify this we have to work even harder than we did for the even case. For simplicity, we will assume that $n \geq 9$, although the statement is true in the case $n = 7$ also.

By Theorem 4, to see that any even permutation can be obtained, we simply need to show that any product of two elementary transpositions can be obtained. Actually something stronger is true: if we can obtain the product of any two *consecutive* transpositions, i.e., a product of the form $(i, i+1)(i+1, i+2)$, then we can obtain any even permutation. (We leave to the reader the verification of this fact.) Carrying out the composition of consecutive transpositions we see that each is an *elementary three cycle*, that is, it permutes three consecutive elements in the cyclic pattern $i \to i + 1 \to i + 2 \to i$. We represent this three cycle as $(i, i + 1, i + 2)$. Again finding such a permutation becomes a matter of playing with the puzzle until the right sequence of moves turns up. Recalling that

$$\sigma = 2, 3, 4, \ldots, n, 1$$
$$\gamma = 4, 3, 2, 1, 5, 6, 7, \ldots, n$$

we can see (Figure 4) that the map

$$\beta = \gamma\sigma\gamma\sigma^{-1}\gamma$$

 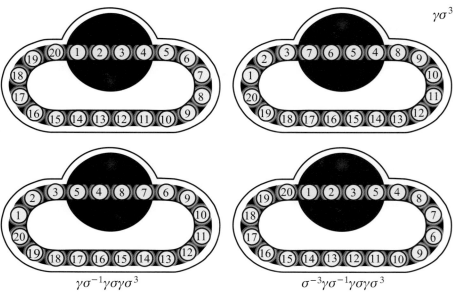

Figure 5. The permutation δ interchanges 4 with 5 and 6 with 8.

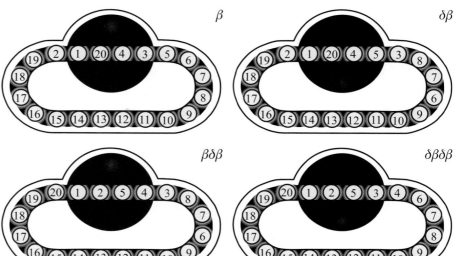

Figure 6. The permutation $\delta\beta\delta\beta$ is the three cycle $(3, 4, 5)$.

is the product of two transpositions $(n, 2)(3, 4)$. Similarly (Figure 5),

$$\delta = \sigma^{-3}\gamma\sigma^{-1}\gamma\sigma\gamma\sigma^3 = (4, 5)(6, 8).$$

A final calculation (Figure 6) tells us

$$\delta\beta\delta\beta = (3, 4, 5).$$

As before, using σ to cycle the numbers around, any elementary three cycle $(a, a + 1, a + 2)$ can be obtained. Thus,

Theorem 5. *The obtainable positions of an n-piece TopSpin puzzle correspond to the even permutations of $\{1,\ldots, n\}$ when n is odd, and correspond to all permutations of $\{1,\ldots, n\}$ when n is even.*

The above discussion can be turned into an algorithm to solve the TopSpin puzzle. Unfortunately, in practice this algorithm is extremely slow. Most readers, however, will find, by just playing around with the TopSpin puzzle, a quick algorithm to solve the puzzle except for the numbers 17 through 20. Thus when I solve the puzzle I typically only worry about two cycles and three cycles at the very end. Curiously, this question of speed leads to an important (and difficult) question in the study of permutation groups. Namely, given a set of permutations (or moves) which *generate* the group S_n, determine which permutation of S_n requires the most such moves to be reached.

At ten I was fascinated by permutation puzzles like the fifteen puzzle. At seventeen, I became enamored of the Rubik's cube, and today I still look for puzzles like these whenever I visit a toy store. For me today, however, the beauty of these puzzles is how easily they lead to deeper mathematics. ■

References

1. J. Gallian, *Contemporary Abstract Algebra*, Houghton Mifflin, New York, 1998.
2. TopSpin is a registered trademark of Binary Arts. Now available as No. Crunch.

Some New Results on Nonattacking Chess Tasks

MARTIN GARDNER

"Chess tasks" is a term for combinatorial problems that involve placing chessmen on square or rectangular boards of size n^2 or $m \times n$ so as to meet certain provisos. They do not include the most common type of problem, that of determining from a given position how to mate in a specified number of moves. Indeed, chess tasks have almost nothing in common with chess except for the use of chess pieces and how they move. (Update: You'll find a column on chess tasks and the entire collection of Gardner's *Scientific American* columns on one CD, *Martin Gardner's Mathematical Games*, MAA, 2005.). Chess tasks include placing sets of pieces on a board to minimize or maximize the number of attacked or unattacked cells, finding the shortest game that ends in stalemate or with all pieces captured, and hundreds of other curious problems of much more interest to mathematicians than to chess players.

A classic family of chess tasks that is now the topic of considerable literature is that of finding the maximum number of pieces, of a given type, that can be placed on a board of given size, so no piece attacks another, and enumerating all possible solutions. I will skim very briefly over such problems before venturing into uncharted waters.

The most analyzed and generalized of all nonattacking tasks involves queens. On the standard 8^2 chessboard there are exactly twelve different ways (rotations and reflections are never considered different) of placing eight queens, obviously the maximum number, so no queen attacks another. Figure 1 shows the twelve patterns. Note that every solution has a queen on a border, four cells from a corner. Pattern 7 is unique and remarkable in that it also solves the problem with the added rule that no *three* queens may be in a straight line, assuming a line can have any orientation, not just along the board's diagonals.

There is a vast literature on nonattacking queens. No formula is known for generating all patterns for n^2 boards, with or without including rotations and reflections. However, computers, at the last count known to me, have found all fundamental solutions for n through 15.

The 4^2 board has one solution, the 5^2 has two, the 6^2 only one (can you find it?), and the 7^2 has six. The 9^2 has 446, and the 10^2 has 92. The order-5 board is of special interest because it permits placing 25 queens (five of each of five colors) so no queen attacks a queen of the same color. Figure 2 shows how this is done.

The task has been generalized to "boards" of more than three dimensions, to toroidal boards that wrap around vertically and horizontally, and to "superqueens" that combine the moves of queens and knights. On the latter task, see Chapter 16 of my *Unexpected Hanging and Other Mathematical Diversions* (1969).

If we pose the same problem for rooks, the maximum number of nonattacking rooks on a chessboard obviously is also eight, but now the number of distinct solutions jumps to 5,282. One simple solution is to put all eight rooks along a main diagonal. If rotations and reflections are included it is easy to show that the number of patterns for an n^2 board is $n!$.

Formulas for the number of basic solutions for nonattacking rooks are complicated and hard to probe. Interested readers will find them explained in "Rooks, Inviolate," by Derek F. Holt, in *Mathematical Gazette,* Vol. 58 (June 1974), pp. 131–134; and "Essentially Different Nonattacking Rook Placements," by Loren C. Larson, in the *Journal of Recreational Mathematics,* Vol. 7 (Summer 1974), pp. 178–183.

The maximum number of nonattacking bishops that will go on a chessboard is 14. (For any n^2 board the number is $2n - 2$.) Figure 3 shows a simple solution. There are 35 others. If rotations and reflections are included, the number of patterns is 2^n. On square boards, if n is even (a number represented by $2k$), the number of distinct solutions is $n^{2k-3} + 2^{k-2}$. If the board's side is odd $(2k + 1)$, the number is $2^{2k-2} + 2^{k-1}$.

For nonattacking knights there are just two solutions on any board: putting the knights on all the black squares, or on all

Reprinted from February 2001, pp. 10–12

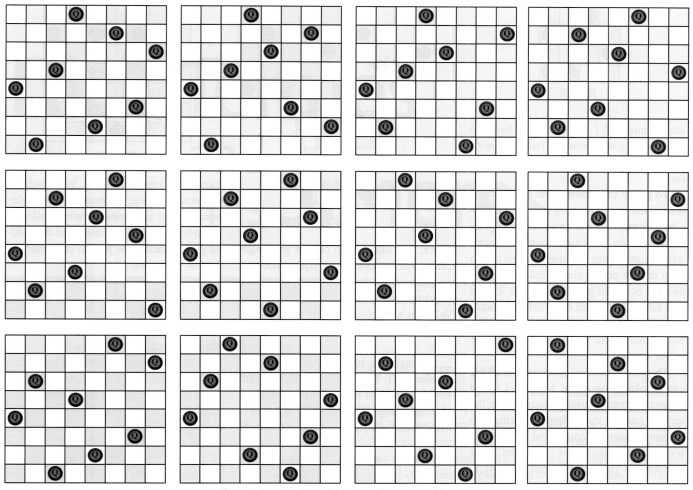

Figure 1. The twelve solutions to the nonattacking queens.

the white squares. For kings, the maximum number on the 8^2 board is 16. One of a very large number of solutions is shown in Figure 4.

We turn now to a new class of nonattacking problems that have received little attention—the task of placing a mixture of different pieces on boards of various sizes so no piece attacks anoth-er. Such problems cease to be trivial when the board is larger than 3×3. On the 4^2 board I found two ways to place two rooks, two bishops, and two knights so no piece attacks another. They are shown in Figure 5. (In all illustrations, N stands for knight.)

Mario Velucchi is a computer scientist and writer who lives at Via Emilia, 106, I-56121, Pisa, Italy. I am much in his debt for aid he has given me on this article. He is skillful in devising fast algorithms for solving all sorts of combinatorial problems including chess tasks. He assures me that the patterns shown in Figure 5 are the only solutions to the task.

Can a queen, king, rook, bishop, and knight be placed on a 4^2 board so no piece

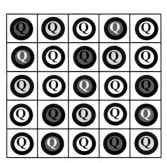

Figure 2

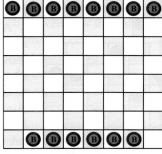

Figure 3

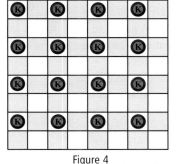

Figure 4

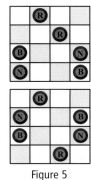

Figure 5

vanishing point above head

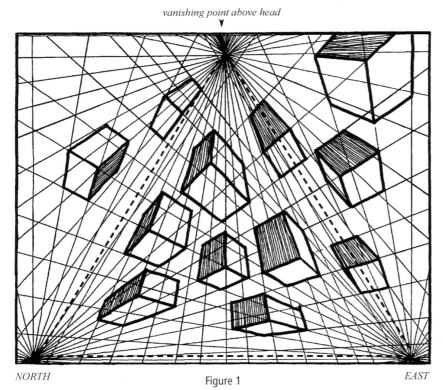

NORTH   Figure 1   EAST

Consider the worldly cube. It has three sets of parallel lines. If the lines of one of those sets meet at a vanishing point ahead of you, as a pair of railroad tracks might, you get one-point perspective. Artists got to this stage in the Renaissance. Two-point perspective takes two different sets of parallel lines on the cube to two vanishing points, say to the east and north of you (90° apart). In 3-point perspective, each of the three sets of parallel lines on the cube has a vanishing point. If, on a page, we draw converging parallel lines from all three directions of the cube, we obtain a grid containing three vanishing points, which perhaps represent the zenith (above you) at the top of the page, and the north and the east at the lower corners of the page. (See Figure 1.)

What in the world can 4-point perspective mean? Termes suggests two ways to grip this concept. For example, you can have vanishing points at the east, west, zenith and nadir (below you)—as if you were hanging onto the middle floor of a sky-scraper and peering above your head and below your feet. (If you *do* grip this perspective, you will be gripping hard!) In this perspective, something strange happens when you attempt to reproduce what you see onto a flat piece of paper. The zenith and nadir vanishing lines bulge in the middle of the paper (as you register nearby objects, such as the flagpole you are clutching) and they taper at the top and bottom (as you scan down the building or upwards to the rescuing helicopter) Thus the up-down perspective lines on your grid curve out like the lines on a football (Figure 2). Termes calls this the first curved-line per-

spective; it is a perspective that can give you vertigo!

There's a different way to think about 4-point perspective. In Termes's "continuous" 4-point perspective, the zenith-nadir (up-down) lines stay parallel, but the east, north, west and south directions all become vanishing points. Imagine trying to reproduce a room around you by sitting with a drawing pad in a swivel chair in the middle of the room. As you face a picture on the north wall, it appears to bulge out, just as the east-west lines leading to the corners of the room in your peripheral vision converge. But as you swivel around in your chair and face an adjacent wall, it's now the north-south directions with the vanishing points. Wait a minute! A vanishing point for this wall is smack in the middle of the previous wall. Obviously, the drawing must suppress some of this intrusion of vanishing points. In fact, the way to approach continuous 4-point perspective is to start with a long strip of paper on which north, east, south, west, north vanishing points are lined up along the equator of the paper. Elliptical grid lines for east-west vanishing points intersect with the elliptical north-south vanishing lines. When the page with its drawing is rolled into a cylinder, the walls and objects in the room look big and small like a visual Doppler effect. (See Fig-

Figure 2

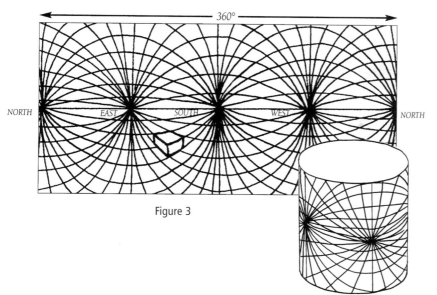

Figure 3

is *your* eye!) view of the cat. He has also painted many cathedrals and other buildings onto spheres. Termes did his preliminary sketches on location at such places as the steps of the *Paris Opera,* the lurking grounds of the phantom of the opera. He sometimes uses his own patented camera, which he mounts on the sides of regular solids, such as dodecahedra. Photographs taken from all the sides of the solid are matched up and reproduced onto actual dodecahedra which adults and children can reconstruct from their flattened forms.

Looking for Order

It may help to understand Termes's 6-point perspective by considering how he mass-produces some of his spheres. A sphere is composed of two hemispherical polyethylene light fixtures which are lightly glued together. A scene is painted on the sphere with acrylic paints, and the sphere is popped apart into its two original hemispheres, which are then put into an oven. The paints melt to the plastic, and as the plastic melts into the disks, the distortions of 5-point perspective are perfectly maintained. The flat paintings on the disks are now reproduced as silk-screens and the process is put into reverse. The silk-screens are printed onto plastic disks which are returned to the oven to be blown back into hemispheres which are then glued together to be reborn as Termespheres. On the other hand, the creation of a *one-of-a-kind* Termesphere may

ure 3.) Squeezing inside the cylinder and looking around produces the truest view. In fact, you don't need to be sitting in the center of the room: you can be located anywhere within the room and the perspective grid will still work.

Confused? Try the wonderful training exercises, abundantly illustrated, in Termes's book, *New Perspective Systems.* Now the plot thickens. To understand 5-point perspective, think of looking directly through a transparent hemisphere which has been formed by slicing a sphere vertically. Facing into it, paint the world around you onto the inside of the hemisphere. Observe the vanishing points east and west, zenith and nadir, and north—the points to your left and right, above and below you, and straight in front of you. Stand back, so the hemisphere looks flat like a disk, and you will find the center of the disk (which was straight ahead) is in fact the north vanishing point. This is 5-point perspective. (See Figure 4.)

Perhaps you now see 6-point perspective quite literally coming around the corner. Just add the vanishing point to the south, that is, behind you. You now see two disk-like drawings—one for the view ahead, and one for the view behind you. Or if you are Dick Termes, you turn around and around and reproduce the view onto a sphere. Now any couches or television cabinets in a room that had disappeared at the edges of disks in 5-point perspective, continue on. Of course, you are seeing the picture from *outside* the sphere. But to Termes, translating a concave view (from inside the sphere) to a convex one (on the outside) is a simple and natural exercise, one which definitely adds to our interest. Incidentally, if these six vanishing points are making you think of the six faces of the cube or the six vertices of its dual, the octahedron, your thinking is on the right track.

Is this life in a fishbowl? Well, yes. Termes has painted life from the inside of a fishbowl in his painting *Fishbowl.* (See page 185.) Fish are swimming around you, and you have a fisheye (this

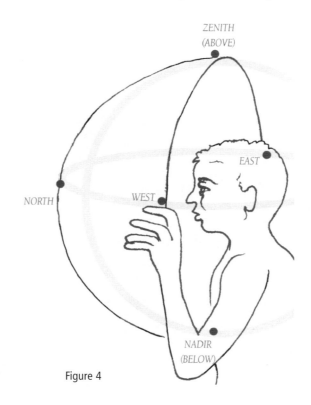

Figure 4

Dick Termes at work on "Platonic Relationships."

take as long as a year.

Is there life beyond 6-point perspective? Yes, says Termes. For example, frequently in a large building like the Notre Dame Cathedral in Paris, the tiling of the floors is diagonal to the main lines of the building. Add a chair or a table "catty-wompus," and you introduce more vanishing points. Edges of objects all lie on great circles; unfortunately, too many of these additions create "an ugly picture." Termes believes that order makes things beautiful, and our eyes can take only so much in the way of disorder.

The largest Termesphere was commissioned by the Law Enforcement Academy in Douglas, Wyoming. This 7-1/2 foot diameter ball was originally destined to become a giant orange, rotating Union 76 gas station sign before it was hijacked by Termes in its transparent state. He explains his painting in this way "You have orderly and disorderly people, and they must live together." Two groups populate the sphere: orderly people, recognizable by their conical hats, *construct* polyhedra and disorderly people in inverted conical hats *chip them away*. This outdoor Termesphere is painted from the inside and like its genetic parent, the Union 76 sign, it lights up and rotates.

Looking for Order is the name of one of the latest Termespheres. This sphere is illustrated with portraits of Einstein at different ages of his life. Reflecting on Einstein's quote that if you could look far enough in one direction you could see the back of your head, one of the artist's many translucent cubes in the portrait emanates from Einstein's eye, circles the sphere, and returns. This rotating piece, which dedicates itself to dimension and time, might have received a nod from Einstein himself. Or perhaps it *is* a nod from Einstein?

One feature of Termespheres is their ability to inspire a good deal of contemplative study as we view our world literally turned inside out. The creation *North is South* repeats the objects on the north and south walls of a room, but the view on east and west walls reveals a subtle difference: a cat seen on the east wall appears at a different angle and size on the west wall. Other, transparent Termespheres allow one to obtain a view inside and outside and ...*through* space. Termes had been unaware of the great perspective artist M.C. Escher until graduate school. Escher once painted himself reflected in a mirrored ball; in admiration of that self-portrait, Termes created a sphere from the point of view of the ball, detailing Escher in his surrounding room!

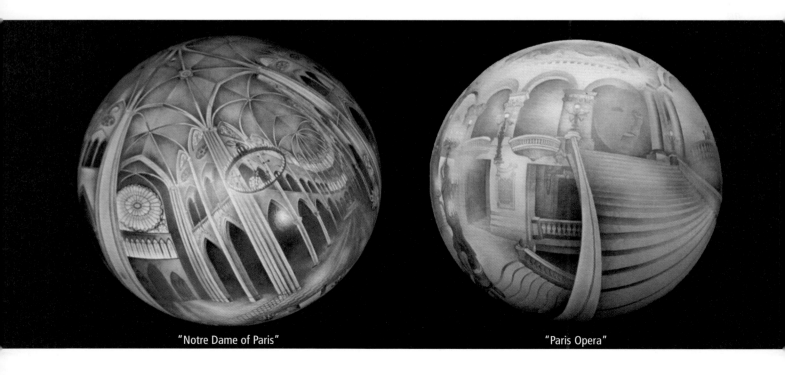

"Notre Dame of Paris"   "Paris Opera"

"When North is South and East is West"

Dick Termes likes nothing better than to get up in the morning and wander about his studio, studying polyhedral structures, reading books on art and geometry, and contemplating the visual world of nature. His love of explaining puts him in demand for lectures and workshops, for example, with the Lakota-Sioux children in South Dakota. Using fondue sticks and styrofoam balls or numerous paper cutouts of triangles and pentagons melded together with tape, the children figure out the five Platonic solids in about one hour and then move on to Archimedean solids. With a new set of toy magnets his wife bought him, Termes has lately been reveling in the beauty of patterns which no mathematician's geometry book has seen. After all, he comments, "mathematics is about patterns, and it's as beautiful as any art you've ever seen." It's no wonder that Termespheres have caught the eye of so many people, from a Jungian society interested in dream worlds, to international art museums, and lately, to the world of mathematicians. Termes presents to amazed audiences at meetings of the Mathematical Association of America. It's no wonder he has a following of students who are charmed by his pleasant and relaxed manner and by his willingness to teach us how to see the world—and life in it —from entirely new perspectives. ■

If you would like to turn your own world inside out, check out www.termesphere.com.

"Fishbowl"

The Edge of the Universe

Noneuclidean Wallpaper

FRANK A. FARRIS
Santa Clara University

Perhaps, like me, you heard the following argument as a child on the playground: "The universe could not possibly have an edge, because if it did you could go there and put your hand through, and that new place would have to be part of the universe too."

If only I had known hyperbolic geometry, I might have refuted this seemingly unassailable argument, in the following manner: "What if you, and all the matter that makes up your measuring instruments, shrink as you approach the edge, and shrink in such a way that you could put your ruler end to end infinitely many

There is nothing below the line!

Figure 1

times and still never reach the edge? After all, what evidence do you have that you do not change size as you move about the world?"

A simple mathematical model of a two-dimensional universe, called the *Poincaré Upper Halfplane*, illustrates the possibility of a universe with an unattainable edge. In this article, I describe this model —a famous example of a noneuclidean geometry—and explain how conversations with an analytic number theorist led me to create wallpaper patterns for its inhabitants. These are interesting not only for their high "Gee whiz!" factor, but also as a window for observing the features of this unusual geometry.

The World of the Shrinking Ruler

To describe the unusual universe we will study in this article, we must stand outside it. Imagine yourself looking down on the ordinary Cartesian plane; the model world consists of those points that lie above the *x*-axis, as if points on or below the *x*-axis have been declared off-limits to inhabitants of our model world, henceforth dubbed the Poincarites. In fact, for the Poincarites, that axis is infinitely far from any point. From our omniscient point of view, the inhabitants' rulers shrink in a particular way as they approach the *x*-axis.

I will later give a precise mathematical rule to describe how this shrinking occurs, but first let us consider how the inhabitants of this world travel if they attempt to move along a straight path. Because physical particles travel in the straightest way possible, in the absence of other forces, understanding how to move without curving is essential to life as a Poincarite.

Any vertical line in the plane, or at least the portion of it that belongs to our new universe, is straight for two reasons. First, the strange change in the size of matter occurs only when moving up and down, not side to side; therefore, if a Poincarite walks up this line with hands out on either side, each hand travels exactly the same distance—a reasonable criterion for straightness. Second, observe that transforming this universe by flipping the plane about that line does nothing to change the size of any measurements, and doing so leaves that line invariant: if the straightest path diverged to the left then, by symmetry, it would also have to diverge to the right; presumably, there is only one way to go straight in any given direction, so that vertical line must be straight.

Heading in a horizontal direction, what path would be the straightest? If a Poincarite maintains a fixed *y*-coordinate and walks to the left, the hand with the lower *y*-coordinate travels farther than the other hand,

Reprinted from September 2001, pp. 16–23

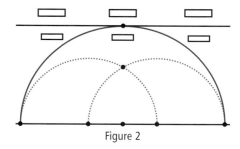

Figure 2

since that lower path is measured with shrunken rulers. Such a person is actually curving to the left! It turns out (we still have not given any rigorous definitions here) that the straightest path for the Poincarite who begins in this direction is a portion of a Euclidean circle whose center is on the *x*-axis, which is the edge of the universe.

The proof is beyond the scope of our discussion, but it turns out that these straightest paths also give the shortest way to connect any two points. Also, between any two points of the Poincarites' world, there is a unique hyperbolic line connecting them.

The space we have described is called the Poincaré Upper Halfplane or the hyperbolic plane. These straightest paths, whether vertical lines or portions of Eu-

clidean circles that meet the boundary at right angles, are called hyperbolic lines. This conceptual universe played an important role in the historical development of noneuclidean geometry. All the axioms of Euclidean geometry are satisfied, except the crucial parallel postulate (can you see how the lines in Figure 2 contradict Euclidean assumptions?). Therefore, this model shows that postulate to be truly independent of the rest of Euclid's system. There are higher dimensional hyperbolic spaces, but we will stick with the two-dimensional version.

To construct our wallpaper it's crucial to understand the *isometries* of the Poincaré Upper Halfplane, that is, the transformations that leave all measurements unchanged. We assume an intuitive familiarity with the Euclidean isometries of translation, reflection, and rotation, and proceed to study the hyperbolic analogues of these.

Reflections

The reflection noted above, flipping the plane about a vertical line, gives a simple example of a hyperbolic isometry. In

Cartesian coordinates, assuming that the vertical line in the figure is the *y*-axis, it would be expressed by the equation $F(x, y) = (-x, y)$. Since this transformation does nothing to change any measurements of figures, this is an isometry of the hyperbolic plane.

What about reflections across other hyperbolic lines? In order to fit with physicists' ideas about empty space, we demand a property of *homogeneity*; space should look essentially the same everywhere, and therefore all hyperbolic lines should behave in the same way. In particular, the two sides of any hyperbolic line should be interchangeable.

Reflection about nonvertical hyperbolic lines is accomplished by a classical process called inversion in a circle, which you may have studied in other contexts. This is a beautiful way to interchange the inside and outside of a circle, leaving points on the circle fixed. If *P* is any point other than the center of the circle, which we'll call *C*, then the image of *P* is the point *P'* on the ray from *C* through *P* so that the product of the distances *CP* and *CP'* is the square of the radius of the circle. Figure 4 illustrates inversion in a

Figure 3. Euclidean wallpaper with many symmetries

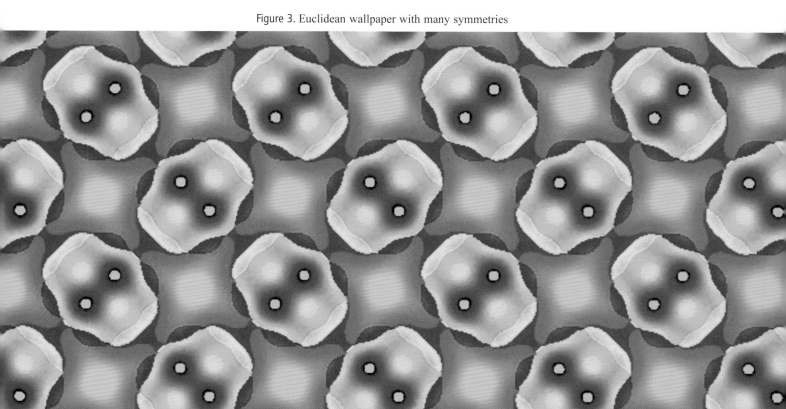

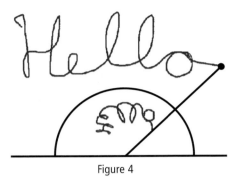

Figure 4

particularly simple semicircle (hyperbolic straight line). The formula for the illustrated inversion is

$$I(x,y) = \left(\frac{x}{x^2 + y^2}, \frac{y}{x^2 + y^2} \right).$$

Notice that in the figure, orientation has been reversed; for this reason, reflections are called *indirect isometries*. If we're willing to use complex coordinates for the Upper Halfplane, then the formula for $I(z)$ becomes appealingly simple

$$I(z) = 1/\overline{z}.$$

If you want to look ahead to Figure 5, you will see a pattern invariant under reflections across certain vertical lines. It takes some imagination to see, but this one is also invariant under the inversion described above. It might help to know that the peacock fans that seem to be largest—for the Poincarites they are all exactly the same horizontal distance across—touch the x-axis at integer points.

Translations

There are two analogues of translations in the hyperbolic plane. The first is a rather obvious shift to the right or left, given in coordinates by

$$P(x, y) = (x + a, y) \text{ or } P(z) = z + a,$$

where a is any real number. Since we don't move anything up or down, P preserves all distances.

The next translation analogue is somewhat suprising, and brings me to say exactly what we mean by distances measured by a shrinking ruler. If we dilate the plane relative to any point on the x-axis, hyperbolic distances, the ones measured by shrinking rulers, are preserved. Suppose $\gamma(t) = (x(t), y(t))$, for $a \le t \le b$ is any parametric curve with $y(t) > 0$. Its hyperbolic length, that is, its length as measured by the Poincarites, is defined to be

$$L = \int_a^b \frac{\sqrt{\left(\frac{dx}{dt}\right)^2 + \left(\frac{dy}{dt}\right)^2}}{y} dt.$$

This is quite similar to the usual formula for arc length, except that the factor of y in the denominator causes an apparently

short piece of arc to count as large in the integral when it is close to the edge. This captures the essence of the shrinking ruler.

Now consider the transformation

$$H(x, y) = (rx, ry) \text{ or } H(z) = rz.$$

Apply this transformation to the curve $\gamma(t)$ and use the integral definition to compute the length of the new curve. It is easy to see that a factor of r cancels from numerator and denominator, leaving the length unchanged; thus, H is an isometry.

Why is H analogous to a translation? Note that the entire y-axis moves along itself under H. Of course, in a Euclidean translation, an entire family of parallel lines slides along itself, but that is not how things work in the hyperbolic plane. We must be content to slide along a single line at a time. While we are on this subject, it is interesting to note that the translations to the right and left, denoted by P above, are analogous to Euclidean translation in that they shift across a family of lines, in this case the family of vertical lines. Surprisingly, there is no line moved along itself by that type of translation.

Rotations and the Rest

To find an example of a rotation, simply compose the two reflections above. Check for yourself that following F by I gives

$$R(x,y) = \left(\frac{-x}{x^2 + y^2}, \frac{y}{x^2 + y^2} \right) \text{ or } R(z) = -\frac{1}{z},$$

which is a hyperbolic rotation of 180 degrees about the point $(0, 1)$.

Again looking ahead to Figure 5, that image is invariant under this half-turn. The fixed point is not labelled, but you can find it in the pale pink area atop one of the large peacock fans. All the noneuclidean wallpapers shown also turn out to be invariant under the transformation

$$R_3(z) = \frac{z-1}{z},$$

which is a rotation through 120° about the point $z = \frac{1}{2} + i\frac{\sqrt{3}}{2}$. Figures 9 and 10 are good places to observe these rotations. In Figure 9, there are centers of two-fold rotation at the points where two lines cross, and centers of three-fold rotation at points where three lines cross.

The collection of isometries of the hyperbolic plane presented so far turns out to be representative of all possibilities. To investigate the totality of these isometries, let us focus on half of them, the set of direct isometries. If you have a feel for the operation of conjugation in the complex plane, you might guess that the formula for any direct isometry will involve only appearances of the variable z, with no $\overline{z}$s required. This is in fact the case.

Using the fact that the composition of two isometries is again an isometry, and looking at the types we have seen so far, it will

Figure 5

now be no surprise that the most general direct isometry of the hyperbolic plane looks like

$$\gamma(z) = \frac{az+b}{cz+d},$$

where a, b, c, and d are real numbers. These are called *fractional linear transformations*.

It takes some algebraic manipulation, but it is not too hard to show that this function takes points with $y > 0$ to other points in the Poincaré Upper Halfplane only when $ad - bc > 0$. Furthermore, note that if $ad - bc = 0$, the numerator and demoninator would have a common factor and γ would degenerate to a constant function, not a candidate for a transformation at all. Suppose we multiply all these coefficients by the same factor; it could be cancelled from numerator and denominator, resulting in the same transformation. Therefore, to avoid redundancies, we assume $ad - bc = 1$.

A helpful shorthand uses a 2 × 2 matrix to keep track of these fractional linear transformations:

$$\gamma(z) = \frac{az+b}{cz+d} = \begin{pmatrix} a & b \\ c & d \end{pmatrix} \circ z.$$

It should be considered a minor miracle that composition of functions corresponds exactly to multiplication of matrices. Try it!

We have now identified the set of direct isometries of the hyperbolic plane with a set of 2 × 2 matrices. If you know a little about groups, you can easily see that this set of matrices forms a group, which is usually called $SL(2, \mathbb{R})$. It is actually subtly different from the group of direct isometries of the hyperbolic plane, in that

$$\begin{pmatrix} a & b \\ c & d \end{pmatrix} \circ z = \begin{pmatrix} -a & -b \\ -c & -d \end{pmatrix} \circ z,$$

so that two different matrices give the same transformation. We leave that distinction to the experts.

Creating Symmetry

As a basic example, consider a process for creating an even function of one variable, that is, a function $f(x)$ invariant with respect to reflection of the real line about the origin:

$$f(-x) = f(x).$$

Given any base function $g(x)$, we can *symmetrize* $g(x)$ to create a new function $f(x)$ by a process of averaging:

$$f(x) = \frac{g(x) + g(-x)}{2}.$$

Clearly, the new function is even. Of course, it is possible that the new function is identically zero, but it is indeed even.

For another easy example, suppose we wanted to create a function of two variables $f(x, y)$ that is invariant under rotation of 120 degrees about the origin. Again, a process of averaging

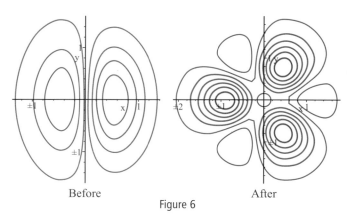

Before   After

Figure 6

may be used, but this time there are three things to be averaged. Suppose $g(x, y)$ is any function of two variables, enjoying whatever properties of continuity or differentiability we wish to impose, and suppose ρ represents this rotation. Define a new symmetrized function by

$$f(x,y) = \frac{g(x,y) + g(\rho(x,y)) + g(\rho^2(x,y))}{3}.$$

Check for yourself that $f(\rho(x, y)) = f(x, y)$, so that f is indeed invariant under the desired rotation. The reason for dividing by three is to make the new function have values in about the same

range as the old. Ponder the level curves in the "before and after" example in Figure 6, where $g(x, y) = (x^2 + 3x)e^{-(2x^2+y^2)}$.

This gives you some idea of a process called *averaging a function over a group action*. In the rotational example, the group consisted of three elements, e, ρ, ρ^2, where e is the identity transformation.

In making wallpaper for the Poincarites, my method was to look for functions that are invariant under a particular set of the isometries described above. Because it is a set much beloved of analytic number theorists, I chose the set of fractional linear transformations where all the coefficients are integers. This set, which also meets the requirements to be a group, is called

$$SL(2,\mathbb{Z}) = \left\{ \begin{pmatrix} a & b \\ c & d \end{pmatrix} \text{ where } a,b,c,d \in \mathbb{Z} \text{ and } ad - bc = 1 \right\}.$$

Surprisingly, my next step amounted (almost) to taking a base function on the plane and averaging it over this infinite group!

But before I describe that averaging, let us imagine what we expect to see. In the swatch of Euclidean wallpaper shown in Figure 3, we can apply a large collection of Euclidean symmetries, reflections, rotations, and translations to the picture and find the pattern left unchanged; of course, we need to imagine that it continues infinitely in all directions, that what we are seeing is

Figure 7

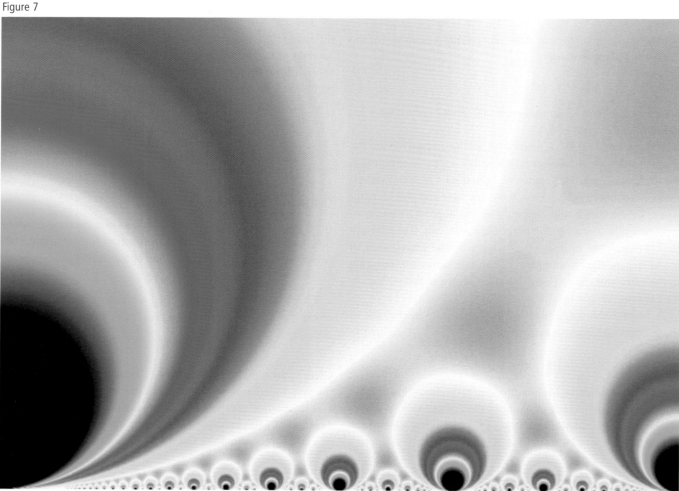

THE EDGE OF THE UNIVERSE

a piece of an infinite pattern. For hyperbolic wallpaper of the type to be constructed, we should be able to apply any of the transformations in $SL(2, \mathbb{Z})$ and find the pattern unchanged.

In particular, consider two families of transformations. The simplest translations we discussed, $P(z) = z + b$, are indeed in $SL(2, \mathbb{Z})$ when b is an integer. (Use $a = d = 1$ and $c = 0$.) Thus, our picture should repeat itself with every unit translation to the right or left; that is reasonably easy to imagine. Call this collection of translations Γ.

This family of transformations has a sort of mirror image, in the set

$$\Gamma' = \left\{ \begin{pmatrix} 1 & 0 \\ n & 1 \end{pmatrix} \text{ where } n \in \mathbb{Z} \right\}.$$

To study a typical element, $\gamma(z) = \frac{1}{nz+1}$, we compute two limits:

$$\lim_{z \to \infty} \gamma(z) = 0$$

$$\lim_{z \to -\frac{1}{n}} \gamma(z) = \infty \ .$$

If the picture is supposed to look the same before and after we apply this transformation, then the behavior as z gets very large should look the same (to the Poincarites, of course) as when z approaches 0 and when z approaches $-1/n$. Of course, z cannot actually *be* 0, or ∞, or $-1/n$, as none of these is a point in our new universe, but the picture should look the same as you approach

any of these points. With some modification, the same argument shows that the pattern must look the same as we approach any rational value on the *x*-axis. Before the first computed image appeared on my screen, I found it hard to imagine such a thing.

Constructing Wallpaper

The group $SL(2, \mathbb{Z})$ is actually too large to perform the averaging we described. To tell the real story, I need to use the language of *cosets* from group theory. If you want to skip this section, imagining the process as one of averaging is a good heuristic.

Recall that our goal is to take some base function $g(z)$ and, by some sort of averaging, produce a new function $f(z)$ which is invariant under every transformation of $SL(2, \mathbb{Z})$. Any function with all those symmetries is called a *modular function*. Modular functions and analogous objects called modular forms are well known to number theorists, so Jeffrey Hoffstein of Brown University was a natural person to ask for help. I told him that I had taken a stab at the construction using naive group averaging, and he showed me how to do it more cleverly. The result was a method I used to make the pictures in this article.

The naive idea would be to start with $g(z)$ and form

$$\sum_{\gamma \in SL(2, \mathbb{Z})} g(\gamma z)$$

Figure 8

hoping that the sum would converge. Unfortunately, it virtually never does. Instead, we start with a special choice of $g(z)$, one that is already invariant under the subgroup, Γ, of integer translations to the left and right. Such a g is easy to invent.

The key idea we need from group theory is that the big group $SL(2, \mathbb{Z})$ can be organized into cosets using the subgroup Γ, where any two elements of a coset differ by an element of Γ. It is a little like organizing the integers into three sets, $\{3n\}$, $\{3n+1\}$, and $\{3n+2\}$, the cosets of the subgroup of integers divisible by 3. To adapt this idea to $SL(2, \mathbb{Z})$, we need to remember that the group operation is matrix multiplication, so γ_1 and γ_2 differ by an element of Γ if $\gamma_1 \gamma_2^{-1} \in \Gamma$.

The following equation shows that any two elements of $SL(2, \mathbb{Z})$ that share the same bottom row belong to the same coset of Γ:

$$\begin{pmatrix} a & b \\ c & d \end{pmatrix} \begin{pmatrix} a' & b' \\ c & d \end{pmatrix}^{-1} = \begin{pmatrix} 1 & a'b - b'a \\ 0 & 1 \end{pmatrix} \in \Gamma.$$

Figure 9

Conversely, multiplying any matrix on the right by an element of Γ does nothing to the bottom row. Thus, each coset of Γ corresponds to a pair of integers c and d. These must be relatively prime (denoted $(c, d) = 1$), as you can see from the equation $ad - bc = 1$.

We now should be able to perform some averaging, but there is one missing ingredient: given a pair of integers c and d, we need to find a top row for our matrix. The Euclidean algorithm comes to the rescue here. Siman Wong of the University of Massachusetts, Amherst, who was also at Brown University at the time, wrote a swatch of code to generate a list of relatively prime c, d pairs and one corresponding pair of values for a and b.

Putting all this together, I was ready to write a program to perform the average:

$$\sum_{(c,d)=1} g(\gamma z)$$

where γ is one of the matrices whose bottom row consists of c and d. Note that we are not dividing by the number of elements in the sum. In the first place, we are summing over an infinite number of elements, and in the second place, the sum converges nicely without doing so, provided we make the right choice for g.

Examples

We are now ready to choose some building blocks and carry out the averaging process to produce images. In light of the previous section, the function we choose as the building block

for averaging must remain the same when you translate to the left and right by integer distances. One type of function that fits this bill is a function that does not depend on x at all. (Recall that we are using complex notation, where the point (x, y) corresponds to the complex number $z = x + iy$.) As an elementary building block, take

$$g(z) = y^s,$$

which is certainly invariant under the translations in Γ. An estimate using an integral test for convergence shows that the sum above will converge as long as s is any complex number with $Re(s) > 1$.

A favorite first example uses

$$g(z) = y^{1.5 + 5i} = y^{1.5}(\cos(5 \ln y) + i\sin(5 \ln y)).$$

This is pictured in Figure 5, with a close-up view of the enticingly complicated part of the image in Figure 7. Note that all the shapes that resemble peacock fans are exactly the same hyperbolic distance across, because any one can be taken into any other by one of our isometries, which the Poincarites see as leaving all distances unchanged. Furthermore, there is one of these fans tangent to the x-axis at every rational number. What a lot of room there is, down near the edge of the universe!

It may be startling to see that the function g, and hence the averaged function f, takes on complex values. How can these be pictured? Space constraints demands that I make a long story short. In the study that led to the article "Vibrating Wallpaper,"

Figure 10

I developed a way to visualize complex-valued functions in the plane using the artist's color wheel. (See www.maa.org/ pubs/amm_complements/complex.html.) Every complex number receives a different color, with white at the center and black out toward infinity; hues are distributed in a circle (Figure 8). When you have a complex-valued function on a domain in the plane, you can color each point of that domain using the color corresponding to the output value for that point. For more information, follow links from my web page. Here, suffice it to say that the black portions of the image are places where the value of the function is very large in magnitude; white spots correspond to places where the function is near zero.

More exciting images are produced using $g(z) = y^s \sin(n\pi x)$, as in Figure 9, or $g(z) = y^s \cos(n\pi x)$. To achieve the required translational invariance, n must be an integer. In this picture, which uses the sine function as a building block, notice the white areas; since $\sin(n\pi) = 0$, this function is zero on a grid of hyperbolic lines. Since $\sin(-x) = -\sin(x)$, it also has an antisymmetry about the y-axis.

It opens a rather enjoyable can of worms to realize that one can superimpose these fundamental building blocks, to produce infinite variations on these pictures. Figure 10 uses a base function that superimposes functions like $y^s \sin(n\pi x)$ and $y^s \cos(n\pi x)$; any linear combination will do. I experimented until I was pleased with the result.

Where To Go From Here

Speedy computers, color monitors, the world wide web, all these give us tools for creating and sharing images that use color to illustrate mathematical ideas in a way not possible even ten years ago, when the computational power necessary to produce the images in this article was simply unavailable. With what you have seen in this article, I hope you will be inspired to create some images of your own. If you want to compute modular functions, and explore the endless variety of possibilities, I have outlined all the steps; there are also animations waiting to be made, showing these wallpapers in vibration. Alternatively, it would be great to see a video game where objects bounce along trajectories that follow hyperbolic lines; one thing to be overcome in that scenario is that a random walk in the hyperbolic plane almost certainly results in your getting lost in that expansive place that the external viewer sees as being down near the edge of the universe.

Tristan Needham's book *Visual Complex Analysis*, Clarendon Press, is an excellent place to learn about fractional linear transformations. To experience the Poincaré Upper Halfplane for yourself, you can use NonEuclid, Java simulation software developed by Joel Castellanos at Rice University. (See cs.unm.edu/~joel/NonEuclid.) With so many possibilities for visualization, the time to study this noneuclidean geometry and explore the edge of the universe is now. ∎

Alfred Bray
Kempe's "Proof"
of the
Four-Color Theorem

TIMOTHY SIPKA
Alma College

Twenty-five years have passed since Wolfgang Haken and Kenneth Appel provided the mathematics community with a proof of the well-known theorem that any map on a plane or surface of a sphere can be colored with at most four colors so that no two adjacent countries have the same color. Their conquest of the four-color theorem came almost a century after the world had accepted the first "proof" of the theorem. In 1879, Alfred B. Kempe published what he and the mathematics community thought was a proof of the four-color theorem. Unfortunately for Kempe, eleven years later P. J. Heawood discovered a flaw. This article will take a close look at Kempe's attempt to prove the four-color theorem. In addition, we will discuss the conjecture's origin and consider Heawood's counterexample that exposed the flaw in Kempe's work.

Origin of the Conjecture

It has been conjectured that early map-makers were the first to notice that four colors would suffice when coloring a map. This claim, though logical and tempting to make, has little evidence to support it. In the early 1960's, Kenneth May reviewed a sample of atlases in the large collection of the Library of Congress and found no tendency by mapmakers to minimize the number of colors. May found very few maps that used only four colors, and those that did usually required only three colors. May concluded, "if cartographers are aware of the four-color conjecture, they have certainly kept the secret well."

So when did the four-color conjecture actually arise? Some believe that Francis Guthrie, a British mathematician, was the first person to make the conjecture. In fact, May believed the conjecture "flashed across the mind of Guthrie while he was coloring a map of England." We do know that in 1852 Francis had a conversation with his brother Frederick in which he stated and attempted a proof of the conjecture. Frederick mentions that Francis "showed me the fact that the greatest necessary number of colours to be used in colouring a map so as to avoid identity of colour in lineally contiguous districts is four." Frederick went on to mention that the proof Francis gave "did not seem altogether satisfactory to himself," which probably explains why Francis never published it. Soon after this conversation, Frederick shared the conjecture with Augustus De Morgan, his mathematics professor at University College London.

Sharing the conjecture with De Morgan was a stroke of luck for the mathematics community. De Morgan immediately began to make inquiries about the problem. In a letter to William R. Hamilton, dated October 23, 1852, we have the first written reference to the conjecture. In that letter, De Morgan asked whether Hamilton had heard of the conjecture. Hamilton promptly replied that he had not, and that he would not likely attempt the problem. We know of two other letters from De Morgan in which he discussed the four-color conjecture. In the first letter, dated December 9, 1853, De Morgan wrote to his former teacher, William Whewell; and in the second, dated June 24, 1854, he wrote to Robert Leslie Ellis.

In addition to spreading news of the conjecture through letters and conversations, De Morgan was also responsible for writing the first article that referred to the conjecture. In a book review of Whewell's *The Philosophy of Discovery*, published in the April 14, 1860 issue of the *Athenaeum*, De Morgan included a paragraph that described the four-color conjecture. The paragraph also contained the comment, "it must have been always known to map-colorers that four different colors are enough" which, quite possibly, initiated the tradition of

Reprinted from November 2002, pp. 21–23, 26

linking the four-color conjecture with cartographers.

Between 1860 and 1878, interest in the four-color problem appeared to wane, and according to Rudolf and Gerda Fritsch, "the problem was not discussed anywhere in the mathematical literature of the time." However, at a meeting of the London Mathematical Society on June 13, 1878, Arthur Cayley appeared to revive interest in the problem when he inquired whether anyone had proved the conjecture. He then published a short paper in which he gave a precise statement of the conjecture and explained where the difficulty was in proving it.

It was not long after Cayley's inquiry that Alfred Bray Kempe, a London barrister and former student of Cayley's, arrived at his now famous and fallacious proof of the four-color conjecture. News of his "proof" was announced in the July 17, 1879 issue of *Nature*, while the full text appeared shortly thereafter in the recently founded *American Journal of Mathematics*.

The "Proof"

There are two observations that should be made when one reads Kempe's paper, observations that may explain why the subtle error in his argument went undetected for eleven years. First, all of his diagrams (there are 16) are relatively simple, and most of them are used to provide examples of the terms he defines. He never provides a nontrivial diagram (map) that demonstrates his argument. Second, the paper is virtually all prose which, though well written, makes it difficult to verify his work.

Though the phrase "mathematical induction" was never mentioned in Kempe's paper, the "patching process" he used made his argument essentially a proof by mathematical induction. Therefore, in presenting Kempe's argument, we will use his vocabulary and basic ideas, but we'll give a more contemporary version of his proof.

As with most induction proofs, the base step is quite obvious: any map containing four or fewer countries can easily be colored with at most four colors. Now, assume that any map containing n countries can be colored with at most four colors, and then let M be a map consisting of $n + 1$ countries. It can then be shown—and Kempe did so—that M must contain at least one country that is adjacent to five or fewer other countries. Let X denote such a country in M; then temporarily disregard X. We are left with a map of n countries, which we'll denote by M-X. Now, color the countries of M-X with at most four colors. Let's use red, blue, green, and yellow as Kempe did.

Kempe actually said "take a piece of paper and cut it out to the same shape" as the country X, and then "fasten this patch to the surface and produce all the boundaries which meet the patch to meet at a point within the patch." In other words, Kempe described a process that physically removed the country X and extended the boundaries of the surrounding countries to meet at a point within the region once covered by X.

In the map M-X, we have colored n countries with at most four colors, and we've left X uncolored. Kempe's goal was to find a way to reduce (if necessary) the number of colors used to color the countries surrounding X so that some color would be "free" for X. He quickly dispensed with the easy cases. First, if X is surrounded by three or fewer countries, then clearly there will be a color available for X. Second, if X is surrounded by four or five countries colored with at most three colors, then there will also be a color available for X. With these cases out of the way, Kempe was left with two cases to consider:

Case 1: X is adjacent to exactly four countries colored with four different colors.

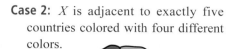

Case 2: X is adjacent to exactly five countries colored with four different colors.

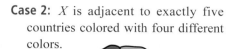

In handling these two cases, Kempe used a technique that today we call "the method of Kempe chains." He first asked that we consider all the countries (he called them districts) in the map which are colored red and green; then he observed that these countries form one or more red-green regions. Kempe's notion of a red-green region was simply a continuous "chain" of countries colored red or green. He then made the important observation that one could interchange the colors in any red-green region, and the map would still remain properly colored. We will now demonstrate, using nontrivial examples, the arguments Kempe gave for the two cases.

For case 1, we first label the countries surrounding X with the letters A, B, C, and D. Kempe then considered two subcases.

Subcase 1.1: Suppose countries A and C belong to different red-green regions.

In Figure 1 we have an example of a map in which countries A and C belong to different red-green regions. In this situation, Kempe observed that "we can interchange the colours of the districts in one of these regions, and the result will be that the districts A and C will be of the same colour, both red or both green." By interchanging the colors in the region containing A, we see in Figure 2 that both A and C are now green, making the color red available for X.

Figure 1

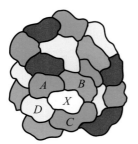

Figure 2

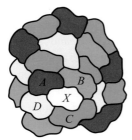

Figure 3

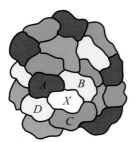

Figure 4

Figure 5

Figure 6

Subcase 1.2: Suppose countries A and C belong to the same red-green region.

In Figure 3 we have an example of a map in which countries A and C belong to the same red-green region. Kempe observed in this case that the red-green region will "form a ring" preventing B and D from belonging to the same blue-yellow region. Therefore, by interchanging the colors in exactly one of these blue-yellow regions, we reduce to three the number of colors surrounding X. In Figure 4 we have interchanged the colors in the blue-yellow region containing B, making the color blue available for X.

For case 2, we label the five countries surrounding X with the letters A, B, C, D, and E. Kempe then considered two subcases.

Subcase 2.1: Suppose we have either countries A and C belonging to different red-yellow regions or countries A and D belonging to different red-green regions.

When one of these alternatives is present in a map, we simply perform an interchange of colors similar to the process used in subcase 1.1. In Figure 5 we have an example of a map in which countries A and C belong to different red-yellow regions. Then, as Kempe claimed, "interchanging the colours in either, A and C become both yellow or both red." If we interchange the red and yellow colors in the region containing A, we obtain the coloring in Figure 6, making red available for X.

Subcase 2.2: Suppose countries A and C belong to the same red-yellow region and countries A and D belong to the same red-green region.

In this, the fourth and final case, Kempe's process for reducing the number of colors surrounding X contained a subtle flaw. In Figure 7 we have an example of a map where countries A and C belong to the same red-yellow region and where countries A and D belong to

Figure 7

the same red-green region. In a case such as this, Kempe correctly observed that "the two regions cut off B from E, so that the blue-green region to which B belongs is different from that to which D and E belong, and the blue-yellow region to which E belongs is different from that to which B and C belong." To reduce the number of colors surrounding X, Kempe then made the claim, "interchanging the colours in the blue-green region to which B belongs, and in the blue-yellow region to which D belongs, B becomes green and E yellow, A, C, and D remaining unchanged." In Figure 8, the interchanges of colors have been performed as Kempe described with the outcome he expected, making the color blue available for X.

Figure 8

Heawood's Counterexample

In the example used in subcase 2.2, Kempe's process worked exactly as he had hoped. By simultaneously interchanging the colors in the blue-green region containing B and the blue-yellow region containing E, the number of colors surrounding X was reduced to three. Unfortunately for Kempe, this proces would not work for all maps satisfying the conditions of subcase 2.2.

In 1890, Percy J. Heawood produced a map for which Kempe's process would fail. Heawood's example revealed a subtlety that had escaped detection by the rest of the mathematics community. And that subtlety was the possibility that the blue-green region containing B and the blue-yellow region containing E might "touch." When this happens, Heawood observed, "Either transposition prevents the other from being of any avail."

In Figure 9 we see the map Heawood used to expose the flaw in Kempe's process for reducing the number of colors in subcase 2.2. Notice that the blue-green region containing B and the blue-yellow region containing E share a boundary. If we interchange the colors in both regions,

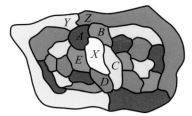

Figure 9

the two countries sharing this boundary, Y and Z, would both receive the color blue. Thus, as Heawood remarked, "Mr. Kempe's proof does not hold unless some modifications can be introduced into it to meet this case of failure."

Kempe certainly tried to fix this "case of failure," but neither he nor any of his contemporaries could do so. The modifications that were needed would require many years of work by many individuals.

Conclusion

The importance of Kempe's work cannot be overlooked. His basic ideas provided the starting point for what would be a century of effort culminating with Appel and Haken's proof. In 1989, as a tribute to Kempe, Appel and Haken declared: "Kempe's argument was extremely clever, and although his "proof" turned out not to be complete, it contained most of the basic ideas that eventually led to the correct proof one century later." ■

For Further Reading

Interested readers will find a detailed history of the four-color problem and a thorough list of the relevant literature in *The Four-Color Theorem: History, Topological Foundations, and Idea of Proof* by Rudolf and Gerda Fritsch.

A Tale Both Shocking and Hyperbolic

DOUGLAS DUNHAM
University of Minnesota, Duluth

Dutch artist M. C. Escher wrote in 1958, upon seeing the pattern of Figure 1, that it "gave me quite a shock." This pattern of curvilinear triangles appeared in a paper by the Canadian geometer H.S.M. Coxeter entitled "Crystal Symmetry and Its Generalizations." Coxeter, and most likely other mathematicians before him, drew such patterns by using classical straightedge-and-compass constructions. Exactly how this was done was a "folk art" until recently when it was explained by Chaim Goodman-Strauss.

To explain how Escher came to be shocked, we go back a few years earlier to the 1954 International Congress of Mathematicians, where Coxeter and Escher first met. This led to friendship and correspondence. A couple of years after their first meeting, Coxeter wrote Escher asking for permission to use some of his striking designs in a paper on symmetry, "Crystal Symmetry and Its Generalizations" (published in the *Transactions of the Royal Society of Canada* in 1957). As a courtesy, Coxeter sent Escher a copy of that paper containing a figure with a hyperbolic tessellation just like that in Figure 1 (in addition to Escher's designs). Escher was quite excited by that figure, since it showed him how to solve a problem that he had wondered about for a long time: how to create a repeating pattern within a limiting circle, so that the basic subpatterns or *motifs* became smaller toward the circular boundary. Escher wrote back to Coxeter telling of his "shock" upon seeing the figure, since it showed him at a glance the solution to his long-standing problem.

Escher, no stranger to straightedge-and-compass constructions, was able to reconstruct the circular arcs in Coxeter's figure. He put these constructions to good use in creating his first circle limit pattern, *Circle Limit I,* which he included with his letter to Coxeter. Figure 2 shows a rough computer rendition of that pattern.

It is easy to see that Figures 1 and 2 are related. Here is how Escher might have created *Circle Limit I* from Figure 1. First switch the colors of half the triangles of Figure 1 so that triangles sharing a hypotenuse have the same color. The result is a pattern of "kites," as shown in Figure 3.

Next, remove small triangular pieces from each of the short sides of the orange kites and paste them onto the long sides. Figure 4 shows the result of doing this for just the central kites. This produces the outlines of the blue fish in *Circle Limit I.* The outlines of the white fish are formed by the holes between the blue fish.

Finally, the pattern of *Circle Limit I* can be reconstructed by filling in the interior details such as the eyes and backbones.

Figure 1. A tessellation of the hyperbolic plane by 30-45-90 triangles.

Reprinted from April 2003, pp. 22–26

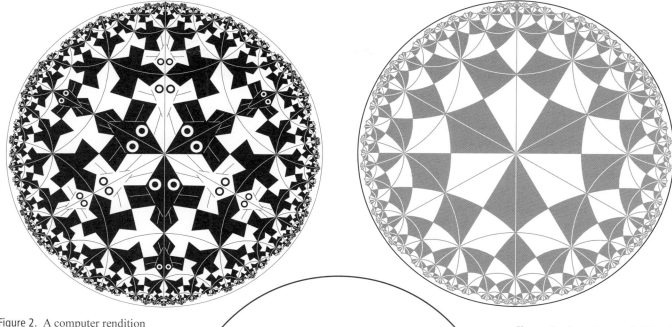

Figure 2. A computer rendition of the pattern in Escher's print *Circle Limit I*.

Figure 3. A pattern of "kites" derived from Figure 1.

A Bit of Hyperbolic Geometry

Mathematicians, and geometers in particular, will recognize the patterns of the figures above as "living in" the Poincaré disk model of hyperbolic geometry. Escher probably knew this, but was not concerned about it since he could use Euclidean constructions to build his patterns. The fact that Poincaré's disk model can be defined in purely Euclidean terms shows that hyperbolic geometry is just as consistent as Euclidean geometry. But that is another story.

The points of the Poincaré disk model of hyperbolic geometry are the interior points of a bounding circle in the Euclidean plane. In this model, hyperbolic lines are represented by circular arcs that are perpendicular to the bounding circle, including diameters. Figures 1 and 2 show examples of these perpendicular circular arcs. Equal hyperbolic distances are represented by ever smaller Euclidean distances as one approaches the bounding circle. For example, all the triangles in Figure 1 are the same hyperbolic size, as are all the blue fish (or white fish) of Figure 2, and the kites of Figure 3. The

Figure 4. The outlines of the central fish formed from the kites of Figure 3.

patterns of Figures 1, 2, and 3 are closely related to the regular hyperbolic tessellation $\{6,4\}$ shown in Figure 5. In general, $\{p,q\}$ denotes the regular tessellation by regular p-sided polygons with q of them meeting at each vertex.

Escher's Criticisms of *Circle Limit I*

Escher had several criticisms of his *Circle Limit I* pattern. First, the fish are "rectilinear," instead of having the curved outlines of real fish. Also, there is no "traffic flow" along the backbone lines - the fish change directions after two fish, and the fish change colors along lines of fish. Another criticism, which Escher didn't make, is the pattern does not have color symmetry since the blue and white fish are not congruent. Before reading further, look back at Figure 2 and try to see why this is true. There are several differences in the shapes of the blue and white fish; the most obvious is the difference in their nose angles.

Some of Escher's criticisms could be overcome by basing the fish pattern on the $\{6,6\}$ tessellation, as shown in Figure 6. In fact, Figure 6 can be recolored in three colors to give it color symmetry, which means that every symmetry (rotation, reflection, etc.) of the uncolored pattern exactly permutes the colors of the fish in the colored pattern. Figure 7 shows that three-colored pattern, which addresses all of Escher's criticisms except for the rectilinearity of the fish.

Escher's Solution: *Circle Limit III*

Escher could have used the methods above to overcome his criticisms, but he didn't. Escher took a different route, which led to his beautiful print *Circle Limit III*. Figure 8 shows an approximate computer-generated version of the *Circle Limit III* pattern.

Escher never publicly explained how he designed *Circle Limit III* but here is how he might have gone about it. From his correspondence with Coxeter, Escher knew that regular hyperbolic tessellations $\{p,q\}$ existed for any p and q satisfying $(p-2)(q-2) > 4$. In particular, he had used the $\{8,3\}$ tessellation as the basis for his second hyperbolic pattern, *Circle Limit II*, and he decided to use that tessellation again for *Circle Limit III*. The $\{8,3\}$ tessellation is shown by the heavy red lines in Figure 9.

Here is one way to get from the $\{8,3\}$ tessellation to *Circle Limit III*. First, connect alternate vertices of the octagons with slightly curved arcs, which are shown as black arcs in Figure 9. This divides up the hyperbolic plane into "squares" and "equilateral" triangles.

Then if we orient the arcs by putting arrowheads on one end, we get the paths of the fish in *Circle Limit III*. This is shown in Figure 10.

A number of years ago when I was trying to figure out how to encode the color symmetry of *Circle Limit III* in my computer program, I drew a pattern of colored arrows as in Figure 10. Later, in 1998, it was my turn to be "shocked" at the Centennial Exhibition of Escher's works when I saw a colored sketch of arrows by Escher just like mine! He had used his drawing in preparation for *Circle Limit III*.

A Bit More Hyperbolic Geometry

It is tempting to guess that the white backbone lines in Figure 8 are hyperbolic lines (i.e. circular arcs perpendicular to the bounding circle). But careful measurements of *Cir-*

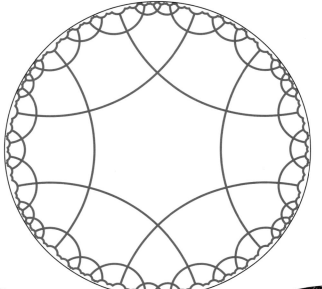

Figure 5. The regular tessellation $\{6,4\}$ of the hyperbolic plane.

Figure 6. A pattern of rectilinear fish based on the $\{6,6\}$ tessellation.   Figure 7. A pattern of rectilinear fish with 3-color symmetry.

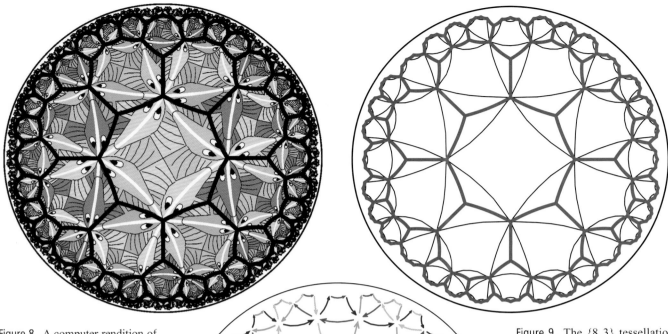

Figure 8. A computer rendition of the *Circle Limit III* pattern and the {8, 3} tessellation (black arcs) upon which it is based.

Figure 9. The {8,3} tessellation (heavy red lines) together with "squares" and triangles (black lines).

Figure 10. The paths of the fish in *Circle Limit III*.

cle Limit III show that all the white arcs make angles of approximately 80 degrees with the bounding circle. This is as it should be, since the backbone arcs are not hyperbolic lines, but equidistant curves, each point of which is an equal hyperbolic distance from a hyperbolic line.

In the Poincaré model, equidistant curves are represented by circular arcs that intersect the bounding circle in acute (or obtuse) angles. Points on such arcs are an equal hyperbolic distance from the hyperbolic line with the same endpoints on the bounding circle. For any acute angle and hyperbolic line, there are two equidistant curves ("branches"), one on each side of the line, making that angle with the bounding circle. Equidistant curves are the hyperbolic analog of small circles in spherical geometry. For example, every point on a small circle of latitude is an equal distance from the equatorial great circle; and there is another small circle in the opposite hemisphere the same distance from the equator.

Each of the backbone arcs in *Circle Limit III* makes the same angle Ω with the bounding circle. Coxeter used hyper-

bolic trigonometry to show that Ω is given by the following expression:

$$\cos\Omega = \sqrt{\frac{3\sqrt{2}-4}{8}}$$

The value of Ω is about 79.97 degrees, which Escher accurately constructed to high precision.

The 2003 MAM Poster Pattern

Much as Escher was inspired by Coxeter's figure, I was inspired by Escher's "Circle Limit" patterns to create a program that could draw them. More than 20 years ago two students, David Witte and John Lindgren, and I succeeded in writing such a program. Having gone to all the trouble to design a program that was more general than we needed to accomplish our goal, we put it to other uses. *Circle Limit III* is certainly Escher's most stunning hyperbolic pattern, so we thought it would be interesting to find related patterns.

Here is my analysis of *Circle Limit III* fish patterns: one can imagine a three parameter family (k, l, m) in which k right fins, l left fins, and m noses meet, where m must be odd so that the fish swim head to tail. The pattern would be hyperbolic, Euclidean, or spherical depending on whether $1/k + 1/l + 1/m$ is less than, equal to, or greater than 1. *Circle Limit III* would be denoted $(4, 3, 3)$ in this system. Escher created a Euclidean pattern in this family, his notebook drawing number 123, denoted $(3, 3, 3)$, in which each fish swims in one of three directions. The pattern on the 2003 Math Awareness Month poster is $(5, 3, 3)$ in this system, and is shown in Figure 11.

Summary

Over a period of five decades, a series of mathematical inspirations and "shocks" have led from Coxeter's figure to the 2003 Math Awareness Month poster. Many people have been inspired by Escher's work, including the authors of articles in the recent book *M.C. Escher's Legacy*. My article and electronic file on the CDRom that accompanies that book contain a number of other examples of computer-generated hyperbolic tessellations inspired by Escher's art. I only hope that the reader has as many enjoyable inspirations and "shocks" in his or her mathematical investigations. ■

For Further Reading

There are illuminating quotes from Escher's correspondence with H. S. M. Coxeter in Coxeter's paper "The non-Euclidean symmetry of Escher's Picture 'Circle Limit III,'" *Leonardo* 12 (1979), 19–25, 32, which also shows Coxeter's calculation of the angle of intersection of the white arcs with the bounding circle in *Circle Limit III*. Read about artists who have been inspired by Escher and are currently creating new mathematical "Escher" art in the book *M. C. Escher's Legacy: A Centennial Celebration*, Doris Schattschneider and Michele Emmer, editors, Springer Verlag, 2003. *Euclidean and Non-Euclidean Geometries*, 3rd Edition, Marvin Greenberg, W. H. Freeman and Co., 1993, has a good account of the history of hyperbolic geometry and the Poincaré disk model. If you want to construct your own hyperbolic tessellation by classical methods, see "Compass and straightedge in the Poincaré disk," Chaim Goodman-Strauss, *American Mathematical Monthly*, 108 (2001), no. 1, 38–49; to do it by computer, see "Hyperbolic symmetry," Douglas Dunham, *Computers and Mathematics with Applications*, Part B 12 (1986), no. 1–2, 139–153.

Acknowledgement

I would like to thank Doris Schattschneider for her considerable help, especially with the history of Escher and Coxeter's correspondence.

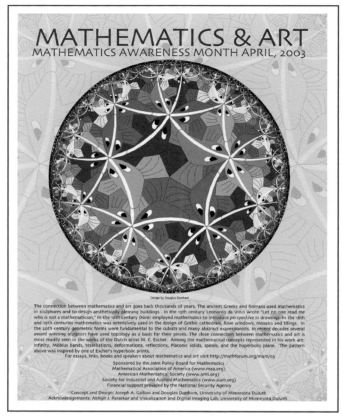

Figure 11. The Math Awareness Month poster.

Symbols of Power

STEPHEN KENNEDY
Carleton College

Visitors to the Centre d'Art Georges Pompidou in Paris who wander into the gallery where Bernar Venet's paintings hang encounter something more shocking than a crucifix in a jar of urine, more horrifying than an animal dung Madonna — mathematical equations. Modern artists have gone to such extremes in defiling venerated cultural icons that doing so is losing the power to rile us up. Now here comes Bernar Venet with a wholly different message, the symbols of mathematics are relevant to modern art, and museums in Brazil and Paris and New York are taking him seriously enough to hang his work. I don't know how the average museum-goer, or the serious avant-garde art aficionado, reacts to Venet's work, but as a mathematician I knew I certainly had a lot of questions.

Who *is* this Guy?

Bernar Venet was born in the Alps in the south of France in 1941. He committed himself to art at a very young age and had paintings on exhibit in Paris when he was eleven. At age fourteen he had his first one-man show and by twenty-two was hanging in the Museum of Modern Art in Paris. His work at this time was already very conceptual and with hindsight we can see a nascent interest in mathematics and science: lengths of unadorned cardboard or plastic tubes; large canvases uniformly covered with a layer of tar; canvases divided into rectangles and a schedule designed (by artist or buyer) that specifies which rectangle is to be painted which color, at the rate of one rectangle per month. (The program to produce the painting is the artwork as much as the final painting itself.)

What's he up to? The artist and critic Thierry Kuntzel explains:

> [T]he search [is] for a kind of *painting degree zero,* the most obvious characteristic of which would be its "absence of style." ... [T]he work has no *subject,* the symbolic codes are thwarted; it has no style, the *aesthetic* codes are immobilized; it is *literal,* the only possible reading being a denotative one.

By the late sixties Venet had firmly grabbed onto mathematics and science as his artistic media. It was his way of removing style and emotion, he was pursuing "the rational image" as opposed to the expressive one. "Look," he tells me, "it's not about *expression.* Everybody's always expressing himself, you do, I do, all the time, by being angry, by being happy. We do art to advance art — that's the point!" His work from this period certainly contains no expression. There are black and white images of the graphs of lines and parabolas; photographic enlargements of

---

C'est ça

Des points sans volume,
Des ondes psi subjectives,
Des nombres (entiers) associés à des propriétés abstraites,
C'est nous,
C'est le contenu matériel de l'univers.

— Bernar Venet

It's this

Points without volume,
Subjective psi waves,
(Whole) numbers associated with abstract properties,
It's us,
It's the material content of the universe.

Bernar Venet

Courtesy Bernar Venet.

Reprinted from April 2001, pp. 17–20

203

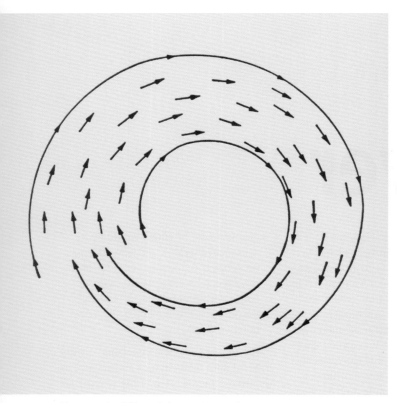

Parametric differential equation in the plane, with spiral integral curves

Undetermined Line, 1990
Collection of Mr. and Mrs. Elliott K. Wolk, New York

Related to Singular Homology Theory, 2000.

$$\cdots \leftarrow H_q(X, X_2; G) \overset{i'_*}{\leftarrow} H_q(X_1, X_1 \cap X_2; G) \overset{\partial}{\leftarrow} H_{q+1}(X; X_1, X_2; G)$$

$$\overset{j'_*}{\leftarrow} H_{q+1}(X, X_2; G) \leftarrow \cdots$$

$$\cdots \rightarrow H^q(X; X_2; G) \overset{i'^*}{\rightarrow} H^q(X_1, X_1 \cap X_2; G) \overset{\delta}{\rightarrow} H^{q+1}(X; X_1, X_2; G)$$

$$\overset{j'^*}{\rightarrow} H^{q+1}(X, X_2; G) \rightarrow \cdots$$

stellar spectra and pages from textbooks and research journals; even whole books. One work is a copy of Azriel Rosenfeld's *Introduction to Algebraic Structures* on a stand open to a random page with blown-up copies of the title and content pages mounted on the wall behind. He explains to me that Rosenfeld's book isn't the object here, the mathematics is. When Rosenfeld is superceded by a newer, updated exposition/understanding of algebraic structures, the exhibitor or owner of this piece of art should throw out Rosenfeld and replace it by the latest thing. In 1968 when the avant-garde Judson Church Theater in New York invited him to give a performance he appeared at the opening with three friends: a mathematician and two physicists each of whom delivered a lecture on his latest research. That was the entire performance. When telling me this story Venet does not reveal how much of the modern art crowd hung around until the lectures were over.

In 1971, at the age of thirty, Venet, proclaiming that he had nothing left to say, quit art; by 1976 he was back. He now, somewhat sheepishly, confesses to youthful arrogance. This is when he turned to sculpture and began making the huge *Undetermined Lines* for which he is best known — these are on display all over the world: China, New Zealand, the U.S., all over Europe. Lately he has turned back to painting, but such paintings! Enormous equations, formulae, and mathematical symbols painted in black on brightly-colored backgrounds. One can't help wondering what he's thinking.

Why Mathematics?

Venet is a devotee of semiotician Jacques Bertin's theory of meaning. Bertin said that a regular representational painting, a picture of a field of flowers say, operates on several different levels of meaning: there's an aesthetic statement, and the literal meaning, and maybe painterly statements about light and composition and motion, not to mention the experiences and reac-

©2001 by Sidney Harris

tions the viewer brings. Bertin calls such images *polysemic*, "poly" meaning many and "semic" from the Greek word for meaning (same root as semantic). A non-representational painting, by contrast is *pansemic*, it can mean anything. Consider a big canvas painted bright red, it could evoke blood, or rage, or innocence (an apple), or nearly anything. Where can one find *monosemy*, images that carry only one level of meaning? Bertin and Venet say only in mathematics. Venet has written, "the picture avoids the expressive and aesthetic because what structures it is a mathematical code which can be in no way invested with values, whose only value is its functioning."

But art is about communication, art is about one person showing us his vision, his passion, and thereby giving us a new way to look at the world or new insight into the larger human condition. Isn't it? If an artist creates art with only one meaning and that one meaning is completely opaque to nearly everyone who sees the painting, what communication can possibly take place?

What is the Purpose of Art?

"The activity of an artist should be to serve art. In other words to do art is to extend the knowledge that we have of art. We know what art has been up until today and we need to keep coming up with new definitions of art. That is what the goal of the artist should be." Venet believes that his use of mathematical symbols is bringing to art a whole new language, allowing artists to create "rational images" as opposed to "expressive images." An

Performance, *May 27–28, 1968, Judson Church Theater, New York.*

Courtesy Giampaolo Prearo Editore, Milano.

$$[\varphi_\alpha(x), \varphi_\beta(y)] = -i\Delta_{\alpha\beta}(x - y).$$

$$U[\sigma, \sigma_0] = I - i\int_{\sigma_0}^{\sigma} \mathcal{H}'(x')U[\sigma', \sigma_0]\, dx'.$$

Courtesy Bernar Venet.

Related to "The Visualization of the Covariant Volume Integral," 1999.

expressive image is one in which the representation of human emotion prevails over the solution of purely artistic problems. An artist who uses art to express his "emotions, his personal feelings, his pathological miseries, his philosophical, metaphysical, or social conceptions ... no longer serves painting: he paints to serve himself." Venet is scornful of such self-indulgence.

"Look," he tells me, "Monet is not important to the history of art because he made beautiful pictures, with a nice blue next to a lovely yellow, any artist can do that. Monet is important because he made a change in our understanding of art. He created a new system of composition and changed radically the subject matter of art. He worked in a very systematic way with, for example, the cathedral of Rouen and by painting it over and over again at different times in different lights he introduced the notion of seriality to art. People look at a Monet or a Cezanne and they say, 'Oh, it is beautiful.' But that is not the point for an artist. In the twentieth century people like Kandinsky and Mondrian introduced a new concept to art— instead of purely representational images they created painting that was only about itself. In a similar way, with my introduction of monosemy, this sounds pretentious, but I think I have enlarged our knowledge of art."

An analogy strikes me as we talk; he sounds a lot like a pure mathematician trying to justify his existence. The same thought occurs to him, "I am very much like a scientist, I always want to do something new. I do not want to go back and redo what I have already done." Later he enlarges the analogy, "I paint with reason, not passion, in that way I am like a scientist."

The Question of Meaning

There's an old story about an American visitor who is on a guided tour of the English countryside made famous by Wordsworth's romantic descriptions of the landscape. He said to his guide, "I can't see much in your scenery here." The guide replied, "Ah,

Courtesy Giampaolo Prearo Editore, Milano.

Undetermined Line. Collection Runnymeade Sculpture Farm, Woodside, California.

don't you wish you could, sir?" It strikes me that something similar must go on between Venet's work and its viewers. I ask him what purpose the viewer of a piece of art serves and if it matters that most of his viewers will not understand the meaning coded in his symbols. "The purpose of the viewer is to *learn something about art*," he tells me. Museum visitors usually assume that his mathematical paintings are studies for the large sculptures for which he is well known. Visitors to his home, where the living room is decorated in huge exact sequences and differential equations, sometimes try to pretend they don't notice.

I suppose it is possible for someone to learn something about art by looking at a painting covered with monosemic symbols that they don't understand, but I wonder about the mathematical viewer; what does Venet imagine is the experience of a mathematician viewing his pictures? "I have been thinking about this; let me make a comparison. Suppose I were to take a beautiful nineteenth-century representational picture of a cow by a very good artist and show it to a farmer who lives in the country and has never been to a museum. Is he going to say, 'Ummm, nice composition, interesting brushwork

and texture.' NO! He is going to say, 'Yes, this cow I know, she is from this country and she is twenty years old and....' He will give you things that the artist did not intend to paint. This is a parallel; I do not understand the meaning of the symbols I paint, but if you do, you take from the picture something that I did not know was there."

One of the things I've been wondering is if Venet is attempting to "avoid the expressive and aesthetic," why put the formulae on such shockingly bright backgrounds? He explains that when he was young and reproducing pages from mathematics texts he worked strictly in black and white. "I tried to be totally rational, like a computer producing art with absolutely no emotion, but now I am older and more human. I'm more liberated. I choose a formula not for its importance in the mathematical context. I choose it for its beauty. There is a visual aesthetic to mathematical symbols that has never been exploited in art. Many people do not see this beauty, but I accept it. It is a new possibility in composition. Why add color? I did black and white when I was younger, restraining myself; now I am liberated, it is going to have more impact on people on a bright yellow background."

As a mathematician who has struggled to invent symbols and good notation, I can believe there is an aesthetic quality to such symbols, but really there is a much deeper beauty in the *ideas* encoded in the symbols. It is a shame that most viewers of Venet's work will never even get a glimpse of the wonders behind the symbols. "Look," he says, "consider $E = mc^2$. This is pure beauty. Here is the story of the universe coded in so few symbols. There is power in those symbols."

For centuries, until very recently, western art was dominated by representations of Christian symbols. There was, at that time, supreme power in those symbols. Now, one could argue, technology and mathematics, its language, are the strongest forces controlling events around us and most people have about as much control over them as your average peasant had over a Renaissance pope. Mysterious important symbols whose meaning is understood only by a few adepts but which ostensibly explain something profound about the world and our place in it — am I talking about mathematics or religion? Maybe we should not be surprised that mathematics is appearing in painting and literature and theater these days. There is power in those symbols. ■

The Conquest of the Kepler Conjecture

DINOJ SURENDRAN
University of Chicago

In August 1998, a message was sent from an Internet cafe in Munich that stunned the mathematical community. It stated that the writer had found "a solution to the Kepler Conjecture, the oldest problem in discrete geometry... the proof relies extensively on methods from the theory of global optimization, linear programming and interval arithmetic... well over 250 pages... computer files require over three gigabytes of space for storage..."

Nearly three years later, a panel of twelve referees is still checking the solution provided by Tom Hales, a mathematician at the University of Michigan, and his graduate student Samuel Ferguson. No serious flaws have been found and none are expected to be found. This article gives a brief overview of their work.

The history of the problem goes back to the early 1600s, when the astronomer-mathematician Johannes Kepler wrote about the problem of packing spheres in space. He asserted that no packing could be better than the Face Centered Cubic (FCC) packing. This is the natural one that arises from packing spheres in a pyramid, as shown in Figure 1.

However, a proof of this simple statement is strangely elusive. A long standing result of Gauss affirms the conjecture for all lattice packings, i.e., all packings where the centers of the spheres form a

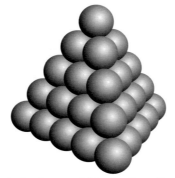

Figure 1. A pyramid in the Face Centered Cubic sphere packing

---

It looks simple at first sight, but reveals its subtle horrors to those who try to solve it.

— S. Singh

---

lattice, but nothing says that a best packing should be of this form. There is even a 1964 conjecture of Claude Rogers that for the corresponding problem in a high enough dimension, the best packing will *not* be a lattice packing. Even worse, there are known non-lattice packings that are as good as FCC.

We state the problem more precisely. An n-dimensional unit sphere in R^n is composed of all points within one unit

from a given center. Kepler's question deals with three dimensions, but we shall often resort to two dimensions for illustrative purposes. A sphere packing is a collection of non-overlapping spheres in R^n. All the packings we deal with will be assumed to be saturated, i.e., have no room for additional spheres. The *density of a packing in a finite region* is the fraction of the region occupied by spheres. The *density of a packing of space* is defined as the limit as $r \to \infty$ of the density of the packing when restricted to a sphere of radius r.

The Kepler Conjecture. *The density of any sphere packing in three-dimensional space is at most $\pi/\sqrt{18}$, which is the density of the FCC packing.*

Related Problems

This question can be generalized in several directions; we give two. First, for what three-dimensional solids can optimum packing densities be found? Second, what is the optimum sphere packing density in n dimensions?

Surprisingly little is known about the first question. Trivial cases like cuboids aside, the first result of this sort was only found in 1990, when András Bezdek showed that the optimum way to pack infinitely long cylinders is in parallel columns in hexagonal fashion. The first re-

Reprinted from April 2001, pp. 8–11, 16

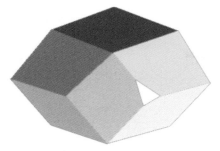

Figure 2. A chipped rhombic dodecahedron.

sult for a bounded solid was by his brother Karoly in 1994. He showed that the best packing involving a rhombic dodecahedron with a corner chipped off (see Figure 2) is the regular tiling of space by the original dodecahedron.

The second question has been studied a lot more. The first result was by the Norwegian mathematician Axel Thue in 1890 for $n = 2$. Here is an adaptation of his proof.

Theorem 1. *The optimum packing of circles in the plane has density* $\pi/\sqrt{12}$, *which is that of the hexagonal packing.*

Proof. Start with an arbitrary packing of unit circles in the plane. Around each circle draw a concentric circle of radius $2/\sqrt{3}$. Where a pair of larger circles overlap, join their points of intersection and use this as the base of two isosceles triangles. The centers of each circle form the third vertex of each triangle. This induces a partition of the plane into three regions, as depicted in Figure 3.

1. The Isosceles Triangles: If the top angle of the triangle is θ radians, then its area is $2/3 \sin\theta$ and the area of the sector is $\theta/2$, so that the packing density is

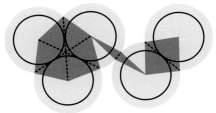

Figure 3. The plane partition induced by a sample circle packing.

$$\frac{3\theta}{4\sin\theta}.$$

θ ranges between 0 and $\pi/3$ radians and the maximum value of the density is $\pi/\sqrt{12}$, attained at $\theta = \pi/3$.

2. Regions of the larger circles not in a triangle: Here the regions are sectors of a pair of concentric circles of radius 1 and $2/\sqrt{3}$ so the density is

$$\left(\frac{1}{2/\sqrt{3}}\right)^2 = \frac{3}{4} < \frac{\pi}{\sqrt{12}}.$$

3. Regions not in any circle: Here the density is zero.

Thus the density of each region of space is at most $\pi/\sqrt{12}$. A packing can only be optimum when it causes space to be divided into equilateral triangles. This only happens for the hexagonal packing of Figure 4.

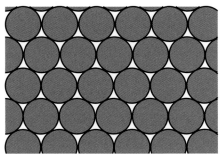

Figure 4. The hexagonal circle packing of the plane.

Unfortunately, this proof cannot be generalized to three dimensions. There are several sphere packings where the density in places is over $\pi/\sqrt{18}$. Experimental evidence, both from real life and computer simulations, indicates that this would be offset by lower densities in other places. The Kepler Conjecture asks whether the offset is enough.

A Finite Calculation

Major progress towards finding a proof of the Kepler Conjecture was made in 1953 when László Fejes Tóth showed how to reduce the problem from a global optimization involving infinitely many variables to a finite calculation.

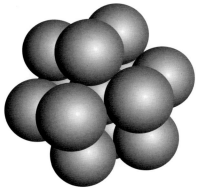

Figure 5. A cluster in the FCC sphere packing.

His method requires a painstaking study of all possible local configurations, called *clusters,* that can arise in a sphere packing. A cluster around a point λ consists of a base sphere centered at λ and non-overlapping spheres with centers within $\tau = 2\sqrt{2}$ units of λ. The reason for the value of τ is technical and can be ignored for now. As an example, the cluster in the FCC packing is shown in Figure 5.

It is possible to define a notion of distance between two clusters with the same number of spheres. The idea is that if two clusters can be obtained from each other by moving spheres around the base one, then there is a finite distance between them. The less the amount of work to convert one cluster to another, the shorter the distance. The set of clusters $\mathcal{C}$ becomes a topological space with this metric. Clusters involving a different number of spheres are in different components of the topological space.

The next step is to find a "correction function" f defined on $\mathcal{C}$ that deals with the problem of certain clusters being denser than the FCC cluster. The function f should be continuous, so that similar clusters will have similar values. We shall also demand that f be *transient,* i.e., that for any packing, if Λ_r is the set of centers within r units of the origin,

$$\sum_{\lambda \in \Lambda_r} f(C_\lambda) = o(r^3). \quad (1)$$

This constraint keeps f from being too large a correction.

Decomposing Space with the Voronoi method

Once all clusters have been studied, attention can be turned to the task of putting them together to form larger and larger configurations that eventually fill up all of space. Building up configurations is analogous to breaking down space, and it is therefore necessary to consider different ways of performing the latter.

The partition of space suggested by Fejes Tóth is the *Voronoi decomposition.* Given a sphere packing, define the *Voronoi cell* of a sphere to be the set of points closer to it than to any other sphere. In a three- (two-) dimensional packing, the Voronoi cells are polyhedra (polygons) containing unit spheres (circles). However, not all such objects are candidates for Voronoi cells—no polyhedra with over sixty faces is, for example. Figure 6 gives an example of a Voronoi decomposition in two dimensions.

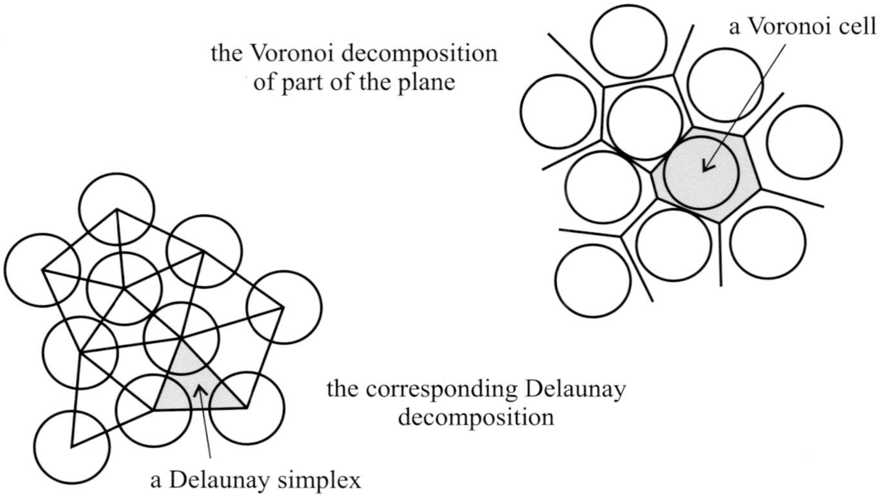

The density of a cell is the ratio of the volume of a unit sphere to its own volume. For example, the rhombic dodecahedron is the only Voronoi cell that occurs in the FCC packing. Its volume is $V_{fcc} = 4\sqrt{2}$ and its density is $\pi/\sqrt{18} \approx 0.7405$.

In two dimensions it turns out that the densest Voronoi cell is the regular hexagon. Since this tiles the plane, the optimum packing consists of circles fitted in each hexagon of the tiling. This is of course the same packing that Thue showed was optimal.

In three dimensions the densest Voronoi cell is the regular dodecahedron, which has a density of about 0.755. (This 1943 conjecture of Fejes Tóth was proved in 1998 by Sean McLaughlin, then an undergraduate at Michigan.) There are several other Voronoi cells that have densities higher than 0.7405. Like the regular dodecahedron, none of them partition space — which is a pity, because then the problem as a whole would be much easier!

Since the density of a packing is a function of the density of its cells, the Kepler Conjecture can be rephrased as the prob-

the Voronoi decomposition
of part of the plane

a Voronoi cell

the corresponding Delaunay
decomposition

a Delaunay simplex

Figure 6. Two decompositions of the plane.

lem of showing that there are no decompositions based primarily on such highly dense cells.

The Correction Function

The use of the correction function can now be explained in more detail. For a given packing let Λ be its set of centers. There is a Voronoi cell V_λ at each point $\lambda \in \Lambda$. Thus the Voronoi Decomposition is

$$\bigcup_{\lambda \in \Lambda} V_\lambda = R^3.$$

It's one of those problems that tells us that we are not as smart as we think we are.

—D. J. Muder

Now we demand that our correction function f be *FCC-compatible,* i.e., that

$$V_{fcc} \le \text{Volume}(V_\lambda) + f(C_\lambda), \quad \forall \lambda \in \Lambda. \quad (2)$$

There are several continuous transient functions known. But finding one that is also FCC-compatible is very difficult since these functions tend to involve well over a hundred variables! However, its existence would settle the Kepler Conjecture, as we now show. If we sum (2)

over Λ_r, we have

$$|\Lambda_r| V_{fcc} \le \sum_{\lambda \in \Lambda_r} \text{Volume}(V_\lambda) + \sum f(C_\lambda)$$

$$\le \frac{4}{3}\pi(r+c)^3 + o(r^3).$$

The second inequality follows from the fact that $\bigcup_{\lambda \in \Lambda_r} V_\lambda$ is contained in a sphere of radius $r + c$ for some constant c. It should be noted that we are making use of the maximality of our packings here; the statement is certainly not true for non-saturated packings, which can have cells of infinite size.

Dividing by $r^3 V_{fcc}$, we have:

$$\frac{|\Lambda_r|}{r^3} \le \frac{4\pi}{3V_{fcc}}\left(1 + \frac{c}{r}\right)^3 + \frac{o(r^3)}{r^3}$$

and sending $r \to \infty$ we find that the density

$$\lim_{r \to \infty} \frac{|\Lambda_r|}{r^3} \le \frac{4\pi(1)^3}{3V_{fcc}} = \frac{4\pi}{3 \cdot 4\sqrt{2}} = \frac{\pi}{\sqrt{18}}.$$

Alternative Decompositions of Space

Fejes Tóth did not pursue the Voronoi decomposition approach further since there was not enough computational power in his day to make it feasible. When Hales took it up in the 1990s, he found it was far too complicated to deal with in certain cases. He decided to try its dual, called the *Delaunay decomposition.*

Consider a network of vertices and edges, where the vertices are the centers of spheres and centers are joined by an edge if and only if their corresponding Voronoi cells have a common face. The resulting partition into simplexes is the Delaunay decomposition. An example in two dimensions is shown in Figure 6.

After some months of work, the Delaunay approach was also becoming too complex to deal with and Hales put the whole problem aside for a while. Insight came in November 1994 when he realized that instead of trying to dogmatically stick to one type of decomposition, it might be better to try a hybrid approach: in other words, partition each region of space into Delaunay simplexes or Voronoi cells depending on what is most convenient.

The good thing about hybrid decompositions is that there are an infinite number of them. As the months went by, Hales and Ferguson found that when the going got tough, it was quicker to change the decomposition and the correction function a bit rather than plough through with the original one. This arbitrariness certainly raised some eyebrows! What makes it permissible is the infinite-dimensionality of the problem.

Two years later, with the deadline for Ferguson's thesis submission looming, the two workers decided to stop adjusting f and the decomposition and push the proof through to the bitter end…

Their final decomposition is a little tricky to define; it contains many, many different Delaunay simplices. The main point is that it is fixed and draws from the best features of the Voronoi and Delaunay decompositions. The correction function f is also fixed and the next step is to confirm that f is indeed FCC-compatible, i.e., that

$$f(C_\lambda) \geq V_{fcc} - \text{Volume}(V_\lambda) \quad \forall \lambda \in \Lambda. \quad (3)$$

Associated Maps and Graphs

Equation (3) represents an optimization problem on f. Unfortunately the space $\mathcal{C}$ of clusters, on which f is defined, is very

Figure 7. The cluster, spherical map and plane graph for the Pentahedral Prism.

complicated and can only be dealt with indirectly. To every cluster $C \in \mathcal{C}$ Hales and Ferguson associate a *spherical map* M_C, which is just a collection of points and edges joining the points in 3-space. An example is shown in Figure 7 for the cluster called the pentahedral prism.

n, for largest n-sided face in map	Number of maps
≤ 3	0
4	1749
5	2459
6	429
7	413
8	44
≥ 9	0
	5094

For 2-dimensional disks this problem has been solved by Thue and Fejes Tóth,… However, the corresponding problem in 3 dimensions remains unsolved. This is a scandalous situation since the (presumably) correct answer has been known since the time of Gauss.

— J. Milnor

Lower bounds for $f(C)$ can be found by studying its associated spherical map. In many cases it is even sufficient to study G_C, which is the plane graph obtained from M_C. (G_C has a lot less information than M_C, which keeps track of geometric details like the position of the spheres relative to each other in space.) There are an infinite number of clusters for which this is not the case, but they can be grouped into a *finite* number of classes according to their maps.

This is certainly the most labor-intensive stage of the problem. To each map is associated a non-linear optimization problem involving up to 160 variables and over 1000 inequalities. At first sight things look hopeless, until it is realized that most constraints are linear and hence standard methods from linear programming can be applied to cut down the number of cases to a more manageable 189. To better appreciate the amount of human toil that went into this stage, an excerpt from a report by Hales in 1996 is in order:

Of the 537… [remaining] maps with only triangles and quadrilaterals, 531 can be discarded by linear programming methods. Five more can be discarded by a refinement of the linear programming methods. This leaves one case. This remaining case is still giving a bound of about 8.25 points and needs to be nudged a bit more before we are finished with this case.

[Footnote, some months later:] We were down to one case, but we continue to fiddle with the scoring rules and inequalities. We are temporarily back up to 33 cases…

The one remaining map referred to above is the pentahedral prism of Figure

7. It turned out to be unbelievably difficult and became the primary focus of Ferguson's doctoral thesis.

Better methods of linear programming bring the number of cases down to 88, at which stage ad hoc computer programs can be written to deal with each. The entire solution contains about 100,000 such calculations.

The End of the Adventure

Ferguson completed his doctorate in August 1997 and returned to Ann Arbor for three months in mid 1998 to help Hales bring the project to its completion:

Theorem 2 (The Kepler Theorem). *The density of any sphere packing in three-dimensional space is at most $\pi/\sqrt{18}$, which is the density of the FCC packing.*

Before the end of this article it should be made clear that there are other packings which are as good as the FCC. One of these is the Hexagonal Closed Packed (HCP) one. This is made of the same layers as the FCC, but the layers are arranged differently, as shown in Figure 8. By fitting layers in different ways it is possible to produce infinitely many FCC-like packings with the same density.

There are other packings, lattice and non-lattice, which are as good as these. Hales and Ferguson were able to determine that all such packings have a local

FCC Packing   HCP Packing

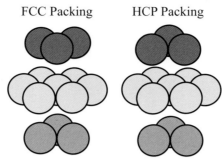

Figure 8. The difference between the FCC and HCP packings.

FCC-type structure for a sufficiently large percentage of space.

What's Next?

There are several interesting open problems in discrete geometry that involve packing unit spheres. Here are a few.

1. It was proved by Van der Waerden and Leech in 1956 that the maximum number of spheres that a sphere could touch in three dimensions is 12. What is the answer in 4 dimensions?

2. What is the minimum volume that can be contained by N planes tangent to a sphere in three dimensions?

3. The Sausage Conjecture is an open question asked by Fejes Tóth in 1975. The problem is that of packing m spheres in R^n, $n \geq 5$ so that the volume of the convex hull generated by the spheres is minimized. The conjecture is that the optimal packing is of spheres next to each other in a straight line, i.e., 'a sausage'. It was confirmed by Betke and Henk for $n \geq 42$ in 1998.

4. The Sausage Catastrophe deals with the above problem for $n = 3, 4$. Wills suggested in 1983 that the optimal solution for small m is indeed a one-dimensional sausage, but for m larger than a certain critical value k_n the optimal configurations are n-dimensional arrangements. He conjectured that $k_3 \approx 56$ and $k_4 \approx 75000$. ∎

For Further Reading

Professor Hales's own account of his discovery, Cannonballs and Honeycombs, can be found in *Notices of the American Mathematical Society*, Vol. 47, No. 4. It is somewhat more technical than this article. Updated information is available at Hales's website www.math.pitt.edu/~thales/kepler06/ countdown including the full 265-page solution. A good introduction to the general topic is *Sphere Packings*, C. Zong, Springer-Verlag, 1999.

Acknowledgements

The author would like to thank Professor Thomas Hales and Dr. Samuel Ferguson for their valuable comments, advice and help with pictures.

A Match

Made in

Mathematics

OLIVIA M. CARDUCCI
Muhlenberg College

Amelia Hopkins is sitting on pins and needles. She is about to graduate from the University of Connecticut's Medical School, she desperately wants to obtain a residency in pediatric medicine, and her fate is in the gentle hands of mathematics. Amelia's anxiety is shared by 20,000 other med school students as they await the results of the National Resident Matching Program (NRMP). The NRMP assigns graduating medical students to hospital residency positions, and these assignments affect the careers of the country's newest doctors and the hospitals that employ them. The driving force in this drama is the algorithm used to make assignments. We turn our attention to the mathematical concepts that underlie this algorithm.

Background on the NRMP

After completing medical school, a doctor must undergo a year of residency training at a hospital in order to be licensed to practice medicine. The process of securing a residency begins much like the process of securing any type of job. Hospitals advertise their positions; medical students in their final year apply for residencies and (if lucky) receive invitations for site visits at some hospitals.

The logical next step would be for the hospitals to extend offers to their top candidates, but the NRMP interrupts this process. After the interviews, hospitals submit ordered lists of favored applicants to the NRMP and applicants submit ordered lists of favored positions. The list that the hospital submits includes only those applicants the hospital would be willing to hire. Similarly, each student's list includes only those positions the student would be willing to accept. The NRMP will not match a hospital and student unless the hospital has included the student on its list and the student has included the hospital on her list.

This process (with some modifications to the algorithm) has been in place since 1951. Prior to that there was fierce competition among hospitals offering residency positions (at that time called internships) for an inadequate supply of medical school graduates. Hospitals tried various tactics to gain an advantage over their competitors, including making offers to students early in their medical school studies or making what came to be known as exploding offers. Exploding offers had to be accepted before a given deadline or they were withdrawn. Students were often given as little as twelve hours to consider an offer.

No one was happy with the situation: not the hospitals, not the medical schools, and certainly not the students. Thus, the medical schools and the hospitals agreed to develop a voluntary centralized matching service to assign students to residency positions. After a false start, the NRMP was born. Today, the NRMP attempts to fill approximately 20,000 positions annually.

The Basic Algorithm

The NRMP collects rank-ordered lists (ROL) from all students and hospitals participating in the match. Each student lists all hospitals she would be willing to work for in order of preference; most preferred hospital first, next most preferred second, etc. Students list an average of 7.5 hospitals, although the entry fee entitles each student to list 15. (Students can list more hospitals for an additional fee.) Similarly, each hospital lists all students it would be willing to hire in order of preference. Different residency positions at the same hospital have different ROLs and can appear on the same student's ROL, although in practice, students typically list only pediatric residencies or only surgical residencies, etc. Sample ROLs for a very small problem are given in Figure 1. First, the lists are preprocessed by removing from the students' lists all hospitals that did not include the student on their lists and by removing from the hospitals' lists all students who did not include the hospital on their lists. If hospital h and student s are on each others' ROL, then (h, s) is an *acceptable* assignment. Thus, at the start of the algorithm every entry in each ROL is part of an acceptable assignment. (The lists in Figure 1 are already preprocessed.)

Students' ROLs				Hospitals' ROLs			
s_1	s_2	s_3	s_4	h_1	h_2	h_3	h_4
h_1	h_2	h_2	h_4	s_2	s_2	s_4	s_1
h_3	h_1	h_3	h_3	s_4	s_3	s_1	s_4
h_2		h_1	h_1	s_3	s_1	s_3	
h_4				s_1			

Figure 1

The basic NRMP algorithm is a *deferred acceptance* algorithm. In this algorithm, each hospital and student is either available or assigned. Hospitals may alternate between being available or assigned, but once a student is assigned, she is always assigned, although the assignment may change (to a more preferred assignment). An available hospital makes offers down its ROL until it makes an offer to a student who tentatively accepts its offer (because the hospital's offer is the best offer the student has received so far). A student who receives an offer accepts it if either the student is available or the student prefers the newly offered assignment to her tentative assignment. If the student prefers her tentative assignment to the newly offered assignment she rejects the offer. The algorithm continues until there are no available hospitals remaining. (Figure 2 gives the algorithm in pseudocode.)

To illustrate how the deferred acceptance algorithm works, consider the ROLs in Figure 1. First, h_1 makes an offer to s_2 which s_2 tentatively accepts. Then h_2 offers s_2 its position causing s_2 to reject h_1's offer and tentatively accept h_2's offer. Hospital 1 is available again. (The order in which the hospitals make offers does not affect the final outcome.) Next h_3 makes an offer to s_4 which s_4 tentatively accepts. Then h_4 makes an offer to s_1 which s_1 tentatively accepts. Remember h_1 is available, so h_1 offers its position to s_4 and is rejected. Finally h_1 makes an offer to s_3 which is tentatively accepted. Since no hospitals are available, the assignments are: h_1 to s_3, h_2 to s_2, h_3 to s_4, and h_4 to s_1.

This algorithm always yields an assignment that forms what mathematicians call a *stable match*. An assignment is stable if no hospital and student are *both* willing to switch from their assignment to each other. Our assignment

$$h_1 \quad h_2 \quad h_3 \quad h_4$$
$$s_3 \quad s_2 \quad s_4 \quad s_1$$

is stable. Although h_1 would rather hire s_2 than s_3 as assigned, s_2 is happier with her current assignment, h_2. Similarly, h_1 would rather hire s_4 than s_3, but again s_4 is happier working for h_3 as assigned. Note that h_1 made offers to both s_2 and s_4 which were rejected (eventually). Hospitals 2, 3, and 4 are happy that their most preferred students will be working for them. The students may be less happy (only s_2 received her most desired assignment), but they cannot convince the hospitals to change who they've hired. Student 1 would prefer to work at h_1, h_2, or h_3 but h_1 is happier with s_3, h_2 is happier with s_2, and h_3 is happier with s_4. Student 3 would prefer to work at h_2 or h_3, but they are both happier with their current assignments. Similarly s_4 would prefer to work at h_4, but h_4 is happier with s_1. Note that none of the unhappy students received an offer from a hospital at which they would have preferred to work.

Table 1 gives all acceptable assignments for the ROLs in Figure 1 and indi-

Table 1

Assignment				Stable?
h_1	h_2	h_3	h_4	
s_1	s_2	s_3	s_4	stable #1
s_2	s_1	s_3	s_4	unstable — s_1 and h_3
s_2	s_3	s_1	s_4	unstable — s_2 and h_2
s_2	s_3	s_4	s_1	unstable — s_2 and h_2
s_3	s_2	s_1	s_4	stable #2
s_3	s_2	s_4	s_1	stable #3
s_4	s_2	s_3	s_1	unstable — s_1 and h_3

cates whether or not the assignment is stable. If an assignment is not stable, Table 1 indicates one reason why not. (There may be others.) For example, the assignment

$$h_1 \qquad h_2 \qquad h_3 \qquad h_4$$
$$s_2 \qquad s_1 \qquad s_3 \qquad s_4$$

is unstable because s_1 would rather work for h_3 than h_2 as assigned *and* h_3 would rather hire s_1 than s_3 as assigned. One phone call from s_1 to h_3 tempts both to break their commitments to their assigned partners in favor of each other. When voluntary matching services like the NRMP allow such instabilities they quickly fall into disuse.

It is not hard to convince yourself that the algorithm described in Figure 2 leads to a stable matching. If a hospital h prefers student s to its assigned student, then the hospital must have made an offer to s and had its offer rejected for a position that s prefers. Thus, s is not willing to switch its assignment to h. If a student s prefers hospital h to its assignment, then h must not have made an offer to s, so h must be assigned a student it prefers to s. Thus, h is not willing to switch and the assignment must be stable. We have the following theorem.

Theorem. *Every set of ROLs admits a stable matching and the deferred acceptance algorithm produces one.*

This theorem tells us a stable matching exists. The next question for any mathematician is: can there be more than one stable matching? A quick look at Table 1 reveals that the example in Figure 1 has three stable matchings, so a stable matching is not unique. The algorithm in Figure 2 yields a stable matching with the surprising property that all the hospitals

```
assign each student and hospital to be available;
while some hospital h is available do
begin
        s := the most preferred student on h's ROL that
               h has not made an offer to;
        if s is available then
              assign s and h to each other
        else
              if s prefers h to her current assignment
                    h' then assign s and h and h' becomes
                    available
              else
                    s rejects h (and h remains available)
end;
output the assignments;
```

Figure 2. The Deferred Acceptance Algorithm

Illustration by John Johnson of Teapot Graphics

agree that no other *stable* matching would be better. This assignment is called the *hospital-optimal* matching. For the ROLs in Figure 1, stable #3 (see Table 1) is the hospital-optimal matching. Hospital 1 is assigned either s_1 or s_3 in a stable matching. Hospital 1 prefers s_3 to s_1, so h_1 prefers stable matchings #2 and #3 to stable matching #1. Hospital 1 is indifferent between stable #2 and stable #3. Hospital 2 is assigned s_2 in all stable matchings, so h_2 has no preference. Hospital 3 is assigned its first choice, s_4, only by stable #3, so h_3 prefers it to the others. Hospital 4 also prefers stable #3. Thus no hospital is better off with stable #1 or stable #2 than it is with stable #3 and some hospitals are better off with stable #3.

There is also a *student-optimal* stable matching with the property that all the students agree no other stable assignment would be better. For the ROLs in Figure 1, stable #1 is the student-optimal stable matching. (Why?) Applying the deferred acceptance algorithm with the students making the offers yields the student-optimal matching. Thus, when we restrict

our attention to stable matchings, our notion of who is competing with whom changes. It is natural to think of the students competing with each other for positions and of the hospitals competing with each other for students. This was the case before the centralized matching service was implemented. But if we restrict ourselves to stable matchings, all the hospitals are united in pursuit of the hospital-optimal matching and all the students are united in pursuit of the student-optimal matching. Thus the competition is between the students and the hospitals over which stable matching to choose.

In fact, the hospital-optimal matching is the worst stable matching for the students because no student is worse off in any other stable matching. The hospital-optimal matching is sometimes referred to as *student-pessimal*. Similarly, the student-optimal assignment is sometimes referred to as *hospital-pessimal*. It is possible to rank all stable matchings in this way, not just the hospital-optimal and student-optimal. If all hospitals agree that no matching in set *A* is better than any match-

ing in set *B*, then all students would agree that no matching in set *B* is better than any matching in set *A*. (In the language of set theory, the hospitals' preferences for the stable matchings form a partial order on the set of stable matchings and the students' preferences are the inverse partial order. Moreover, with an appropriate meet and join, the set of stable matchings forms a distributive lattice.)

Even more surprising than the existence of the hospital-optimal and student-optimal matches is the fact that if a hospital or student is unmatched in one stable matching, the hospital or student is unmatched in all stable matchings. Medical school graduates often want to work at major hospitals, so rural hospitals can have difficulty filling their positions. In the past, the NRMP has argued that the hospital-optimal match gives these hospitals the best opportunity to fill their positions; however, mathematical analysis indicates that the choice of stable matchings does not affect which positions are filled. One would expect the NRMP to take comfort in the fact that although they must choose a particular stable matching, their choice does not affect who receives an assignment and who does not.

Complications

The original NRMP algorithm served well for many years, but real world considerations have recently forced the NRMP to make adjustments. The most significant of these occurred in 1983. In 1951 when the NRMP was first instituted, there were few married couples looking for internships together, but by the early 1980's the number of such couples had increased significantly. Originally, these couples were matched by a "couples algorithm" applied near the end of the regular NRMP procedure. Couples submitted individual preference lists along with a set of positions that they considered to be in the same geographic area and an indication of the "preferred partner." Members of couples entered the standard algorithm as individuals with the one caveat that individuals who were members of couples were allowed to hold multiple offers. Near the end of the process, the set of offers being held by the members of a couple were examined to determine which, if any, were acceptably close together. If more than one pair were acceptably close together, the preferred partner's preference determined which pair the couple was assigned.

Many couples were unhappy with this process and a relatively high percentage of them chose not to participate in the NRMP. This did not threaten the existence of the NRMP when the number of couples was small; however as the number of couples increased, the NRMP felt it was important to attract couples to the program. Thus beginning in 1983 they allowed couples to submit joint preference lists where the couple presents a single ROL ranking pairs of positions. For the first time couples were able to decide that assigning him to position a in New York and her to position b in New York was the most desirable assignment, but assigning him to position c in Chicago and her to position d in Chicago was more desirable than assigning him to position a in New York and her to position z in New York. If he were the preferred partner and

he ranked position a first, the old algorithm would assign them to New York if she could be assigned *any* position in New York.

At the same time, hospitals were becoming dissatisfied with having to specify what residency positions were available so far in advance. A hospital might want to have four residents in orthopedic surgery and six in general surgery but, if it could only recruit three in orthopedic surgery, would prefer to have seven in general surgery. As medicine became increasingly specialized, such considerations grew more important. Thus the NRMP now allows hospitals to designate certain residencies that can revert to another position. The new post must be of a type that the hospital already offers, so that the new post appears on some students' ROLs.

Most positions assigned by the NRMP are intended to immediately follow medical school (first-year positions), but some hospitals are now offering positions through the NRMP that require a year of residency training before the applicant is qualified to assume the position (second-year positions). The hospital offering the second-year position does not dictate which first-year position the applicant completes. For each second-year position on a student's ROL, the student submits an associated ROL of first-year positions. Thus, if a student who had temporarily accepted a second-year position in Chicago receives a preferred offer of a second-year position in New York, the student's ROL of first-year positions changes. Another minor complication imposed by a small number of hospitals is that they be assigned an even number of residents. (These positions involve two assignments; half the group has assignment A first while the other half has assignment B first. Then they switch. With an even number of residents each assignment has the same number of residents for the entire year.) The NRMP has modified their algorithm to accommodate these changes.

Until the 1998 match, the algorithm the NRMP used was a modified deferred acceptance algorithm with the hospitals making offers to the students. The modi-

fications were necessary to accommodate the real world considerations described above. For example, if a member of a couple received an offer, the other individual in the couple would make offers to the hospitals near the partner's offer to try to make a match. If the couple was successful in making a match, then the couple would hold both offers; if not the individual originally receiving the offer would reject the offer. Thus, although primarily the hospitals make offers, under some circumstances the students make offers.

Not much is known about matching problems with these complications, except that even the existence of a stable matching is no longer guaranteed and that it is possible to have more than one stable matching and for these matchings to leave different numbers of unmatched students. If more than one stable matching exists, it is not known which the NRMP will select. No one knows an algorithm that either produces a stable matching or indicates that none exists. On the bright side, the NRMP has been checking its results for stability since the late 1970s and each year the algorithm has succeeded in producing a stable match.

Recent Controversy — Mathematics Under Fire

Over the years there has been criticism of the NRMP. Often the NRMP is criticized for using the hospital-optimal match. Once couples were allowed to submit joint preference lists, it was no longer true that the algorithms produced the hospital-optimal match, although it was believed to do so. And it does seem likely that the modified algorithm favors the group that initially makes the offers. This criticism came to a head in 1995 with an article in *Academic Medicine* arguing that the NRMP algorithm should be modified to favor the students rather than the hospitals. This article, and the responses to it, prompted the American Medical Student Association to petition for a change. Although participation rates of both students and hospitals remain high, the NRMP was concerned that the criticism would diminish participation, especially among students.

The NRMP commissioned a study in which a new algorithm was constructed which reversed the roles of the students and hospitals. This was more difficult than simply having the students make offers because of the existence of couples, reverting positions, and the other complications described above. The new algorithm was run using ROLs from previous years and the match that resulted was compared with the actual match from that year. Approximately 0.1% of the applicants' positions differed between the two matches. In spite of the very small differences, the NRMP decided to implement the new algorithm starting in 1998.

The stable matching problem with its blend of theoretical depth, computational complexity and immediate real-world consequences continues to inspire research by mathematicians, computer scientists, and economists for the best possible solution. Amelia Hopkins just wishes they would hurry up. ∎

For Further Reading

A readable discussion of the stable marriage problem and its application to college admissions can be found in College Admissions and the Stability of Marriage, *The American Mathematical Monthly*, David Gale and L. S. Shapley, 1962. A good place to start a deeper investigation into this problem is *Two-Sided Matching: A Study in Game-Theoretic Modelling and Analysis*, Alvin Roth and Marilda A. Oliveira Sotomayor, Cambridge U. Press, 1990 (a mathematical economics approach) or *The Stable Marriage Problem: Structure and Algorithms*, Dan Gusfield and Robert W. Irving, MIT Press, 1989 (a computer science approach). A series of articles in the June 1995 issue of *Academic Medicine* provides insight into the NRMP algorithm and surrounding issues.

Acknowledgement

The author wishes to thank William Dunham for reading early versions and for his very helpful comments. His help greatly improved the quality of this paper.

How Many **Women Mathematicians** Can You **Name?**

JUDY GREEN
Marymount University

Until my last semester as an undergraduate student in 1964, my answer to the question of the title would have been "One: Emmy Noether, the German algebraist." That semester a woman mathematician, Yvonne Choquet-Bruhat, was a visiting professor at my undergraduate institution, Cornell, so my list increased to two! If you restrict your answer to those women who were active by the middle of the twentieth century, you are unlikely to be able to name more than seven: Hypatia (c. 370–415); Gabrielle-Émilie Le Tonnelier de Breteuil, Marquise du Châtelet (1706–1749); Maria Gaetana Agnesi (1718–1799); Sophie Germain (1776–1831); Mary Somerville (1780–1872); Sofia Kovalevskaia (1850–1891), and Emmy Noether (1882–1935).

By the time I got my PhD in 1972, my list of women mathematicians active by mid-century had increased by one, Dorothy Maharam Stone, whom I met when she was visiting Yale. However, there were a number of women on the mathematics faculty of my doctoral institution, the University of Maryland, so I could name a number of women who had become mathematicians in the 1950s and 1960s, including my dissertation advisor, Carol Karp. In the late 1970s I became interested in the history

of women in mathematics and I have been working in that field ever since, collaborating with Jeanne LaDuke of DePaul University. What we learned is that women have been mathematicians for longer, and in greater numbers, than most people, even most mathematicians, realize. The second half of this paper summarizes and updates a paper Jeanne LaDuke and I wrote in 1987, "Women in the American mathematical community: The pre-1940 PhDs." (*Mathematical Intelligencer* 9 no 1: 11–23); it also relies on another of our papers, "Contributors to American mathematics: An overview and selection" (in G. Kass-Simon and Patricia Farnes (eds.), *Women of Science: Righting the Record*, Bloomington: Indiana University Press, 1990).

Rather than repeat the often-told tales of the seven famous women mathematicians, I will start in the late nineteenth century when English-speaking women had already had access to training in mathematics beyond arithmetic for about half a century and some were beginning to receive real training as mathematicians. While it was still unusual for women to receive higher education of any sort, it was not a secret that when they did, they studied mathematics. In fact, in 1894 George Bernard

Shaw wrote about it in his play, *Mrs. Warren's Profession*. In the first act, the following dialogue takes place between a middle-aged gentleman, Mr. Praed, and Mrs. Warren's twenty-two-year-old daughter, Vivie. Vivie has just taken the mathematical tripos, the honors examination in mathematics at Cambridge, and has achieved the same score as the third wrangler, that is the male candidate with the third highest score. Mr. Praed is quite impressed with Vivie's accomplishment but Vivie tells him that she "wouldn't do it again for the same money," explaining that

> Mrs. Latham, my tutor at Newnham, told my mother that I could distinguish myself in the mathematical tripos if I went in for it in earnest. The papers were full just then of Philippa Summers beating the senior wrangler… and nothing would please my mother but that I should do the same thing. I said flatly it was not worth my while to face the grind since I was not going in for teaching; but I offered to try for fourth wrangler or thereabouts for £50. She closed with me at that, after a little grumbling; and I was better than my bargain. But I wouldn't do it again for that. £200 would have been nearer the mark.

Reprinted from November 2001, pp. 9–14

One can surmise that Shaw chose to have Vivie Warren study mathematics at Newnham and made reference to Philippa Summers, who is supposed to have beaten the senior or top-scoring wrangler in a previous year, because of a real incident. In 1890 a student at Newnham, Philippa Fawcett, who regularly spent six hours a day studying for the mathematical tripos, did indeed score above the senior wrangler. In fact, she scored thirteen percent higher than the man who received the title of senior wrangler. Her accomplishment was discussed in the English newspapers and even the *New York Times* ran an article, "Miss Fawcett's Honor: The sort of girl this lady Senior Wrangler is," describing her success and the significance it had for the higher education of women.

Although Fawcett's achievement was unprecedented, women from Newnham and Girton, the first two women's colleges at Cambridge, had been formally competing in the tripos since 1881. Previous to that a woman had to obtain special permission to sit for the tripos and had to find an examiner to grade her paper, and even then, her score was officially ignored. The change came because of an incident in 1880 when Charlotte Angas Scott of Girton scored between the seventh and eighth wranglers. Since women were not mentioned at the awards ceremony, the undergraduates who were present shouted her name and cheered her as the true eighth wrangler. In subsequent years the rankings of the women were given separately but relative to the men's rankings and the successful women were awarded a special certificate, though not a degree. Women did not receive degrees from Cambridge until 1948!

Neither the fictitious Vivie Warren nor the real Philippa Fawcett pursued an academic career in mathematics; Vivie, in the play, becomes an actuary and Philippa Fawcett, after about ten years as a Lecturer at Newnham, went to South Africa and helped develop a system of

Charlotte Angas Scott
Courtesy of the Bryn Mawr Library.

farm schools in the Transvaal. She returned to England in 1905 and spent the remainder of her career working for the London County Council first as Assistant to the Director of Education and later as Assistant Education Officer for Higher Education. On the other hand, by the time Philippa Fawcett had distinguished herself in the tripos, Charlotte Scott had come to the United States to Bryn Mawr College in Pennsylvania.

Charlotte Angas Scott was born in 1858 and was educated by private tutors until she enrolled at Girton College, Cambridge, in 1876. Girton had opened seven years earlier as England's first college for women and was located three miles from the Cambridge University. Even the supporters of higher education for women were against their receiving degrees. One such supporter was quoted as saying: "If given the BA, they must next have the MA... [and e]ven the BA would enable them to take 5 books at a time out of the University Library..." Despite the fact that Cambridge was not to grant degrees to women for another seventy-two years, even in 1876 most of the Cambridge professors allowed women to listen to their lectures. Thus the women of Girton had opportunities to obtain a real education and Scott took

advantage of this education. Although she received no official recognition for her achievements at Cambridge, in 1882 the University of London opened all degrees, prizes, and honors to women and Scott received a BSc by examination that year. She served as Lecturer at Girton and continued her mathematical studies at Cambridge, where her main interest was algebraic geometry. In 1885 she received a DSc from the University of London, again based on examinations. She was hired by Bryn Mawr, then a newly founded women's college, to head its mathematics department, and remained there for forty years, retiring in 1925, six years before her death.

Scott's influence on American mathematics was publicly acknowledged in many ways. In 1906 she was ranked by her peers as fourteenth among the top ninety-three mathematicians of the period. That same year Scott served as vice-president of the American Mathematical Society, being the first woman to hold that office and the only one to do so until seventy years later when Mary Gray, one of the founders of the Association for Women in Mathematics (AWM), was elected.

Another Girton-educated Englishwoman, Grace Chisholm Young, also has an important place in the history of women in mathematics. Grace Chisholm was born in 1868, the year before Girton was founded. Like Scott, she did not receive any formal education until she entered Girton in 1889. At the end of her first year she heard Philippa Fawcett of Newnham College announced as "Above the Senior Wrangler." When Chisholm sat for the mathematical tripos two years later, she placed between the twenty-third and twenty-fourth Wranglers. She accepted a challenge to become the first woman to sit for the Oxford Honour exams and received First Class Honours with the highest score on the exam.

Chisholm returned to Girton the following year and then went to Göttingen to be one of three women first officially

Grace Chisholm Young and her son Frank

admitted to study at the university there. She wrote a dissertation and, in 1895, became the first of the three pioneering women students to receive her PhD. As such, she was the first woman to receive a doctorate as a regularly enrolled student in a university administered by the Prussian government. Although Sofia Kovalevskaia had received a PhD *in absentia* from Göttingen in 1874, she had never taken classes or been enrolled at any German university.

Chisholm returned to England with a doctorate but without a job and started a mathematical collaboration with William Henry Young, one of her Girton tutors. A year later they were married. The following year their first child was born and they moved to Göttingen where they both spent their time doing mathematical research. Although the Youngs were doing very well mathematically at Göttingen, neither of them was earning money, so Will Young eventually had to resume his tutoring duties at Cambridge. He did this part-time while Grace Young and their children remained on the continent. Based on his published work to that time, Will Young received a DSc from Cambridge in 1903 but was still only able to get part-time jobs. Meanwhile, in Göttingen, Grace Young was

studying medicine, raising four children, and doing mathematics. In 1908 the family, now with six children, moved to Switzerland. Will Young didn't get his first regular appointment until five years later, in 1913, and then it was in Calcutta. Grace Young and the children continued to live in Switzerland and finally, in 1919, Will Young got a job closer to his family, in Wales.

Early in their career, the Youngs produced many mathematical works together, but almost all of the papers were published under Will Young's name alone. The first of the works appearing under both their names was a 1905 geometry book intended for elementary school children. In 1914, Grace Chisholm Young started publishing under her own name again, this time on the foundations of the differential calculus.

At the outbreak of World War II Grace Young was visiting in England and Will Young was stranded alone in Switzerland, where he died in 1942 without having seen his family again. Grace Young died two years later in England. The Young's mathematical heritage includes their work as well as a daughter, a son, and a granddaughter who became mathematicians.

As mentioned earlier, Chisholm was one of the three first women to be officially admitted to study at Göttingen. The other two women were Americans, Margaret Maltby, who came to study physics, and Mary Frances Winston, who also came to study mathematics and who received her PhD two years after Chisholm. It is not a coincidence that none of these three women were German and that they had come to Göttingen to study mathematics and physics. At about this time, early in the 1890s, there had been discussion in Germany concerning admission of women to the universities. While the Prussian Minister of Culture was not unsympathetic to the idea, the overseer of the University at Göttingen was firmly against it.

In spite of that, it was decided that foreign women should be admitted to study mathematics. Felix Klein, the mathematician responsible for bringing Chisholm and Winston to Göttingen, explained later,

> Mathematics had here rendered a pioneering service to the other disciplines. With it matters are, indeed, most straightforward. In mathematics, deception as to whether real understanding is present or not, is least possible.

In the summer of 1893 Klein came to the United States with mathematical models to be displayed at the Columbian Exposition in Chicago and to speak at the International Mathematical Congress held in conjunction with the Exposition. In Chicago Klein met Mary Winston, a graduate student at the University of Chicago whose undergraduate degree was from the University of Wisconsin. After teaching for two years in Milwaukee she studied with Charlotte Scott at Bryn Mawr before coming to the University of Chicago in its inaugural year, 1892. Klein agreed to sponsor her admission to the university but could not provide her with financial support.

Five of Felix Klein's students, including Mary Winston and Grace Chisholm, at Göttingen
Courtesy Sylvia Wiegand

Although Winston applied for a European fellowship from the Association for Collegiate Alumnae, she did not receive it and was able to go to Germany only because of the generosity of a woman mathematician, Christine Ladd-Franklin, who personally provided her with a $500 stipend. Mary Winston arrived in Göttingen in the fall of 1893 and waited for Klein to clear the way for her admission to the university. A few weeks after her arrival, Winston wrote her family that the people in Göttingen were very skeptical as to her chances for admission; they were wrong.

Two years after coming to Germany, Winston published a short paper in a German mathematical journal. In a book entitled *A History of Mathematics in America before 1900* (Chicago: Mathematical Association of America, 1934), the authors, David Eugene Smith and Jekuthiel Ginsburg, note that this particular journal contains fifteen articles published by Americans between 1893 and 1897. They then list the authors of fourteen of these articles, omitting only the name Mary Winston. Winston's paper was based on a talk she had given in the mathematics seminar at Göttingen within months of her arrival in Germany. That talk was the first such given by a woman and she wrote her family that the presentation "went off reasonably well.... I do not think that anyone will draw the conclusion from it that women cannot learn Mathematics."

Upon her return to the United States in 1896, Mary Winston took a job teaching high school in Missouri. The following year she received her PhD from Göttingen and became Professor of Mathematics at Kansas State Agricultural College, now Kansas State University. Three years later she resigned and married Henry Byron Newson, a mathematician at the University of Kansas. Henry Byron and Mary Winston Newson had three children born in 1901, 1903, and 1909. Mary Winston was widowed in 1910 when her youngest child was just

Christine Ladd-Franklin
Courtesy of Christine Ladd-Franklin Papers, Rare Book and Manuscript Library, Columbia University

three months old. She moved in with her parents, who were then living in Lawrence. She returned to teaching, but not to mathematical research, a few years later at Washburn College in Topeka, Kansas. Her son reported that she took that job because Topeka was within commuting distance of Lawrence and her parents could care for the children during the week. Newson remained at Washburn until 1921; she spent the rest of her career at Eureka College in Illinois, retiring in 1942.

While Mary Winston Newson was the first American woman to receive a degree from a foreign university, by the time her degree was awarded in 1897 eight American women had been awarded PhDs in mathematics in the United States. The first American woman to earn a PhD in mathematics was Christine Ladd, who is the Christine Ladd-Franklin who provided the $500 that allowed Mary Winston to go to Germany. Ladd graduated from Vassar College in 1869 and during the next ten years she taught school and began publishing mathematics, including several articles that appeared in an American journal and at least twenty mathematical

questions or solutions to questions in a British periodical, *The Educational Times*. She also attended classes at Harvard and, in 1878, she applied to Johns Hopkins University to study mathematics at the graduate level. Since Hopkins was not open to women, her admission was far from routine. The head of the mathematics department, J. J. Sylvester, had read some of her published work and wrote on her behalf both for admission and for the granting of a fellowship. Ladd was admitted, but under the condition that she was to attend only Sylvester's lectures. Although she was not required to pay tuition, she was not given a fellowship. After her first year of attendance, 1878–79, she was voted the $500 stipend of a fellowship for the following year. Despite this, her name was not included on the actual list of fellows but appeared in a footnote. Thus, like Philippa Fawcett and Charlotte Scott, Christine Ladd received the recognition she deserved but not the title.

During her four-year stay at Johns Hopkins, Ladd did not confine her studies to those she could pursue with Sylvester, but also attended classes given by other members of the mathematics faculty. She also continued her contributions to mathematical journals. By 1882 she had written a dissertation under the direction of the logician Charles S. Peirce. However, she did not receive the PhD she had earned simply because Hopkins was unwilling to grant degrees to women. The following year she married a member of the Johns Hopkins mathematics faculty, Fabian Franklin; they had two children, a son who died in infancy and a daughter who was born in 1884. Starting about 1887 Ladd-Franklin began a second research career in the physiological optics of color vision. She later served as a lecturer in logic and psychology at Hopkins and then at Columbia. Actually Ladd had originally wanted to study physics but switched to mathematics, what she called "the next best subject," because

Wellesley Mathematics Department faculty
1927–1928: (Merrill, Smith, Comygs, Young,
Stark, Curtis, Copeland)
Courtesy of Wellesley College Archives.

physics laboratories were not open to women.

In 1926, Johns Hopkins offered Ladd-Franklin, then seventy-eight years old, an honorary doctorate in recognition of her work in color vision. Ladd-Franklin convinced Hopkins to award her the PhD in mathematics that she had earned forty-four years earlier. *The New York Times* reported that the ovation she was accorded was "one of the outstanding features of the day."

In 1886 Winifred Edgerton became the first American woman to be *awarded* a PhD in mathematics when Columbia University granted her the degree. In order for Edgerton, an 1883 graduate of Wellesley, to study at Columbia, which, like Hopkins, admitted only men, the trustees had to approve her request to study mathematics and astronomy. While the request was eventually granted, it took several meetings of the trustees. In 1982 one of her sons reported "that a condition of her admission was to dust the astronomical [instruments] and so comport herself as not to disturb the men students…. When working alone in the observatory she would arrange dolls around the room to keep her company. If she heard someone coming she hid them in a window box."

By the spring of 1886 Edgerton had written a dissertation and the trustees voted to award her the degree of Doctor of Philosophy *cum laude*. Her degree was both the first PhD in mathematics awarded to an American woman and the first degree of any kind that Columbia awarded to a woman.

For two years after the receipt of her doctorate, Edgerton taught at a school for girls in New York. In 1887 she married Frederick James Hamilton Merrill, a geologist who did not approve of her involvement in the movement to increase the availability of education for women. They soon had four children. Although she participated in the founding of Barnard College, her husband objected to her attendance at meetings that were held in a man's office and she resigned her position on Barnard's original Board of Trustees. In 1890 she moved to Albany, New York, where her husband worked for the State of New York. In 1904 the family returned to New York City and two years later she founded a school for girls in Greenwich, Connecticut, where she taught for 20 years. She was honored in 1933 when Columbia hung her portrait in what was then the woman's graduate clubroom in Philosophy Hall; its inscription reads: "She opened the door."

Although they were the first two American women to be awarded PhDs in mathematics, neither Ladd nor Edgerton emerged from an intellectual vacuum. Ladd had studied with the astronomer Maria Mitchell at Vassar, and Edgerton was a product of the entirely female mathematics department of Wellesley College. Wellesley's mathematics faculty stayed entirely female for many years and later most of the faculty had doctorates from among the best known schools in the country: Harvard, Johns Hopkins, the University of Chicago, the University of Pennsylvania, and Yale.

With many women PhDs on the mathematics faculty at Wellesley, it is clear that Ladd and Edgerton did not lack for successors, although it did take another seven years, until 1893, for the second PhD in mathematics to be awarded to a woman in this country. However, by the end of the nineteenth century Edgerton and Winston had been joined by eight women who had received PhDs in mathematics: three from Cornell, three from Yale, and two from Bryn Mawr College, both under the direction of Charlotte Scott. Even counting Christine Ladd-Franklin, the number eleven sounds quite small by modern standards. However, one must bear in mind that the PhD was a far-less-common degree then than it is now and only about 150 American men had received PhDs in mathematics by the turn of the century.

After the turn of the century, the number of American women entering the field of mathematics began to increase and by the mid 1930s was sufficiently established to motivate a study of the history of American women in mathematics by Helen Owens, a 1910 Cornell PhD in mathematics. The percentage of women receiving PhDs also increased from the 1890s through to the 1930s.

Courtesy of Wellesley College Archives. Photo by Pach Bros.

Winifred Edgerton

Approximate Percentage of US Mathematics PhDs Earned by Women

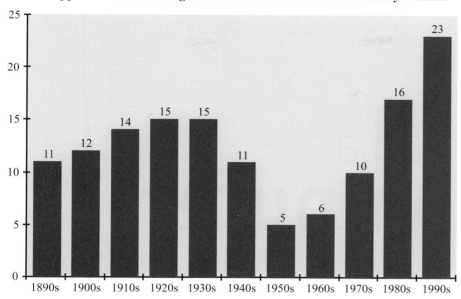

However, starting in the mid-1930s the percentages began to drop and by the mid-1940s they had dropped below ten percent, lower than at any time in the first four decades of the century. Although the drop ended in the post-Sputnik boom, it took until the 1980s to reach the level of the 1930s and it took until 1991 to surpass twenty percent.

While the percentages of women were dropping during the late 1930s through around 1960, the numbers dropped only from the mid-1930s through the mid-1940s. Furthermore, the increase from the 1960s to the 1970s in the percentage of women getting PhDs does not show the dramatic growth in the 1970s in the numbers of women getting PhDs. This growth continued into the 1980s, but at a slower rate. Since the 1990s, the numbers have been consistently above 200 PhDs being granted annually to women by schools in the United States. In 1999 the number exceeded 300 for the first time when 318 women received PhDs in mathematics in the United States; these women made up twenty-eight percent of all the PhDs granted in this country. If this growth continues, the visibility of women in mathematics should increase to the point where it will soon seem absurd to ask how many women mathematicians you can name.

If you would like to find more information about women in mathematics, consult the section on "Women in Mathematics" in *The History of Mathematics from Antiquity to the Present: A Selective Bibliography, edited by Joseph W. Dauben*, revised edition on CD-ROM edited by Albert C. Lewis, in cooperation with the International Commission on the History of Mathematics (Providence, RI: American Mathematical Society, 2000). This bibliography lists specific articles and books on most of the women referred to here. For more information on American women mathematicians see the articles cited in the second paragraph; for information on Fawcett and the education of women in England see *Philippa Fawcett and the Mathematical Tripos* by Stephen Sikos (Cambridge: Newnham College, 1990). ∎

Number of Mathematics PhDs Earned by Women at US Schools

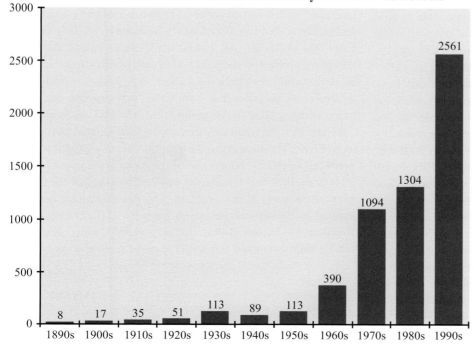

If
Pascal
had a
Computer

THADDEUS N. SELDEN and
BRUCE F. TORRENCE
Randolph-Macon College

The year is 1654. The young French scientist and mathematician Blaise Pascal has just been presented with a deceptively simple mathematical question, a question whose solution would in short time mark the birth of the theory of probability. The question was posed by Pascal's friend, the gentleman, soldier, socialite, and notorious gambler Antoine Gombaud, Chevalier de Méré. It involves a game of chance between two players, whom we imagine as tossing a coin several times in succession. Each time a head is tossed the first player scores a point, and each time a tail is tossed the second player scores a point. The first player to reach a predetermined number of points wins a pot of cash. The problem posed to Pascal later became known as the Problem of the Points. It goes like this: Suppose this game of chance is interrupted before either player has reached the number of points required to win. For instance, suppose the game is stopped when the first player is 3 points shy of winning, while the second player is 2 points shy of winning. How can the money in the pot be fairly divided between the two players?

Now you can dismiss this question on several grounds. First, you might object that the game itself is too simpleminded, too boring, too *blasé*. But very little imagination is needed to find examples of real games modeled this way. In a tennis match, for instance, the match is won by the first player to win three sets. Winning a set in tennis (between players of equal skill) is analogous to winning a point in our game of chance. Second, you might object to the idea of dividing the stakes between players when neither has yet won. However, this notion, it appears, is truly related to the realities of gambling, where one player may lose his nerve and move to abandon the game. The pot of cash is typically comprised of equal amounts of money put forth by the two gamers. If one player wants out before the game has ended (presumably because he is losing), he must compensate the other player with a fair division of the stakes. The question is simply: how should the stakes be divided?

The question was provocative for Pascal. A child prodigy of frail health, he made sporadic contributions to mathematics and science during his short life of 39 years. In the summer of 1654, when he devised his solution to the problem of the points, he was 31. He would remain mathematically active for only another few months, at which point a profound religious experience would lead him to abandon mathematics, physics, and all worldly endeavors for most of the remainder of his life.[1] His work on the problem of the points would mark a mathematical pinnacle in a short and controversial career. The progress of the work is documented by a correspondence between Pascal and his esteemed countryman, Pierre de Fermat. Fermat was 22 years senior to Pascal, and had a well-deserved reputation as both a superb mathematician and a gracious and modest correspondent. The two exchanged a series of letters, some (but not all) of which have survived. It is clear from this exchange that Pascal and Fermat independently solved the problem of the points, with each using different techniques. Pascal later published a short book, *Traité du triangle arithmétique*, an influential work which is responsible for his name being applied to the arithmetical triangle, i.e., Pascal's Triangle. In it, he rigorously derives the fundamental relationships among the numbers in the triangle, and, as an application produces a complete solution to the problem of the points.

It is worth noting that Pascal and Fermat were not the first to try their hands at this problem. Specific cases were documented in Italian manuscripts as early as 1380. During the Renaissance, Pacioli (1494), Tartaglia (1556), and Forestani (1603) each attempted but failed to solve particular cases of

---

[1] Pascal returned only once more to mathematics when in 1658 he reportedly "beguiled a persistent toothache by meditating on the problem of the cycloid." Leibniz later credited this work of Pascal's as directly contributing to his discovery of calculus.

Reprinted from November 2001, pp. 15–20

the problem. The most thorough attempt came from the prominent Italian mathematician Girolamo Cardano in 1539, although the solution he proposed was incorrect. By the year 1654, known results in probability amounted to "the idea of the exhaustive enumeration of the fundamental probability set, which had already been given by Galileo" [David]. It is fair to say that the modern theory of probability was born with the treatment given this problem by Pascal.

In this article, we will examine the two strikingly different approaches to the problem of the points developed by Fermat and Pascal. We will then demonstrate how a computer algebra system such as *Maple* or *Mathematica* can be harnessed to suggest a general solution for each mathematician's approach, and to show that the two approaches do indeed produce equivalent solutions. It is of interest to note that Pascal did have a computer of sorts; he developed and produced a calculating machine a decade earlier, the first of its kind. It is safe to say that it was not quite up to the task of symbolic computation, although the ideas that went into its development may have led in part to the recursive thinking Pascal would later employ. Pascal also bears the dubious honor of having a computer language named after him—some of our older readers may have heard of it.

Fermat's Approach

Throughout the discourse, Fermat sticks to a single principle for calculating probabilities. It is simply stated: enumerate every possible way for the game to end if play were allowed to continue. Find the likelihood of each of these outcomes, and sum those for which a particular player would win the game. This sum represents that player's likelihood of winning, so this is the proportion of the stakes that should be awarded to him.

An example will suffice to illustrate this method. (The approach taken in this example is the same as that outlined in Fermat's letter of September 25, 1654. That example involves a game with three players; we have employed the same reasoning to suit the current context.) Suppose a game is played in which the stakes will be awarded to the first player to win three points. If player A has two points, player B one, and the game is interrupted, how do we fairly divide the stakes? Under this scenario, at most two additional tosses of the coin would be required for a winner to emerge. If A wins the first toss, he is the winner; there is a $\frac{1}{2}$ probability that this will occur. If A loses this toss, but wins the next, he still wins. And as there are four equally likely outcomes in two tosses of a coin, and in just one of these does A lose the first toss and win the second, he has probability $\frac{1}{4}$ of winning the game in two tosses. Putting these together, we see that since there is a $\frac{1}{2}$ chance of A winning on the first toss, and a $\frac{1}{4}$ chance of his winning on the second, A has a $\frac{3}{4}$ probability of winning the game. He is therefore entitled to $\frac{3}{4}$ of the stakes. It is important to observe that this solution applies not only to a game where three points are required to win, but to any game where A lacks one point and B lacks two, regardless of the total number of points required to win.

Fermat concludes "And this rule is sound and applicable to all cases," and closes with his usual signature "I am, with all my heart, Sir, your, etc. FERMAT."

Pascal's Approach

Fermat, residing in Toulouse, sent his solution to Pascal in Paris. Upon receiving it Pascal replies, "Your method is very sound and is the one which first came to my mind in this research; but because the labor of the combinations is excessive, I have found a short cut and indeed another method which is much quicker and neater, which I would like to tell you here in a few words: for henceforth I would like to open my heart to you, if I may, as I am so overjoyed with our agreement. I see that the truth is the same in Toulouse as in Paris." Pascal is correct in noting that the labor inherent in Fermat's approach is formidable. For instance, if A lacks four games, and B lacks five, the proportion of the stakes going to A is $163/256$. We encourage the reader to perform this calculation, as it illustrates the complexity inherent in enumerating all the combinations.

Implicit in the correspondence between Pascal and Fermat is an understanding of the concept of mathematical expectation (a concept whose formal definition would be given three years later by Christiaan Huygens). Pascal and Fermat refer to the value of a game as the likelihood of winning the game times the amount of the stakes to be awarded the winner. For instance, if $60 were offered if one could roll a six on a fair die, the value of the game would be $1/6$ of $60, or $10.

Pascal, like Fermat, presents his solution in the form of an example. He employs what would today be called a recursive algorithm. He is strikingly modern in his approach. He supposes in his example that a total of three games are to be won. As in the preceeding example, the first player has scored two points, and the second player only one. He supposes in addition that each player has contributed 32 pistoles to the stakes. If the first player wins the next point, he takes the entire 64 pistoles. If he loses the next POINT, then each player has won two games, and so the stakes should be evenly divided. Pascal explains: "Then consider, Sir, if the first man wins he gets 64 pistoles, if he loses he gets 32. Thus if they do not wish to risk this last game, but wish to separate without playing it, the first man must say: 'I am certain to get 32 pistoles, even if I lose I still get them; but as for the other 32, perhaps I will get them, perhaps you will get them, the chances are equal. Let us divide those 32 pistoles in half and give one half to me as well as my 32 which are mine for sure'. He will then have 48 pistoles and the other 16." This is the same as the example outlined in the previous section: A gets $3/4$ of the stakes if play is halted when he lacks 1 game, and his opponent lacks 2.

We take this opportunity to introduce a convenient notation. Let $E[m, n]$ denote the amount of the stakes that A should be awarded in the case where A lacks m points and B lacks n. This is the expected value of A's winnings if play were allowed to continue. We thus have established that $E[1, 2] = 3/4$ of the total amount of the stakes, or 48.

Pascal continues. Note that he refers to each point played as a 'game.' "Let us suppose now that the first had won two games and the other had won none, and they begin to play a new game. The conditions of this new game are such that if the first man wins it, he takes all the money, 64 pistoles; if the other wins it they are in the same position as in the preceding case, when the first man had won two games and the other *one*.

"Now, we have already shown in this case, that 48 pistoles are due to the one who has two games: thus if they do not wish to play that new game, he must say: 'If I win it I will have all the stakes, that is 64; if I lose it, 48 will legitimately be mine; then give me the 48 which I have in any case, even if I lose, and let us share the other 16 in half, since there is as good a chance for you to win them as for me.' Thus he will have 48 and 8, which is 56 pistoles." Pascal is demonstrating a recurrence relation, in our notation,

$$E[1,3] = E[1,2] + \frac{1}{2}\big(E[0,3] - E[1,2]\big)$$

$$= \frac{1}{2}E[0,3] + \frac{1}{2}E[1,2].$$

Pascal continues: "Let us suppose, finally, that the first had won one game, and the other none. You see, Sir, that if they begin a new game, the conditions of it are such that, if the first man wins it he will have *two* games to *none*, and thus by the preceding case 56 belong to him; if he loses it, they each have one game, then 32 pistoles belong to him. So he must say: 'If you do not wish to play, give me 32 pistoles which are mine in any case, let us take half each of the remainder taken from 56". Thus the first player gets $32 + (1/2)24 = 44$ pistoles, or in our notation, $E[2,3] = 44$.

Pascal closes his letter in a manner that underscores his respect and admiration for Fermat: "This is poor recognition of the honor you do me in suffering my wearisome discourse for so long. I intended to say no more than two words to you and I have not yet told you what is closest to my heart, and that is the more I know you the more I admire and honor you, and if you could see how much that is you would find a place in your affections for him who is, Sir, your etc. PASCAL."

In this discourse, Pascal has introduced a recurrence relation for an equitable division of the stakes. In our notation, this relation takes the form

$$E[m,n] = \frac{1}{2}E[m-1,n] + \frac{1}{2}E[m,n-1]$$

where m and n are positive integers. Since these values depend upon the total amount of the stakes, it is simpler to deal with the underlying *proportion* of the stakes belonging to each player. We denote by $P[m, n]$ the proportion of the stakes that fairly belong to player A when he lacks m points and his opponent lacks n. Equivalently, $P[m, n]$ is the probablilty that A will win the game under these circumstances. $P[m, n]$ and $E[m, n]$ are then proportional, the product of $P[m, n]$ with the amount of the stakes yields $E[m, n]$. Thus the same recurrence relation holds for the probabilities:

$$P[m,n] = \frac{1}{2}P[m-1,n] + \frac{1}{2}P[m,n-1].$$

The base cases are $P[0, n] = 1$ for positive n (if A needs no games to win, he is certain to win), and $P[m, m] = \frac{1}{2}$ for positive m (if each player lacks the same number of points, A has a $\frac{1}{2}$ chance of winning). All other cases are then computed from these base cases recursively. The longwinded example provided above by Pascal can be dealt with efficiently with this notation as follows:

$$P[2,3] = \frac{1}{2}P[1,3] + \frac{1}{2}P[2,2]$$
$$= \frac{1}{2}\left(\frac{1}{2}P[0,3] + \frac{1}{2}P[1,2]\right) + \frac{1}{2}\cdot\frac{1}{2}$$
$$= \frac{1}{2}\left(\frac{1}{2} + \frac{1}{2}\left(\frac{1}{2}P[0,2] + \frac{1}{2}P[1,1]\right)\right) + \frac{1}{4}$$
$$= \frac{1}{2}\left(\frac{1}{2} + \frac{1}{2}\left(\frac{1}{2} + \frac{1}{2}\cdot\frac{1}{2}\right)\right) + \frac{1}{4}$$
$$= \frac{11}{16}.$$

Unlike Fermat, Pascal did not end with a vague statement about the generality of his method. *Au contraire*, Pascal probed further and used his recursion relation to deduce an explicit formula for $P[m, n]$. This deduction was rigorously presented in his *Traité du triangle arithmétique*. It is, however, easily discovered (and in a more general form, for we do not assume that each player has an equal chance of winning a point) with the aid of a computer algebra system. This approach also leads to a simple proof that the approaches used by Pascal and Fermat lead to equivalent solutions.

Using *Mathematica*, we first program Pascal's recurrence relation and initial conditions. We find it convenient to denote the respective probabilities of players A and B winning a single point as $a/_{a+b}$ and $b/_{a+b}$, as these quantities clearly sum to 1. (We also replaced Pascal's boundary condition $P[m,m] = \frac{1}{2}$ with the condition $P[m, 0] = 0$, which says simply that A cannot win when B requires no games to win.) We then use *Mathematica* to calculate recursively a table of values for $P[m, n]$. We do this with the hope that a clear pattern will emerge. The table below shows $P[m, n]$, with rows corresponding to $m = 1$, $m = 2, \ldots, m = 5$, and columns running from $n = 1$ to $n = 4$. Thus $P[2, 3]$ appears in row 2 and column 3; if $a = b = \frac{1}{2}$, it reduces to $\frac{11}{16}$ and agrees with our earlier calculation.

$$\frac{a}{(a+b)^1} \qquad \frac{a(a+2b)}{(a+b)^2} \qquad \frac{a(a^2+3ab+3b^2)}{(a+b)^3} \qquad \frac{a(a^3+4a^2b+6b^2+4b^3)}{(a+b)^4}$$

$$\frac{a^2}{(a+b)^2} \qquad \frac{a^2(a+3b)}{(a+b)^3} \qquad \frac{a^2(a^2+4ab+6b^2)}{(a+b)^4} \qquad \frac{a^2(a^3+5a^2b+10b^2+10b^3)}{(a+b)^5}$$

$$\frac{a^3}{(a+b)^3} \qquad \frac{a^3(a+4b)}{(a+b)^4} \qquad \frac{a^3(a^2+5ab+10b^2)}{(a+b)^5} \qquad \frac{a^3(a^3+6a^2b+15b^2+20b^3)}{(a+b)^6}$$

$$\frac{a^4}{(a+b)^4} \qquad \frac{a^4(a+5b)}{(a+b)^5} \qquad \frac{a^4(a^2+6ab+15b^2)}{(a+b)^6} \qquad \frac{a^4(a^3+7a^2b+21b^2+35b^3)}{(a+b)^7}$$

$$\frac{a^5}{(a+b)^5} \qquad \frac{a^5(a+6b)}{(a+b)^6} \qquad \frac{a^5(a^2+7ab+21b^2)}{(a+b)^7} \qquad \frac{a^5(a^3+8a^2b+28b^2+56b^3)}{(a+b)^8}$$

Can you conjecture a general expression for $P[m, n]$ on the basis of this evidence? Finding a suitable expression for the coefficients of the powers of a and b is the tough part. We turn to Pascal's Triangle for clues.

The Arithmetical Triangle

Pascal's treatise on the Arithmetical Triangle was written in 1654, presumably to aid in the formal solution of this problem. Pascal begins the treatise with a recursive definition of the triangle. The arithmetical triangle consists of a table of nonnegative integers, with rows and columns numbered from 0 to ∞. The entries in row 0 and in column 0 are all 1's. To find an entry whose position is in the interior of the table, simply add the two numbers that appear above and to the left of that position. The beginning of the triangle then looks like this:

1	1	1	1	1	1	1	1
1	2	3	4	5	6	7	
1	3	6	10	15	21		
1	4	10	20	35			
1	5	15	35				
1	6	21					
1	7						
1							

The number in row t and column s is denoted $\binom{s+t}{s}$. Pascal was the first to present the now standard combinatorial argument demonstrating that $\binom{s+t}{s}$ represents the number of ways to choose s positions in a row of $s + t$ positions. For instance, if $s + t = 7$ people are arranged in a row, there are $\binom{7}{3} = 35$ ways to choose precisely 3 of them. Pascal, in this treatise, was also among the first mathematicians to formally employ the technique of mathematical induction.[2] It is widely acknowl-

---

[2] At present, the first known use of induction was by Levi ben Gerson in the year 1321, for none other than proving a formula for numbers in the arithmetical triangle. Earlier, Pascal was given credit for first implementing mathematical induction, and earlier still John Bernoulli was credited. The term "induction" was not yet in use; it was coined by Augustus DeMorgan two centuries after the time of Pascal.

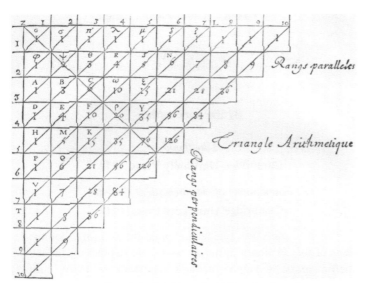

Pascal's original Arithmetical Triangle, from the
Traité du triangle arithmétique (1665)

edged that virtually all of the results about the triangle contained in Pascal's treatise were known at the time it was written. The arithmetical triangle was known to be an ancient invention at the time Pascal's treatise appeared. His work provided, however, a comprehensive and modern treatment that became widely cited by those who followed. Pascal was also "a little forgetful about his sources" [Edwards], and failed in his treatise to credit others with results that were surely not his own. It is for these reasons that his name has become associated with the arithmetical triangle. A key result concerning the numbers in the triangle appears as the conclusion to part I of the treatise: The binomial coefficient $\binom{s+t}{s}$ can be directly computed via the formula

$$\binom{s+t}{s} = \frac{(s+t)!}{s!\,t!}$$

where $n!$ denotes the factorial of n.

The Solution to the Problem of the Points

Having the arithmetical triangle at our disposal, a careful look at the table of values we created for $P[m, n]$ is in order. The pattern is now unmistakable; a comparison of the coefficients appearing in the table with the numbers in the arithmetical triangle leads to the following conjecture:

$$P[m,n] = \sum_{i=0}^{n-1} \binom{m+n-1}{i} \frac{a^{m+n-1-i}b^i}{(a+b)^{m+n-1}}$$
$$= \sum_{i=0}^{n-1} \binom{m+n-1}{i} p^{m+n-1-i}q^i \tag{1}$$

where $p = a/_{a+b}$ and $q = b/_{a+b}$. When $p = q = \frac{1}{2}$, this is precisely Pascal's solution to the problem of the points. This equation

can be proven from the recursion relation for $P[m, n]$ and the boundary conditions $P[m, 0] = 0$ and $P[0, n] = 1$, by using induction on $m + n$. The inductive step makes use of the defining relation for the numbers in the arithmetical triangle. We encourage the reader to reconstruct this proof. When $p = q = \frac{1}{2}$, this is essentially the proof that Pascal used in his treatise on the arithmetical triangle. Pascal noted that he came to the formula by observing that the recursion relation for $P[m, n]$ differed from that of the arithmetical triangle only by a factor of $\frac{1}{2}$, and by the boundary conditions.

Equation 1 above is of course more general than Pascal's solution, as it allows for players having an unequal chance of scoring a point. It provides a mild justification for our use of a computer algebra system, albeit a rather weak one. This general formula was first established by John Bernoulli in the year 1710 in a letter to Montmort. Bernoulli and Montmort exchanged a series of letters at the time after having read the already historic letters of Pascal and Fermat. Bernoulli noted that the expression for $P[m, n]$ above can be thought of as the first n summands in the standard expansion of the binomial $(p+q)^{m+n-1}$.

A better justification for the employment of a computer algebra system is to find another useful form for Equation 1. The entries in the table of values for $P[m, n]$ take on a very different look when we ask our computer algebra system to expand, rather than simplify, them. To save space, we only show the columns $n = 3$ and $n = 4$.

$\dfrac{ab^2}{(a+b)^3} + \dfrac{ab}{(a+b)^2} + \dfrac{a}{(a+b)^1}$	$\dfrac{ab^3}{(a+b)^4} + \dfrac{ab^2}{(a+b)^3} + \dfrac{ab}{(a+b)^2} + \dfrac{a}{(a+b)^1}$
$\dfrac{3a^2b^2}{(a+b)^4} + \dfrac{2a^2b}{(a+b)^3} + \dfrac{a^2}{(a+b)^2}$	$\dfrac{4a^2b^3}{(a+b)^5} + \dfrac{3a^2b^2}{(a+b)^4} + \dfrac{2a^2b}{(a+b)^3} + \dfrac{a^2}{(a+b)^2}$
$\dfrac{6a^3b^2}{(a+b)^5} + \dfrac{3a^3b}{(a+b)^4} + \dfrac{a^3}{(a+b)^3}$	$\dfrac{10a^3b^3}{(a+b)^6} + \dfrac{6a^3b^2}{(a+b)^5} + \dfrac{3a^3b}{(a+b)^4} + \dfrac{a^3}{(a+b)^3}$
$\dfrac{10a^4b^2}{(a+b)^6} + \dfrac{4a^4b}{(a+b)^5} + \dfrac{a^4}{(a+b)^4}$	$\dfrac{20a^4b^3}{(a+b)^7} + \dfrac{10a^4b^2}{(a+b)^6} + \dfrac{4a^4b}{(a+b)^5} + \dfrac{a^4}{(a+b)^4}$
$\dfrac{15a^5b^2}{(a+b)^7} + \dfrac{5a^5b}{(a+b)^6} + \dfrac{a^5}{(a+b)^5}$	$\dfrac{35a^5b^3}{(a+b)^8} + \dfrac{15a^5b^2}{(a+b)^7} + \dfrac{5a^5b}{(a+b)^6} + \dfrac{a^5}{(a+b)^5}$

A careful look at the results leads to the conjecture:

$$P[m,n] = \sum_{i=0}^{n-1} \binom{m-1+i}{m-1} \left(\frac{a}{a+b}\right)^m \left(\frac{b}{a+b}\right)^i$$
$$= \sum_{i=0}^{n-1} \binom{m-1+i}{m-1} p^m q^i. \tag{2}$$

This equation is also easily established by an induction argument nearly identical to the one which establishes Equation 1. Since the recursion relation that defines $P[m,n]$ admits a single solution with the given boundary conditions, this proves that the two solutions are indeed equivalent. This equivalence was first noted by Montmort in 1713, using direct algebraic methods on a few examples. Montmort did not provide a general proof.

However, no formula of this ilk was ever presented by either Pascal or Fermat. If Pascal had a computer, perhaps he would have discovered it; it's certainly fun to realize that undergraduates today have at their disposal tools capable of revealing patterns that the masters of yesteryear overlooked. What is perhaps more interesting, however, is that an elementary proof of Equation 2 exists that was within the grasp of Pascal in 1654; that is to say, this proof uses only ideas that are found in his treatise on the arithmetical triangle. It is even more interesting that this direct proof is itself a straightforward generalization of Fermat's approach to the problem. Perhaps this was reason enough for Pascal to fail to pursue it.

To outline the direct proof of Equation 2 let us first introduce a useful notation for representing a sequence of consecutive points played during a game: we use an ordered tuple of a's and b's. For instance, if player A wins a point, and player B wins the next two points, we denote this as (a, b, b). We refer to any such sequence as a "point-sequence." Suppose now that A has probability p of winning each point, and B has probability $q = 1 - p$ of doing the same (in the correspondence between Pascal and Fermat it was always assumed that $p = q = \frac{1}{2}$). We note that any sequence of s a's and t b's has likelihood $p^s q^t$. Thus, for example, the likelihood of one win for A, and 2 for B, in any order, is $3pq^2$. Why 3? One pq^2 is needed for each of the sequences (a, b, b), (b, a, b), and (b, b, a). The coefficient 3 is the number of ways to choose one position for the a from the three that are available. Recall that $\binom{s+t}{s}$ represents the number of ways to choose s positions (for the a's) from all available positions in a point-sequence of length $s + t$. So the likelihood of s wins for A and t for B, in any order, is $\binom{s+t}{s}p^s q^t$.

We are now ready to establish Equation 2 directly. We argue in a manner not unlike Fermat's approach to the problem of the points. Suppose that A needs m points to win and B needs n when play is interrupted. We wish to find $P[m,n]$, the probability that A would win if play were to continue. But this is the probability that A scores m points *before* B scores n. Our task reduces to enumerating the possible games in which A wins, and summing the probabilities of these games. Note that every possible game in which A wins is represented by a point-sequence that contains exactly m a's, fewer than n b's, AND has an a in the final position (player A's winning point). Now organize these sequences by their length. The least number of points that need to be played if A is to win is m, and there is but one winning sequence in this case: the m-tuple $(a, a,..., a)$, which has probability p^m. If $m + 1$ points are played, the winning games for A are those $(m + 1)$-tuples comprised of exactly m a's and 1 b, with an a in the final position. The probability of such a game is $\binom{m}{m-1}p^m q$, the coefficient $\binom{m}{m-1}$ representing the number of ways to choose $m - 1$ positions among the first m for the a's (the final position must contain the last a, player A's winning point). One continues in this fashion for games requiring more and more points, noting that at most $m + n - 1$ points will need to be played (for in any $(m + n)$-tuple, there must be either at least m a's or at least n b's, and hence a winner). An expression for $P[m, n]$ results:

$$P[m,n] = \sum_{i=0}^{n-1} \binom{m-1+i}{m-1}p^m q^i.$$

And this is precisely the result we are after. ■

For Further Reading

There are several good, readable accounts of the Fermat-Pascal correspondence, in particular: *Pascal's Arithmetical Triangle*, A. W. F. Edwards, Charles Griffin & Company LTD, London, 1987 and *A History of Probability and Statistics and their Applications before 1750*, A. Hald, John Wiley and Sons, New York, 1990. English translations of the surviving letters can be found in *Games, God, and Gambling*, F. N. David, Charles Griffin & Company LTD, London, 1962.

President Garfield and the Pythagorean Theorem

VICTOR E. HILL IV
Williams College

In 1876 a member of the House of Representatives produced a new proof of the Pythagorean Theorem. The proof itself is somewhat unusual in that it is based on a trapezoid, although inspection shows that it bears a resemblance to an argument found in the oldest known Chinese mathematics text. More interesting is the fact that Congressman James Abram Garfield, who later became the twentieth President of the United States, apparent-

ly had a rather mixed ability and background in mathematics, although he exhibited an unusually high level of overall intelligence.

Garfield was born in Ohio on 19 November 1831, the youngest of four children, and was barely two years old when his father succumbed to what is euphemistically recorded as a cold and sore throat, but as Jane Austen's Mrs. Bennet says, "People do not die of little trifling colds." It was almost certainly pneumonia, but the customs of the time did not allow for a grown man to be acknowledged to have been felled by an "unmanly" disease like pneumonia. Garfield's early schooling took place at the usual one-room schoolhouses near his rural home. His 13-year-old sister Mehitabel (named for her Puritan grandmother) carried her five-year-old brother through the snow on her back to attend classes. Eventually Eliza Garfield donated a portion of her land for a schoolhouse nearer their home. Garfield was also inspired by the visits of the local teacher when he boarded around with various neighborhood families. Here we have the

beginning of the classic rags-to-riches story, fulfilling the heroic archetype of the day: the boy from the poor background who did grow up to be President of the United States. Garfield was, in fact, the last United States President to be born in a log cabin.

Garfield displayed rudimentary geometrical talent at an early age. His older brother Thomas sacrificed his own education in order to work for the family. Under his example and tutelage, James learned basic principles of carpentry; several surviving accounts relate that he had a good eye for angles, corners, and how pieces of wood would fit together. When he was about 15, he had a carpentry job building a woodshed for a black-salter; the employer noticed that James could not only read and write, but was "death on figgers." Garfield accepted a job keeping the man's account books and used the time to devour his employer's library after hours.

Garfield subsequently went to Cleveland and was hired by a cousin as a "canal boy." One of his responsibilities was guiding a canal boat as it encountered another boat going in the opposite direction; this operation is tricky in that both boats have ropes attaching them to the mules or horses on shore, and care is required to ensure that the ropes do not tangle. Occasionally the ropes would get

James A. Garfield

Reprinted from February 2002, pp. 9–11, 15

snarled, and Garfield's job was to disentangle them without getting them caught on the bridges overhead; this task required some sense of spatial relations and perhaps, in modern terms, elementary knot theory.

In March 1849 Garfield entered Geauga Seminary in Ohio, having been encouraged to return to school by the cousin who employed him and who was impressed by his knowledge in all subjects. At Geauga he alternated between his studies and stints of teaching in the local school November–February and March–May. In his diary for May he recorded, "The *Algebra* has been nothing but *theory* for the past month. I like it however, very well." Three weeks later he wrote, "Upon this memorable (I guess) day, we have finished our Algebra, and commenced reviewing. I can look back to the time when it seemed a Herculean task to me but I [am] glad to know that *Perseverentia vincit omnia*." How much he learned may, however, be questioned since he noted two months later, "My Algebra was not examined because it was so small, and they had not time." In August, he wrote, "Agreed to take the Mental Arithmetic class reluctantly. Hope for success." These entries do not sound like the record of a particularly successful student of mathematics.

In August 1851 Garfield entered a college preparatory school, the Western Reserve Eclectic Institute (now Hiram College). During October he recorded in his diary, "I have today commenced the study of Geometry alone without class or teacher. I commenced this morning and in about two hours I got the definitions and 8 propositions and the scholiums and corollaries appended." That would suggest very rapid assimilation of material for two hours of work, but we have no further record of any interest or study of geometry in Garfield's diary, which makes his return to the Pythagorean Theorem some 25 years later all the more intriguing.

Garfield wrote to the presidents of Brown, Yale, and Williams Colleges, stating how much he knew and asking how long it would take him to finish his degree at each school. President Mark Hopkins of Williams replied, "If you come here, we shall be glad to do what we can for you," and more directly, "I can only say that if you come, we shall be glad to do for you what the circumstances will admit." Two other factors may have come into play. One is that the ranks of the junior class had just been significantly reduced by the mass expulsion of a bunch of its rowdier members following an incident apparently involving some spiked lemonade; having virtually no endowment or patronage at the time, the College could hardly afford to be overly selective. In addition, this was a time at which Hopkins took an interest in admitting "sincere yet rustic" men from the West to combat what he called "signs of effeteness" in his students. Garfield did decide to attend Williams, where he hoped to complete his studies in one year. On arrival at the beginning of the summer 1854, he was examined orally and was found highly proficient in Greek and Latin but unable to solve even simple geometry problems and was told to bring up his mathematics. Since Garfield was admitted as a junior, he would have had a deficiency in mathematics. The college catalog for those years shows a freshman requirement of algebra to the binomial theorem and Loomis's geometry, and a sophomore year of higher algebra and Loomis's analytical geometry and calculus, with no further mathematics require-

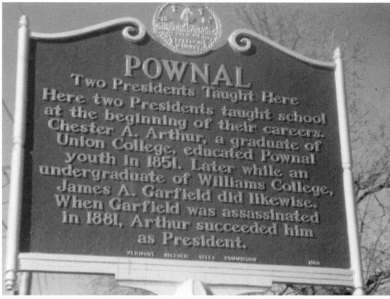

Photo courtesy of Victor E. Hill IV.

ment (or offerings) for upperclass students. Garfield, however, remained cheerful, as he recorded, "I recite…in Analytical Geometry and Conic Sections.… Privately I am bringing up the Trigonometries and Surveying and Navigation.… My mind seems unusually clear and vigorous in Mathematics, and I have considerable hope and faith in the future." Whatever he lacked in accomplishment, he made up for in confidence, an attribute he displayed throughout his life.

Williams marked the first time that Garfield experienced any sort of educational setback. In Ohio, which was really still frontier country, he had been teaching advanced courses, and his studies had always come easily to him. Indeed, at his first teaching appointment, he was accounted a wonder because he was the only instructor in two counties who could solve a problem requiring long division! Thus he arrived at the College with the expectation that he could study for nine weeks and enter as a senior. (It is important to realize that Garfield was 22 at the time, significantly older than most of the other students; his nickname among his peers was "Old Gar.") President Hopkins allowed him to attend the last few classes of the term and Commencement; this experience, along with his oral entrance examination, enabled him to see that he was not as well prepared as he had believed.

Sadly, Garfield's diaries during his Williams years, 1854–56, are sparse, and information about what he learned about mathematics, presumably on his own, while a college student is entirely lacking. Like many students of limited means, he spent the long summer and winter vacations teaching in nearby communities. The winter months, in particular, were times when boys in their mid-teens (who could be spared then from farm work) were given to older male teachers such as Garfield. A historical marker in Pownal, Vermont (just over the Massachusetts state line), commemorates the teaching activity of both Garfield and his successor in the Presidency, Chester A. Arthur, in that community. Garfield was unusual in that, in addition to teaching, he filled in numerous times in the pulpits of churches of the Disciples of Christ and in conducting revival meetings, making him the only preacher to become a U.S. President.

In his second term at Williams, Garfield paid the bills by teaching two writing courses. The first grew out of a lecture engagement he had fulfilled in Pownal. The second took place in Poestenkill, New York, about 21 miles from Williamstown. He was due at college when the second course ended, but he was persuaded to stay an additional week to assist with a revival there. The local Disciples were so pleased that they gave him $20 and some new clothes to make up for his having missed the first three weeks of the term.

Garfield returned to teach at Hiram after graduating from Williams and married his former student Lucretia Rudolph on 11 November 1858. He enlisted in the Civil War and was made an officer, preparing himself by fashioning companies, troops, and soldiers out of wood and using them as physical examples to teach himself and his officers the principles of

maneuvers and tactics—another example of his geometrical intuition at work. He maintained contacts with Williams alumni and, in fact, spent the day before his Presidential Inauguration attending a Williams reunion in Washington at which he also gave a speech—this caused him to stay up until 2:30 A.M. writing his inaugural address. (That inaugural was the first to be attended by the President's own mother.) It is ironic, in view of his longtime college connection, that Garfield was assassinated in 1881 en route to his 25th Williams College reunion.

Aside from the curiosity of the Pythagorean Theorem, Garfield did exercise mathematical skills in Congress. He served on the committee which oversaw the 1880 census, addressing the members of the American Social Science Association on the subject. Garfield also served (at his request) on the Ways and Means Committee, where he furthered his interest in the mathematics of finance. The same boy who had been "death on figgers" was entirely self-taught in these matters and is reported to have made something of a nuisance of himself on the committees with the tenacity of his mathematical interests.

Although Garfield was not a particularly strong student in college, he did have at least one remarkable ability, which he delighted to demonstrate: he could write in Latin with one hand and, simultaneously, in Greek with the other! This curiosity does reflect his extensive background in classical languages; in fact he and his wife routinely spoke Latin and Greek in their home. Another indication of Garfield's intellect is that he was admitted to the Bar while serving in Congress; he argued his first case ever before the Supreme Court—and won.

Garfield took his proof of the Pythagorean Theorem with him when he spoke at the Dartmouth College Chapel on 7 March 1876. He recorded in his diary, "After meeting [I] had a private conference of an hour with Professors

Garfield's Proof of the Pythagorean Theorem

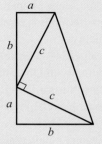

Area of trapezoid

$\frac{1}{2}$ (sum of bases)(altitude)

$$= \tfrac{1}{2}(a+b)(a+b)$$
$$= \tfrac{1}{2}a^2 + ab + \tfrac{1}{2}b^2.$$

Sum of areas of three triangles

$$= \tfrac{1}{2}ab + \tfrac{1}{2}ab + \tfrac{1}{2}c^2$$
$$= ab + \tfrac{1}{2}c^2.$$

So $\tfrac{1}{2}a^2 + \tfrac{1}{2}b^2 = \tfrac{1}{2}c^2$ and $a^2 + b^2 = c^2$.

Quimby and Parker in reference to [ex-Senator James Wallis] Patterson's connection with the Credit Mobilier. I showed my solution of the *pons asinorum* to Professor Quimby, who said it was new and asked for a copy for publication in a Mathematical journal." (*Pons asinorum,* which is Latin for "bridge of fools," usually refers to Euclid's proof that the base angles of an isosceles triangle are equal.) Garfield's solution was published in the *New England Journal of Education* in 1876. The text begins:

In a personal interview with Gen. James A. Garfield, Member of Congress from Ohio, we were shown the following demonstration of the pons asinorum, which he had hit upon in some mathematical amusements and discussions with other M.C.'s. We do not remember to have seen it before, and we think it something on which the members of both houses can unite without distinction of party.

The proof consists of two copies of the triangle, with their legs placed on a common straight line, as shown in the figure. The trapezoid is then completed by joining the remaining vertices of the triangles, computing the area of the trapezoid, and comparing it to the sum of the areas of the three triangles into which it is decomposed.

Adjoining the mirror reflection of this trapezoid, one obtains the square found in the ancient Chinese text, the *Arithmetic Classic of the Gnomon and the Circular Paths of Heaven*; however, there is no indication that Garfield or any of his colleagues in Congress had any knowledge of this or any similar historical proofs; indeed, what we have seen of Garfield's biography shows that his mathematical training, especially in geometry, was limited and largely self-taught. Thus he can be credited with some innate instinct and a worthy degree of geometrical curiosity, perhaps not often found in the halls of legislatures. ■

Acknowledgments

The author acknowledges the kind assistance of Amy Rupert and Linda Hall of the Williams College Special Collections, and of his research assistant, Victoria C. H. Resnick of Indiana University.

For Further Reading

Read Allen Peskin's *Garfield: a Biography*, Kent State University Press, for more details of Garfield's life. *The Pythagorean Proposition* by Elisha S. Loomis, published by the National Council of Teachers of Mathematics, contains hundreds of proofs of the theorem: Garfield's is number 231. You can read about the *Arithmetic Classic of the Gnomon and the Circular Paths of Heaven* in David Burton's *The History of Mathematics: An Introduction*.

Life
and Death
on the
Go Board

PETER SCHUMER
Middlebury College

The ancient Asian board game, Go, has been revered as the ultimate game of intellectual and aesthetic enlightenment for millennia. Originating in China or Tibet about 4000 years ago where it was known as Wei Chi (or Game of Encirclement), the game spread to Korea and then Japan along with other Buddhist practices and beliefs about the year 500 C.E. The game flourished in Japan where it spread from the Buddhist priesthood and the court aristocracy to the samurai class and then to the population at large. Along with music, calligraphy, and painting, Go was considered one of the essential four arts that a well-educated citizen should cultivate.

Today the game is played by more than 30 million enthusiasts worldwide and by a growing number of professional players in Japan, China, and Korea who compete in lucrative tournaments and televised matches to appreciative Asian audiences. In addition to prestigious national titles and championships there are now several annual international championships sponsored by multi-national corporations such as Fujitsu Limited, the Ing Foundation, and the Tongyang Securities Company. There are, of course, a far larger number of strong amateur players with a host of tournaments played throughout the globe—some over the internet. The top amateurs representing over fifty nations compete in the annual World Amateur Go Championship. As you might expect, the individual winner is usually from either China or Japan, but players from the West have been faring better every year.

Like mathematics, Go originates in simple elements that lead naturally to unfathomable complexity. Its basic structure is intellectually pleasing and elegant and it encourages and rewards careful thinking, long-range planning, and deep problem-solving skills. Both pursuits involve unlimited creativity within a rigid structure. In both disciplines, there is a strong sense among practitioners that they are uncovering deep universal truths rather than creating temporary or artificial forms. Both mathematics and Go have a rich and fascinating history peopled with brilliant practitioners whose profound insights serve as inspirations to all those who follow. Furthermore, in both pursuits the joy of discovery and the thrill of self-improvement serve as grand incentives and a source of perpetual rewards.

Fortunately, one does not have to choose between these two beautiful realms. You can be a mathematician *and* a Go player! Okay, so how do you play?

The game is played on a 19 by 19 line grid laid out on a rectangular wooden board. The board starts out empty. Two players (black and white) alternately place identically shaped lenticular "stones" of their own color on the board beginning with black (in these illustrations we use blue for greater visual effect). The object is to safely surround the largest total area. Stones can connect horizontally or vertically with friendly stones of the same color, but can also be captured if they lose all their "breathing spaces" to the opponent's stones. Each competitor's final score is actually the total number of empty intersections surrounded by that player's stones minus the number of his or her stones which were captured. Let us illustrate further.

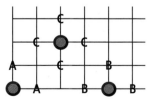

Figure 1. Liberties
The blue stone in the corner has two liberties (or breathing spaces) marked A. Once white has occupied both of them, the blue stone would be removed from the board by white and kept as a prisoner until the end of the game. Similarly the stone on the bottom edge has three liberties (marked B), and the stone in the middle has four liberties (marked C).

Reprinted from February 2002, pp. 16–19

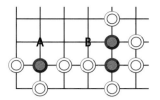

Figure 2. Capture

The single blue stone is in *atari,* which means it has just one remaining liberty. If white plays at A, then the blue stone will be removed. Similarly, the two blue stones on the right are in atari since five out of their six liberties are taken. If white plays at B, the two blue stones are removed together.

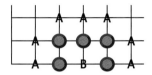

Figure 3. An Eye

The group of blue stones have eight liberties, seven external liberties (marked A) and one internal liberty (marked B). White must play all the points marked A before playing B in order to remove the blue stones. The reason that B must be played last is that "suicide is illegal," no stone can be played where it has no liberties. After white makes all the A moves, a white move at B would be legal since the blue stones would be removed from the board as part of white's last move.

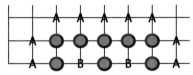

Figure 4. Two Eyes

All the blue stones are part of one group (connected along vertical or horizontal lines). However, unlike the blue group in Figure 3 which has one "eye," this group has two "eyes." Once white has played at all the points marked A, blue still has two internal liberties (B). However, white cannot play at either one since to do so would be an illegal suicide move. That is, white needs to play both in order to remove any blue stones, but cannot do so. Hence the blue stones are alive and blue has two points of territory.

On any given turn, a player may "pass" rather than place a stone if he or she cannot find a constructive move to make. In keeping with the polite etiquette of the game, a Go game ends when both players pass in succession.

The rules are quite simple and elegant, but give you little understanding of how a game is played. Figure 7 shows a sample game on a 9 × 9 board. Play through the game in order (from move 1 to move 42) with paper and pencil or with a real board and stones. Beginners usually learn to play on quarter boards where long-range strategic concepts do not play a significant role.

To improve at the game, one must play hundreds and even thousands of games, carefully replay the games of experts, and spend endless hours studying opening play, tactics and strategy, proper shape and form, and working out localized life and

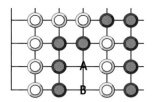

Figure 5. Seki

Blue has four stones surrounded by white stones and white has three stones surrounded by blue stones. If either blue or white plays A or B, the opponent will play the other point and capture. This is a *seki* (or impasse) where neither player will play here. At the end of the game, all the stones are alive, but neither point (A nor B) counts as territory for either player.

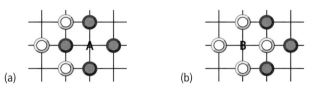

Figure 6. Ko rule

Figure (a) shows a *ko* situation. White can play at A and capture a blue stone. The situation would then look like (b). It would appear that blue could immediately play back at B, capturing a white stone and returning the board to (a). Conceivably this could go on indefinitely. (The word ko is a Buddhist word meaning eternity.) The ko rule states that on the next turn after white first takes at A, blue has to play somewhere else on the board instead of at B. On blue's subsequent turn, blue could then capture white by playing at B if white hasn't filled in that point in the meantime. So the ko rule states that a player must make an intervening move before returning to a particular ko.

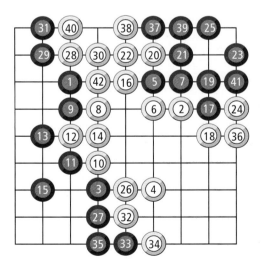

Figure 7. A 9 × 9 game

Blue's plays are odd-numbered and white's are even-numbered. At the end of the game all the blue stones and white stones are alive. White wins with 20 points of territory to blue's 19 points. For blue to place any additional stones on the board would simply subtract from his own territory or add a prisoner for white to take (and similarly for white). So blue passed for move 43 and white passed for move 44, thus ending the game.

death problems (known as *tsume-go* problems in Japan). In one such type of problem, the *status* problem, the reader must determine if a given group of blue stones are dead, alive, or unsettled. A dead group is one where perfect play by both players would result in the capture and removal of the group even if blue were to play first. An alive group is one where perfect play results in a living group even if white plays first. An unsettled group is one where the life or death of the stones depends on whose turn it is to play. In an actual game, players are usually in no rush to play where the stones are completely settled (alive or dead), but rather choose to play where groups are unsettled. Of course, in a real game there are many individual groups and the life or death of some groups might be subtly interconnected. In addition, timing and the proper order of moves is essential. So determining the next best move in a game often involves as much art and intuition as logic and deduction. But in a *tsume-go* problem the answer is generally definitive (though usually not at all obvious).

Here is a standard beginner's problem. What is the status of the blue group below? Think about the answer before reading the subsequent explanation.

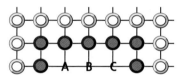

Figure 8. Eye space of three in a row
The blue group is unsettled. A blue play at B would create a live group as in Figure 4. If white gets to play B first, then blue will be dead. (In theory, next white plays at A, blue plays at C capturing two white stones, then white plays at A putting blue in atari, and if blue plays B, then white plays A removing all nine blue stones. Blue cannot improve his outcome by playing at A or C after white's play at B. (Can you see why?) Hence in practice, when white first plays at B, the players leave the situation as is until the end of the game, at which point the seven blue stones are removed as white's prisoners.)

Go proverbs are a large part of the heritage of Go and help the student to gain insights into the game. Figure 9 illustrates, "an eye-space of four in a row is alive" and Figure 10 is summarized by the proverb "rectangular six is alive." Other proverbs serve as more general guidelines. Some examples are, "don't throw good stones after bad," "don't touch stones you want to attack," "the weak player pushes without thinking," and "the opponent's key point is your own." And of

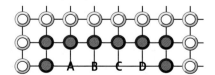

Figure 9. Eye space of four in a row
A white play at either B or C is met with a blue play at C or B respectively (making two safe eyes). So blue is alive with four points of territory.

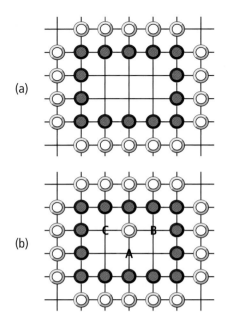

Figure 10. Rectangular six in the middle of the board
In (a), blue has surrounded a 2×3 rectangle consisting of six open points. If white plays in the center as in (b), then blue should respond at A. Now white B is met by blue C and vice versa. So blue's group is alive. (This diagram might take a bit more thought than some of the previous ones.)

course there are more spiritual and enigmatic maxims such as, "watch the fire from the opposite shore" and "close the door to capture the thief."

Just for fun, Figure 11 shows one final life and death problem. What is the status of the blue group?

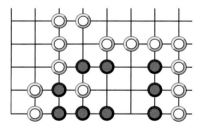

Figure 11. Challenge Problem
White has to find a way to capture all of blue's stones. (See solution on p. 237.)

Almost all devotees of the game of Go develop a philosophical sense of the game and it becomes deeply rooted in their psyche and sense of themselves. Players can often tell when their opponent is somewhat timid, too greedy, or overconfident. From July 21 to 25, 1846 two of the world's greatest players, Kuwahara Shusaku (blue) and Gennan Inseki (white) played a highly anticipated, long, arduous match. Professionals watching the game could not tell who was winning, but a doctor at the event said he knew Shusaku was going to win. When pressed how he knew, he replied that once Shusaku played move 127, Gennan's ears turned red belying his calm demeanor. As predicted, in the end Shusaku was victorious and this game has come to be known as the "ear-reddening" game.

To get some sense of the seriousness and single-minded dedication to the game, an excellent source is *The Master of Go* by the Nobel Prize-winning author Yasunari Kawabata. The book chronicles the actual "retirement" match between an old grand champion and the best of the rising young stars at the time. This one game began in June of 1938 and lasted until December! I'll let you read the account of the game if you wish to find out who won.

The Japanese professional Chizu Kobayashi has said that "the Go board is [her] mirror, when her mind is cloudy and confused, she cannot see [her] face." It is reported that on his deathbed, a wise Chinese philosopher remarked that if he could live his life over again he would dedicate half of it to playing Go. The twentieth-century Japanese Go champion, Kaoru Iwamoto, commented only somewhat facetiously that the philosopher seemed to lack any real enthusiasm for the game by promising to dedicate such a small part of his life to Go. Iwamoto himself participated in a famous game played on the outskirts of Hiroshima on August 6, 1945—the day of the atomic bomb attack. After the shock wave blew out the windows and knocked the board and players on the floor, the game was immediately reassembled and Iwamoto went on to win.

Of course one must decide for oneself how deeply to dedicate your energy, mind, and spirit to either mathematics or the game of Go, but in my view any time invested is time well spent indeed. ■

For Further Reading

An excellent introduction to the rules of the game including a full-board example game is *The Magic of Go* by Cho Chikun. It also has brief but informative essays on professional Go in Japan, China, and Korea. As previously mentioned, Yasunari Kawabata's *The Master of Go* is a superb novel about a famous Japanese Go contest and the struggle between tradition and change. To sharpen your Go technique step-by-step, work through the 1377 problems in Kano Yoshinori's, *Graded Go Problems for Beginners*, Vols. I–IV. The book is divided into subsections such as "How to capture stones," "How to save endangered stones," "Living groups and dead groups," and so on. Another very nice and leisurely way to learn about Go and deepen your understanding of the game is Janice Kim's and Jeong Soo-hyun's, *Learn to Play Go*, Vols. I–IV. To learn more about the game, where you can meet other Go players in your area, and other Go-related resources contact the American Go Association on line at www.usgo.org.

Solution to the Challenge Problem

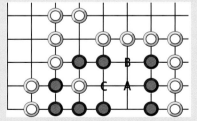

Figure 12.
White must first play at A. Blue is compelled to play at B to keep white's stones separated. But then white plays C and blue cannot form two eyes. If blue ever captures the three white stones, white must immediately play back at C. Thus blue's stones are essentially dead once white plays at A.

In Search of a Practical Map Fold

THOMAS HULL
Merrimack College

In the history of origami there has been occasional mention of how various people folded maps. Examples of ancient folded maps can be found, but modern paper-folders have been searching for "better" map folds than the awkward, fold in half over and over again method. One notable example is the Miura map fold, invented by Professor Koryo Miura of Japan, which can be opened and closed easily merely by separating two opposite corners. Such a trick is not a mere novelty; Miura has used his map fold to design solar panel arrays that can be opened and closed easily in space satellites. Other map folds exist that are sold commercially. I have seen one subway map that opens and closes using an array of pleats along the side.

But I had never thought of other, more practical possibilities for map folds until I received the following email:

Dear Sir:

Is there a way to fold a large map that enables the user to view a conveniently small section of it, but also to view adjacent sections of the map with minor manipulations? This question was raised by one of my fellow medical students. I had shown him an origami tessellation and he was impressed with my folding technique, though I told him I got it off the internet. My friend was in Desert Storm. He was an Army jeep driver for a general. There was another general who had his own jeep driver but this other jeep driver knew this great way to fold maps. He was able to take a large map and fold it so that his general, in the jeep, could look at one section of it, and then with minor manipulations, look at adjacent sections without having to unfold it. My friend claimed that this manner of folding could allow the user to go to sections of the map both North, South, AND East and West. This

was quite a feat, because the map could be used as the jeep was being driven around, with wind blowing. The other jeep driver was very jealous of his map folding and would not share it because it gave him a step up above all the other jeep drivers and made his general more efficient than the other generals.

I asked my friend if there were diagonal folds and he said that he thought so, but he never got closer to the other's maps than two jeeps parked side by side, as generals compared notes. I asked if there were cuts in the maps, he said there might have been but he didn't remember any. He said that maps tended to wear out and break at the creases. My friend suggested that I come up with something because I was obviously interested in origami.

I then called up another friend of mine who was also in Desert Storm.... My second friend had not heard nor seen anything like this map folding description but said that it would give a soldier quite an advantage and maybe it was kept secret. He related what a pain it was to stop, unfold the map and then refold it each time, all the while making a big white target for the enemy to shoot at.

Have you heard of anything like this? Do you have any ideas? I suspect that it might be topologically impossible, but I am not sure.

Sincerely,

Shannon S. Roberts

Needless to say, I was intrigued! This seems like such a logical map-folding problem, yet I had never encountered it before. I had never heard of the requirement that the map be folded up and have it be easy to access other, neighboring parts of the map.

Reprinted from February 2002, pp. 22–24

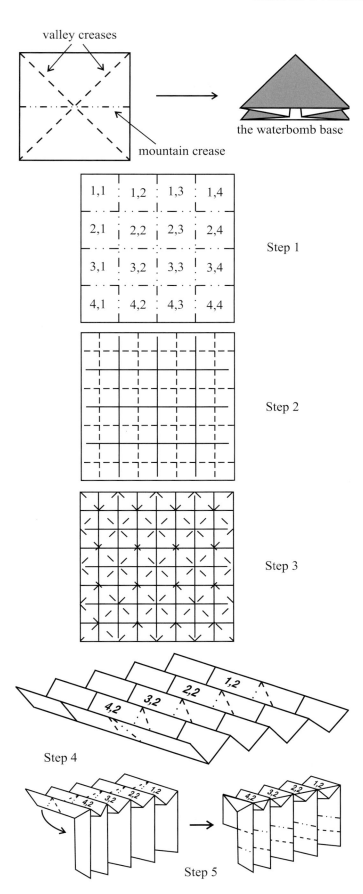

valley creases

the waterbomb base

mountain crease

1,1	1,2	1,3	1,4
2,1	2,2	2,3	2,4
3,1	3,2	3,3	3,4
4,1	4,2	4,3	4,4

Step 1

Step 2

Step 3

Step 4

Step 5

In this article I'd like to share a solution to this practical map folding problem that I devised. So read no further if you'd like to try thinking of a solution on your own.

The Waterbomb Base Solution

Being an origamist, as well as a mathematician, I immediately thought that I could use waterbomb-base collapses to hide parts of the map that I didn't want to see. A waterbomb base is a classic origami sequence which folds the paper along the diagonals using valley creases and along the middle using a mountain crease to collapse the paper into a layered triangle, as shown. (Note the two different types of dashed lines that we use to distinguish valley and mountain creases. These are standard symbols in origami books.)

The thing that attracted me to the waterbomb base for this project was the fact that it tessellates quite well. That is, it is possible to fold a grid of waterbomb bases on one sheet of paper. Take a standard tiling of the plane with squares and imagine putting the waterbomb base crease pattern into each square (you might need a vertical mountain crease in each waterbomb as well). This would be your crease pattern, and amazingly enough, it can collapse into a flat object! Furthermore, these waterbombs would be easy to open and close to reveal different parts of the paper. Thus, if we tried folding a map in this way, we might have a folded map like the one Shannon's friend described.

Now, describing this fold and actually folding it are two different things. What follows is a description of how to fold a square map (though the same technique will work for rectangular maps) in this way, using a 4 × 4 grid of tessellated waterbomb bases.

Step 1: Fold the map (map side up) into fourths horizontally and vertically, making the creases mountain folds. In these pictures I'll label the 16 sections of the map-grid (x, y) where x is the row and y is the column, each going from 1 to 4.

Step 2: Make valley creases at the 1/8th marks, horizontally and vertically.

Step 3: Make a bunch of diagonal valley creases. You're only dividing each diagonal into 8ths, so you won't hit every intersection with these diagonals. Can you see the 16 waterbomb bases?

All the creases are now in place. All that remains is to collapse it! This isn't very easy, and it needs to be done in the right way. (But there is a lot of variability—the map will still work even if some of the layers are put in awkward places.)

Step 4: Accordion pleat the paper along the horizontal creases. I've labeled the column (1, 2) through (4, 2) as a reference mark.

Step 5: The paper to the right of the (1, 2)–(4, 2) column needs to be reversed down. Step 4 has dashed lines indicating where some of the mountain and valley creases are placed. The paper should look like the left side of the figure after this step.

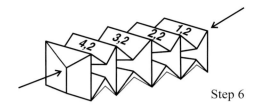

Step 6

Step 6: Reverse the paper to the left of the (1, 2)–(4, 2) column in the same way, as shown in the right picture of Step 5. At this point the paper to the left of the (1, 2)–(4, 2) column is in place and ready to go. The right side needs to be collapsed some more. Make the mountain crease between columns 3 and 4 (as shown in Step 1) and make the valley down the middle of column 4. (This will require opening it all up a bit.) Then waterbomb base collapses can be made in the 3rd column squares… Your model, at this point, might not look exactly like the drawing in Step 6, depending on how you choose to arrange the layers. Persevere and flatten the whole thing into one square. At this point any one of the second column squares can be revealed easily. So, if you open the (3, 2) square you should be able to "book fold" one layer down to reveal the (2, 2) square. Then try folding to the left to get (2, 3)!

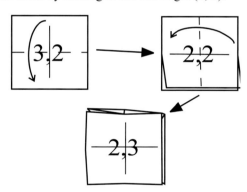

It's not very easy, is it? To get to (2, 3) you'll have to open up one of the waterbomb base collapsed squares. This might be doable with no resistance, or you may have to slide some layers around underneath to see (2, 3) and get it in the proper orientation. This all depends on how you collapsed everything down.

I'm not sure how satisfied Shannon Roberts would be with this solution. It is possible to carefully arrange the layers so that if you're looking at, say, (2, 2) then you can easily go North and South with book folds and go East and West by opening water-bomb collapses. This can easily be done with one hand and would probably serve the needs of army jeep generals.

However, if you get used to the manipulations, you can learn how to quickly go from one square on the grid to any other, as opposed to just going N, S, E, or W. I'm not sure if that's what Shannon claims the Desert Storm soldier could do, and it may be awkward doing this with one hand while bullets are flying around.

But this leads to many wonderful questions and thoughts!

- Can a three-axis version be made with a hexagonal tile grid, instead of a square one? I don't think it can without sacrificing parts of the map to the mechanism, which can then never be viewed.

- Might this have a nice application for storytelling? Imagine having a large version with pictures drawn on every square. A story could be told, showing one square at a time. The audience would be bewildered at how many pictures were hidden in the fold! It might even look like a magic trick if done properly.

There are other solutions to this map folding problem as well. Stephen Canon, a math major at Brown University, came up with a solution that is based on the mechanics of the square twist. Here is a crease pattern that works for folding a square map into a 3 × 3 grid.

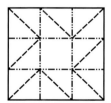

See if you can collapse this! Stephen Canon's method can be expanded to fold a square map into any $n \times n$ grid, but like the waterbomb base method I described above, the layers of paper get very unwieldy as the grid gets bigger.

It remains an open question whether or not the United States Army feels that such map folds are vital to our nation's security. But it does offer a great example of how geometric origami can turn out to be useful in surprising settings. ■

The World's
First Mathematics
Textbook

UNDERWOOD DUDLEY
DePauw University

Let us look back, as far as we can, on how and why mathematics has been taught. Let us look back at the world's first mathematics textbook that we know about. Since its contents are more than four thousand years old, there can't have been many earlier.

It is the Rhind papyrus, named not for its author but for Alexander Henry Rhind (1833–1863), a Scotsman who, for his health, fled British winters. He was in Luxor, Egypt in the winter of 1858–9. Luxor is surrounded by hundreds of miles of desert and, besides watching the Nile go by, there was very little to do there other than investigate antiquities, of which there are many—across the river there is the Valley of the Kings, a vast tomb-field. Rhind was offered, among other items, a roll of papyrus, eighteen feet long and thirteen inches high, that had been dug up by an anonymous Egyptian from beneath the sands that had preserved it for thousands of years. He bought it, and after his very early death (on the way home from another trip to Egypt) it was sold to the British Museum, where it has been to this day.

The papyrus is a copy, made around 1650 B.C., of a work that dates back at least two hundred years earlier, and that work must have had even earlier sources. Evidence of that is that the papyrus has problems about pyramids in it. When the copy was made the great pyramids of Giza were more than one thousand years old, the Egyptians had given up pyramid-building and, most likely, making up problems about them. The problems were as old as the pyramids.

The copyist, whose name is rendered as Ahmes or Ahmose, copied also the grandiose title of the papyrus, which can be translated as

Accurate Rendering
The Entrance into the Knowledge
of All Existing Things
and All Obscure Secrets

We all know mathematics is powerful, but it is not *that* powerful. The hyperbole may be just an example of the way that ancient Egyptians did things—if the past is a foreign country, then the past of more than three millennia ago is *very* foreign, even alien—or it may reflect the respect that they had for mathematics. Ahmes did not title the papyrus, as a modern textbook writer might, *Arithmetic, with Applications to Agriculture*. Mathematics was more exalted than that.

The contents of the papyrus are mainly problems and solutions. Thus was mathematics taught at the dawn of civilization, and thus is it taught today. The teacher presents a problem and shows how it is solved, probably gives some more examples, and then gives the student exercises to do.

The problems were, as could be expected, primitive. The ancient Egyptians were the *first*. Never before in the history of the world had anyone considered such problems. Every civilization after theirs had predecessors to build on, but they had nothing behind them. So it is no surprise that what they did was, to us, crude and limited. To them, it was knowledge of all existing things and all obscure secrets.

The ancient Egyptians wrote numbers in the natural way. That is, one was a mark, a vertical line, |, two was two of them, | |, three was | | |, and four was | | | |. After that there were separate symbols for 5, 6, 7, 8, and 9. Similarly, there were symbols for 10, 20, …, 100, 200, …, 1000, … . Since the Egyptians did not hit on the idea of place-value notation, they could not represent numbers as we do, with just ten digits. Their notation used the same idea as Roman numerals, which also do not have place-value (an X is a ten no matter where it appears). We would find such a notation awkward and problems like CLXIV + MMDXIX would take us a long time to do, but that is because we are used to our numerals and were trained to do additions with pencil and paper:

$$164$$
$$+\,2519$$
$$2683.$$

Ancient Egyptians and Romans did not write numbers down in order to calculate with them: they used some variation of the abacus or counting-board to perform calculations and used their symbols only to record the results.

So, adding and subtracting, even with Egyptian symbols, presents no problem. The Egyptians solved the problem of multiplying by doubling. The two-times table is all that they needed to calculate any product, though they had to do more work than we do when we multiply using pencil and paper.

For example, to multiply 19 by 15, the ancient Egyptian calculator would keep doubling 15:

1	15
2	30
4	60
8	120
16	240.

Since $19 \times 15 = (16 + 2 + 1) \times 15 = 16 \times 15 + 2 \times 15 + 1 \times 15$, summing $240 + 30 + 15 = 285$ gives the product.

This is quite a good method. If we were not blessed with calculators, it wouldn't be a bad method to teach.

Of course, all numbers are not integers and the Egyptians, even as we, sometimes had to deal with fractions. Their crude mathematics had not advanced to fractions with a numerator or a denominator. That idea, which is not an easy one, as witness the difficulty that some children have with it, had not occurred to them. The only fractions they had were reciprocals: $1/2$, $1/3$, $1/4$, $1/5$, and so on. (They had a special symbol for $2/3$, since that fraction arises so often, but that is the only exception.) When we want to multiply $1/7$ by 2, we write $2/7$ and are done with it, but Ahmes could not do that since he did not have the idea of two-sevenths. Since multiplication was done by repeated doubling, it was necessary to know what twice a reciprocal was. The fraction had to be expressed in terms of reciprocals, and that is what the first part of the Rhind papyrus lets calculators do: it is a table showing what two times reciprocals are.

For example, the table says that

$$2\left(\frac{1}{5}\right) = \frac{1}{3} + \frac{1}{15} \quad \text{and} \quad 2\left(\frac{1}{7}\right) = \frac{1}{4} + \frac{1}{28}.$$

It is easy for us to check those but how the Egyptians discovered them is unknown. Trial and error is a sufficient explanation for some examples, but what about

$$2\left(\frac{1}{83}\right) = \frac{1}{60} + \frac{1}{332} + \frac{1}{415} + \frac{1}{498}?$$

Even though the ancient Egyptians did not have the idea of numerators and denominators, they were far from stupid.

To double reciprocals of even numbers, no table is needed, since 2 times $1/8$ is $1/4$, 2 times $1/10$ is $1/5$, and so on. The table in the papyrus thus gives 2 times $1/n$ for $n = 3, 5, \ldots , 101$. Some more examples are

$$\frac{2}{9} = \frac{1}{6} + \frac{1}{18}$$

$$\frac{2}{11} = \frac{1}{6} + \frac{1}{66}$$

$$\vdots$$

$$\frac{2}{97} = \frac{1}{56} + \frac{1}{679} + \frac{1}{776}$$

$$\frac{2}{99} = \frac{1}{66} + \frac{1}{98}$$

$$\frac{2}{101} = \frac{1}{101} + \frac{1}{202} + \frac{1}{303} + \frac{1}{606}.$$

The table was also needed for division, which the Egyptians accomplished by multiplying by the reciprocal of the divisor. (Dividing by 6 and multiplying by $1/6$ amount to the same thing.) So, if the Egyptians needed to calculate $13/10$, they would multiply $1/10$ by 13 using their doubling method:

1	$\dfrac{1}{10}$
2	$\dfrac{1}{5}$
4	$\dfrac{1}{3} + \dfrac{1}{15}$
8	$\dfrac{2}{3} + \dfrac{1}{10} + \dfrac{1}{30}$

To get $13/10$, add 1, 4 and 8 times $1/10$, combining the $1/3 + 2/3$ to get 1 and the $1/10 + 1/10$ to get $1/5$—

$$\frac{1}{10} + \left(\frac{1}{3} + \frac{1}{15}\right) + \left(\frac{2}{3} + \frac{1}{10} + \frac{1}{30}\right) = 1 + \frac{1}{5} + \frac{1}{15} + \frac{1}{30}.$$

That is, in fact, what we would express as $13/10$.

So, the Egyptians could add, subtract, multiply, and divide, though with more difficulty than we have. Though we may find their methods clumsy, they were probably every bit as smart as we are. Problem 9 in the papyrus shows its unknown author showing off his smartness. The problem is to multiply $1/2 + 1/14$ by $1 + 1/2 + 1/4$. Here is what is in the papyrus:

1	$\dfrac{1}{2}$	$\dfrac{1}{14}$
$\dfrac{1}{2}$	$\dfrac{1}{4}$	$\dfrac{1}{28}$
$\dfrac{1}{4}$	$\dfrac{1}{8}$	$\dfrac{1}{56}$

Total 1.

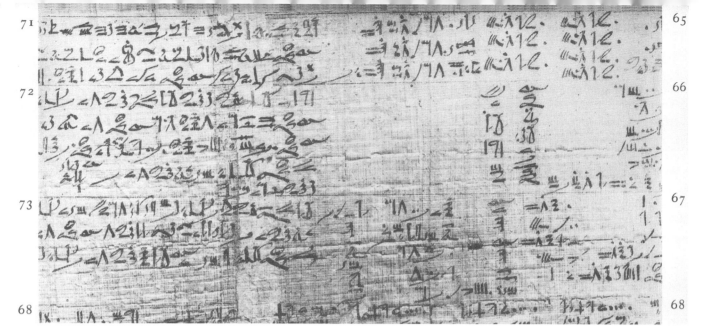

From the Rhind Papyrus

The calculation is correct: $\frac{1}{2} + \frac{1}{14}$ (that is, $\frac{4}{7}$) times $1 + \frac{1}{2} + \frac{1}{4}$ (that is, $\frac{7}{4}$) is 1. You or I would not be able to go from ($\frac{1}{2} + \frac{1}{14}$) + ($\frac{1}{4} + \frac{1}{28}$) + ($\frac{1}{8} + \frac{1}{56}$) to 1 as quickly as the practiced ancient Egyptian calculators could, but the author's smartness is demonstrated by his making up a problem with a neat answer.

The first twenty-three problems in the papyrus are like the last one. Number 24 is different. It asks, what is the quantity so that when $\frac{1}{7}$ of it is added to it, the total is 19. Today, if we had to solve that problem (an unlikely event), we would use algebra:

$$x + \frac{1}{7}x = 19$$

$$\frac{8}{7}x = 19$$

$$x = \frac{7}{8} \cdot 19 = \frac{133}{8} = 16\frac{5}{8}.$$

The ancient Egyptians, who had no algebra, could not solve the problem that way. But they could solve it nevertheless, using a fine method: by guessing an answer and then correcting it. The papyrus says to guess 7. That plus its $\frac{1}{7}$ is 8. That's quite a bit off from 19 but, the papyrus says, see what we have to multiply 8 by to get 19. When we multiply 7 by that multiplier, we'll get what we want.

That is, the papyrus says, find y so that $8y = 19$. Then $7y$ will be the answer. So it will be, since $7y$ plus $\frac{1}{7}$ of it is $7y + y = 8y = 19$. The papyrus calculates that $2 + \frac{1}{4} + \frac{1}{8}$ times 8 is 19, so $2 + \frac{1}{4} + \frac{1}{8}$ times 7, or $16 + \frac{1}{2} + \frac{1}{8}$, is the answer, as it is.

So, were we ancient Egyptian students being taken through the text, we would now be able do all the arithmetical operations and to solve linear equations. In the manner of teachers over the millennia, the papyrus gives a few more problems exactly like problem 24 but with different numbers, and then progresses to a more complicated example. Problem 29 says take a quantity, add $\frac{2}{3}$ of it, then add $\frac{1}{3}$ of the total, then add

$\frac{1}{3}$ of that sum and the grand total is 10. What did you start with? That was the climax of the papyrus's linear equations.

The problem is to solve

$$\left(x + \frac{2}{3}x\right) + \frac{1}{3}\left(x + \frac{2}{3}x\right) + \frac{1}{3}\left(\left(x + \frac{2}{3}x\right) + \frac{1}{3}\left(x + \frac{2}{3}x\right)\right) = 10.$$

Easy for us ($x = \frac{27}{8}$), but not easy to do using the guess-and-correct method.

In the second half of the papyrus the practical problems start. Problem 43 is to find the amount of grain in a cylindrical container, height 6 and diameter 9. Problem 52 is to find the area of a trapezoid, problems 56 to 59 are about the slopes and altitudes of pyramids, and problem 65 is one that could appear in textbooks today: how to divide 100 loaves among 10 men when three of them get double shares. Problem 71 is about pouring off $\frac{1}{4}$ of a container of beer and replacing it with water.

Those are practical problems. The Egyptians were a practical people. Not for them the philosophical speculations of the Greeks. In fact, not for them speculation of any kind. Egyptian mathematics was essentially unchanged for two thousand years, a long time. The Rhind papyrus was a practical text, containing no "theory" at all. Its solutions give no reasons, nor do they cite any underlying principles. They had the form, "Do this. Then do this. Then do this. There is the answer, it is right." There was theory behind the papyrus, but it wasn't written down. It may have been that instructors would pass along to students the reasons for why the solutions were as they were, but it is quite possible that they did not. I suspect that many instructors, in the long, long history of Egyptian mathematics education did not understand the reasons, and I suspect that the majority of Egyptian students did not care about them. Practical people want results, unencumbered by theorizing. Even now, many students will say, "Just show me how to do the problems," with the implication that they will then do them, but with a minimum of thought going into the process and none into why the process is what it is.

However, the papyrus is not entirely practical. Mathematics has power, and not just power over finding the volumes of containers of grain, and solving problems about dividing loaves or watering beer. Mathematics has now, and had then, the power to fascinate minds, quite independent of its uses. That problem about adding $2/3$ and so on to get 10 arose from no practical situation. It is the result of an ancient Egyptian's delight in pure mathematics, in particular of his joy and pride in being able to solve linear equations.

This shows up elsewhere in the papyrus. Problem 67 is to find the number of cattle in a herd if $2/3$ of $1/3$ of them make 70. This is not a practical problem. The owner of the herd knew how many animals were in it: he could count them. This is a problem that was proposed because it was a nice problem and because it would exercise and expand the reasoning powers of those who were forced to try to solve it.

Impracticality is even more obvious in problem 40, which asks how to divide 100 loaves among 5 men so that the shares are in arithmetic progression and that $1/7$ of the largest three shares is equal to the sum of the smallest two. That is a problem that has never had to be solved in the "real world," ours, the ancient Egyptians', or any other possible world. Never! Some ancient Egyptian, one with a lively mind, one day tired of those problems about dividing m loaves among n people, even if some of them got double or triple shares, and thought, "What would happen if I wanted to divide the loaves so that the shares were in an arithmetic progression?" (He did not use those words—"arithmetic progression" was not in the ancient Egyptian vocabulary—but that is the idea that he had.) He then worked on the problem, not because it had any use, but because it was interesting, and fun, and he actually was able to solve it. A discovery! It so impressed the mathematics education establishment of, say, 2000 B.C. that it got into the curriculum, and it, and its offspring, have been there ever since. Even for the eminently practical Egyptians, mathematics was studied for its own sake.

The answer, by the way, is that the smallest share is $1 2/3$ and the difference between shares is $9 1/6$. Hard enough for us, harder still for ancient Egyptians, using ancient Egyptian methods and calculations.

Then there is problem 79, which asks: if there are seven houses, each house with seven cats, seven mice for each cat, seven ears of grain for each mouse, and each ear of grain would produce seven measures of grain if planted, how many items are there altogether? The answer is

$$7 + 49 + 343 + 2401 + 16807, \text{ or } 19607.$$

Talk about adding apples and oranges! This is adding houses and cats and mice and grain. The answer is nonsensical, and the problem is as far from a practical problem as you are likely to get. It is for fun.

By the way, if you want something of yours to go rolling down the centuries, don't bother with anything physical. Make up a new problem. Physical objects decay; problems have more staying power. The seven-cats-and-so-on problem shows up in Fibonacci's *Liber Abaci*, a text that was published in 1202, and survives, in a way, in "As I was going to St. Ives…." It may outlast the pyramids.

There are several lessons to be learned from the Rhind papyrus. The most obvious is that instruction in mathematics has changed hardly at all since the subject was first discovered. Teachers have been passing mathematics on to the next generation in much the same way over the centuries: problem, solution, practice. As all teachers of mathematics know, the method does not always work, but if there is a better way, no one has found it in four thousand years of looking.

Another constant is the emphasis on process and answer. In every age, instructors of mathematics have pointed back to their predecessors with scorn, saying that they taught mathematics by rote, drilling their students to perform mechanically, without understanding. In this age, instructors in every age say, it will be different, we will have our students *understand*. Every age fails. Many students in every age would just as soon not understand. It would be fine with them to learn the processes and let it go at that. Many students have minds that have not developed to the point where they are capable of understanding. It is not their fault any more than it is the fault of third-graders that they can't understand Shakespeare. Faced with resistance and incapacity, if instructors (in any age) are to accomplish anything they must concentrate on rules, formulas, and algorithms. Understanding is always to be striven for, but it is not always to be had. Some students can solve word problems and some cannot, even given the best of intentions by both instructors and students.

The papyrus also shows that even at the dawn of mathematics we find artificial word problems, devoid of any connection with anything practical. They exist because by solving them both mathematics and reasoning are learned. Students, and they are many, who say, "I can't do word problems; just give me the formula and I'm fine" have missed out. They have not gotten what they should have from their mathematics. Of course, the ability to carry out a sequence of calculations correctly, even without understanding, is good to have—it helps when it's time to fill out income tax forms—but they have not developed their reasoning power as much as they could have.

Finally, the papyrus illustrates the immense attractive power of mathematics. It does not have this power over everyone, but those who are susceptible to it are pulled in. They delight in it for its own sake, and they try to share their delight by making up problems about cats and mice, or about loaves in arithmetic progression.

The papyrus shows how far we have come in four thousand years and at the same time how much we are unchanged. ■

The
Instability
of
Democratic
Decisions

PHILIP D. STRAFFIN JR.
Beloit College

In any society, people with different preferences have to decide on common courses of action. In a democratic society, we often do this by voting, usually by majority rule. When there are many possible courses of action, a decision may require a sequence of votes—a legislature might vote on amendments to amendments, then on amendments, then on a final bill. When this happens, it's important to understand how *stable* the result is. We would like to think that the result of democratic decision-making reflects "the will of the people." If the decision were made again tomorrow by the same people with the same preferences, but perhaps by a different sequence of votes, the decision should be the same, or at least approximately the same.

One way that political scientists have approached this question is by building and analyzing mathematical models. I'd like to show you how one class of models—geometric or "spatial" models—deals with the question of the stability of democratic decisions. I think you'll find the insights they offer surprising, and perhaps disturbing.

One-dimensional Spatial Models

The oldest spatial model in politics dates back to the time of the French Revolution, when radicals sat on the left side of the Assembly and royalists sat on the right side. Since then, common political discourse has often represented preferences by placing voters along a line, with "liberal" voters to the left, "conservative" voters to the right, and "centrist" voters in between. For example, Figure 1 shows a one-dimensional spatial model of the U.S. Supreme Court constructed by the *Washington Post* by considering how justices voted in twelve key decisions in 1998. Notice that Justice Kennedy occupies an important position, in that he is the middle, or *median*, voter. He has the swing vote between liberal Justices Souter, Breyer, Ginsburg and Stevens, and conservative Justices O'Connor,

FIGURE 1. A one-dimensional spatial model of the U.S. Supreme Court in 1998.

Rehnquist, Thomas and Scalia. We might expect many decisions to be made in accordance with Kennedy's preferences.

When voters vote on alternative courses of action, those alternatives may be able to be positioned along the same continuum as the voters. For instance, voters might choose among candidates whose platforms might be liberal or conservative, or legislators might vote on bills which have liberal or conservative content. In the 1950s, Anthony Downs proposed that we analyze voting by assuming that voters and alternatives can be represented as points on a line, and that in choosing between alternatives, each voter will vote for the alternative which is closest to him or her. This is the classical "Downsian" spatial model of voting.

Let's see what the Downsian model says about the stability of voting. For simplicity, we'll always assume there are an odd number of voters, so we won't need to worry about ties. Suppose that the voters in Figure 2 choose between alternatives x and y. Notice that which alternative will win is entirely determined by whether the median voter is closer to x or y (so in Figure 2, y will win). In other words, in any pairwise majority vote, *the alternative closer to the median voter will win*. If a series of alternatives are considered, the one closest to the median voter will win, and this will be true regardless of the order in which the alternatives are voted on. The one alterna-

FIGURE 2. The one-dimensional Downsian voting model.

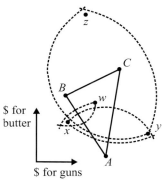

FIGURE 3. Spatial voting in two dimensions.

tive which will beat any other alternative is the exact point occupied by the median voter. This result is known in the political science literature as the *Median Voter Theorem*.

One way to think of this result is to imagine x and y as the positions of two candidates who can change their positions by the statements they make during the campaign. If y starts nearer to the median voter, the first candidate has incentive to move x to the right to get closer to the median voter. Then the second candidate might move y to the left to recapture the lead. We should see the positions of the two candidates converging towards the center, until there is very little to distinguish between them. (Of course, there are some practical political problems. For instance, if a candidate tries to move too far too fast, voters may question his sincerity. Or if the two candidates become indistinguishable, there may be incentive for a third candidate to enter the race taking an extreme position on the left or right.)

Although we might not like having to choose between two almost indistiguishable candidates, in many ways the Median Voter Theorem is reassuring about the nature of democratic decisions. It says they will be stable and they will be centrist—maybe not pleasing everyone but not making anyone too unhappy.

Two-dimensional Spatial Models

Unfortunately, we know that although democratic decisions are sometimes centrist, they are not always so. It is also true that both politicians and voters know that the liberal-conservative continuum is too simplistic to capture preferences over a broad range of issues. A politician, for example, might describe herself as "liberal on social issues, but conservative on economic issues." Mathematically, the natural way to model such preferences would be to represent voters as points not along a line, but in a plane (or a space of three dimensions or n dimensions, but for this article we'll stick to two dimensions). Alternatives—bills or candidates—are represented by points in the same plane, and we retain the Downsian assumption that in a decision between two alternatives, voters will vote for the closer one.

Surprisingly, the simple change from one dimension to two—allowing voter preferences to be slightly more complex—has profound consequences for the stability of democratic decisions. Consider, for example, Figure 3, in which there are just three voters A, B, and C who must decide on a

budget which will spend a certain amount of money for social purposes and a certain amount for military protection. If these voters are asked to decide between bills w and x, voter C will certainly vote for w, but x is closer than w to both A and B. (To make this clearer in the figure, I have shown arcs of circles centered at A and B, passing through w. Notice that x is inside both of these arcs.) Hence A and B will vote for x, which will beat w by two votes to one.

So far so good. But now suppose a new budget y is proposed, and our voters must choose between x and y. With the help of the appropriate circular arcs, you should check that A and C will vote for y, so y beats x by two to one. Finally, if z is paired against y, z will get the votes of B and C and beat y by two to one. Our small democratic society started with w and by a series of decisive majority votes, ended up choosing z. This is a strange outcome, because you can see that the voters would unanimously prefer w to z. My colleagues in the Beloit College Academic Senate find this phenomenon familiar. If you have ever been part of a legislative body, has anything like this happened to you?

McKelvey's Theorem

Let's think more carefully about the example in Figure 3. First of all, since each alternative in the cycle w, x, y, z, w would beat the previous alternative, none of those alternatives is stable. In fact, by starting the cycle in different places, we could make any one of the four alternatives the ultimate winner. The decision of this society seems to have little to do with the voters' preferences, and everything to do with the order in which alternatives are presented. This phenomenon is known in political science as the *agenda effect*. The practical consequence is that "s/he who controls the agenda, controls the outcome." In the U.S. House of Representatives, the most prized committee assignment is to the Rules Committee, which determines the voting agenda.

In Figure 3, a clever agenda controller could lead our voters from w to z. It is natural to ask where else they could be led. Political scientist Richard McKelvey published the startling answer in 1976: they can be led, by a finite sequence of majority votes, to *any point in the plane*! In two dimensions, the agenda effect is absolute.

To prove McKelvey's Theorem for the situation in Figure 3, I'll start by telling you that I didn't choose points x, y, and z at random. The circles through w centered at A and B meet at a second point, outside triangle ABC, which is the reflection of w in the line AB. To get x, I took that reflection and moved it slightly in toward the line AB. Similarly, I got y by reflecting x in the line AC and moving it in slightly. (In politics, this maneuver goes by the name of "splitting the winning coalition.") Finally, z is the reflection of y in the line BC, again moved in slightly. Of course, there is no reason we need to stop here: we could iterate the procedure of reflecting in lines AB, AC, and BC (always remembering to move in slightly) as many times as we wish. To see what happens if we do that, we need a lovely theorem from transformational geometry: *The product of reflections in three sides of a triangle is a glide reflection.* Figure 4 illustrates this result. The glide reflection which is the product of the three reflections first translates the plane along vector **v** and then reflects it across **v**. The effect of iterating the sequence of reflections, then, is to iterate a glide reflection.

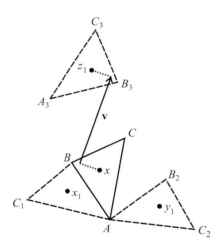

Figure 4. The product of reflections in three sides of a triangle is a glide reflection.

The strategy of our agenda controller is now clear. To lead the voters from w to any chosen point q, first iterate the glide reflection n times to get to a point z_n which is farther from all of the voters than q is. Then propose q as a final option. The voters should adopt q gratefully and unanimously.

The Plott Conditions for Stability

For three voters at the vertices of a triangle, there is no stability. Every alternative can be beaten by other alternatives, and in fact every alternative can be reached from every other alternative by a finite sequence of majority votes. (See how the argument above shows that?) But clearly this is a very special,

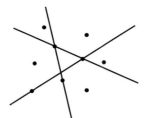 

Figure 5. An unstable voting configuration of nine voters. Figure 6. A stable voting configuration of nine voters.

and simple, configuration. What about other configurations? It turns out that the conclusion of McKelvey's Theorem holds "almost always." Given any configuration of an odd number of voters in the plane, define a *median line* to be a line such that each of the two regions into which it divides the plane contains fewer than half the number of voters. Figures 5 and 6 show some median lines. Notice that any median line must pass through the position of at least one voter.

The key observation is that given any alternative x, reflecting x across any median line and moving in slightly toward the median line gives a new point y which beats x under majority rule. (Can you give a quick proof of this?) Hence if there are three median lines which do not all pass through the same point, i.e., form a triangle as in Figure 5, we can use exactly the argument in the previous section to prove complete instability. Thus the only case in which instability doesn't hold is if *all median lines pass through the same point*, as in Figure 6. In this case, the common point must be the location of some voter—the *median voter M*—and all lines through M must be median lines. If M is the location of a single voter and we imagine a median line rotating around M with one half-plane picking up voters as the other loses voters, we see that whenever a median line contains k points on one side of M, it must contain exactly k points on the other side of M, i.e., the configuration must have the kind of "symmetry" of Figure 6. Configurations of this kind were first identified by Charles Plott in 1967. (If more than one voter is located at M, other configurations are possible.)

A Plott configuration is stable. You can check that, just as in one dimension, in any vote between two alternatives, the one which is closer to the median voter M will win, and an alternative positioned at M will beat any other alternative. In fact, we can think of the one-dimensional situation as a Plott configuration in which all of the alternatives happen to be positioned on one line.

Finally, notice that configurations with more than one median voter or the symmetry of a Plott configuration are extremely rare, in the sense that if we throw an odd number of points randomly on the plane, the probability that such a configuration will result is zero. Hence with probability one, a two-dimensional configuration of an odd number of voters will have complete instability.

Morals of This Story

The results of our foray into spatial models of voting seem discouraging, perhaps even dismaying, for those who believe in majority rule as a method of democratic decision-making. If issues are very simple, in that it makes sense to represent voters' positions along a line, decisions can be stable. But even a small amount of complexity destroys that stability in the most drastic possible way, and gives ultimate power to anyone who can control the voting agenda.

What should we conclude from this? One possible response—and one which has sometimes been made by non-mathematical political scientists—is that this is "only a model," with no relation to complex reality. The problem with this dismissal is that other political scientists recognize real political behavior in what the model portrays: that different procedural rules can produce different outcomes, that minorities can propose amendments which split majorities, that control of the voting agenda can be enormously important.

An extreme response on the other side is that the model has uncovered a fundamental flaw in democratic society. It shows that in any society of even moderate complexity, making decisions by majority rule gives power not to the people, but to professionals who know how to manipulate political procedures to their own advantage.

A more moderate response, and one which has been influential in political science for the past twenty years, comes from focusing on the aspect of the model which produces the instability: the freedom to introduce new alternatives anywhere in the plane. This line of research asks what kinds of "germaneness" rules, parliamentary procedures and voting rules might best promote stability without limiting in too serious a way the freedom of voters to propose alternatives.

For me, a main moral is that mathematical models can focus ideas, bring new insights, and suggest new lines of research in the study of important social questions. ■

For Further Reading

A good general introduction to spatial voting models can be found in J. Enelow and M. Hinich, *The Spatial Theory of Voting* and *Advances in the Spatial Theory of Voting*, Cambridge University Press (1984) and (1990). A nice proof that the product of reflections in three sides of a triangle is a glide reflection can be found in I. M. Yaglom, *Geometric Transformations I*, Mathematical Association of America, 1962. A more detailed discussion of voting instability appears in P. D. Straffin, Power and stability in politics, pp. 1128–1151 in R. Aumann and S. Hart, eds., *Handbook of Game Theory with Economic Applications*, vol. 2, Elsevier (1994).

A Baseball Giant, A Math Giant, and the Epsilon in the Middle

CARL POMERANCE
Dartmouth College

The 28th anniversary of Hank Aaron's 715th major league home run, which eclipsed the career total of the great Babe Ruth, seems a perfect time to tell the tale of a special baseball.

Paul Erdős is famous not only for his theorems, but also for his discovery and development of young mathematicians, with stories of his "epsilons" being legend. I am fortunate beyond words that Erdős played a pivotal role in my career, helping me to develop my talents in combinatorial number theory, eventually leading to my involvement in the analysis of number-theoretical algorithms. However, Paul did not discover me when I was an epsilon. Rather, I had already done my doctorate at Harvard University under John Tate, and was several years into an assistant professorship at the University of Georgia. Unlike most Harvard-trained number theorists, I had not become an expert in algebraic geometry; my true love was elementary number theory. Among the first papers I studied at Georgia was Erdős's 1956 paper "On pseudoprimes and Carmichael numbers" in *Publicationes Mathematicae Debrecen*. Here he found a good upper bound for the distribution of Carmichael numbers and gave his famous heuristic argument on why they should be relatively numerous. More than twenty years after I read this paper, Red Alford, Andrew Granville and I would add some new elements and make this heuristic argument into a proof. We dedicated the paper, which appeared in *Annals of Mathematics*, to Erdős on his eightieth birthday.

But I get ahead of the story. After the Second World War, baseball became integrated with African-American players and also players from Latin America. Among the first American blacks to play for the major leagues were Jackie Robinson and Henry "Hank" Aaron. They are American heroes since not only

On April 8, 1974, in Atlanta, Georgia, Henry Aaron hit his 715th major league home run, thus eclipsing the previous mark of 714 long held by Babe Ruth. This event received so much advance publicity that the numbers 714 and 715 were on millions of lips. Questions like "When do you think he'll get 715?" were perfectly understood, even with no mention of Aaron, Ruth, or home run.

In all of the hubbub it appears certain interesting properties of 714 and 715 were overlooked. Indeed we first note that

$$714 \cdot 715 = 510510 = 2 \cdot 3 \cdot 5 \cdot 7 \cdot 11 \cdot 13 \cdot 17 = P_7$$

where P_k denotes the product of the first k primes. Is this really unusual? That is, are there other pairs of consecutive numbers whose product is P_k for some k? We readily see that $1 \cdot 2 = P_1, 2 \cdot 3 = P_2, 5 \cdot 6 = P_3, 14 \cdot 15 = P_4$. Putting the problem to the CDC 6400 at the University of Georgia (using 5 minutes of computer time), we found that the only P_k which can be expressed as the product of two consecutive numbers in the range $1 \le k \le 3049$ are P_1, P_2, P_3, P_4, and P_7. Hence if there is any other pair of consecutive integers whose product is a P_k, then these integers exceed 10^{6021}. Great as Henry Aaron is, we believe he will never again hit two consecutive home runs whose numbers have their product equal to some P_k. However, on April 26, 1974, Henry Aaron did hit his 15th grand slam home run, breaking the old National League mark of 14, and, of course, $14 \cdot 15 = P_4$.

By Carol Nelson, David E. Penney, and Carl Pomerance. Excerpted with permission from Journal of Recreational Mathematics, Spring 1974.

were they excellent athletes, but, with great personal dignity, they also put up with constant degrading insults and even death threats. It became especially bad for Aaron, when in the spring of 1974 it appeared that he might actually surpass George Herman "Babe" Ruth's supposedly unbeatable career record of 714 home runs. On April 8 of that year, Aaron succeeded, hitting number 715. By his retirement several years later, he had hit a total of 755 home runs, and the address of the baseball field in Atlanta is given with this in mind: it is 755 Hank Aaron Boulevard.

In the spring of 1974, I had still not met Erdős, and I was still trying to find my true character as a mathematician. At the actual moment when the 714 record was broken, being more of a mathematician than a baseball fan, I guess, I started looking at interesting properties of the numbers 714 and 715. The first thing that I noticed was that they factor very easily, and in fact, their product is the product of the first 7 primes. (It is conjectured now that this is the last pair of consecutive integers whose product is the product of the first k primes for some k.) The next day, I challenged my colleague, David Penney, to find an interesting property of 714 and 715. He discovered the same property, but he also challenged a class he was teaching that morning, and one of the students, Jeremy Jordan, discovered that $S(714) = S(715)$, where $S(n)$ is the sum of the prime factors of n taken with multiplicity. (Since 714 and 715 are both square free, one might also have taken a sum of distinct prime divisors, but it is a little simpler to consider the completely additive function $S(n)$.)

Penny and I, together with another student, Carol Nelson, wrote a short, humorous paper [see an excerpt in the box] on our observations....

Say an integer n with $S(n) = S(n + 1)$ is a "Ruth-Aaron number." (We had originally called n an "Aaron number," but in retrospect it seems fairer to honor both baseball greats.) In our paper we wrote: "The numerical data suggest that Aaron numbers are rare. We suspect they have density 0, but we cannot prove this." These words started my life over as a mathematician in the Erdős school. Paul read this article, which was published within a few months of the actual baseball event, and wrote to me that he knew how to prove density 0, and would like to visit me at Georgia to discuss it. This then became the subject of our first joint paper, in which we also discussed the joint distribution of the largest prime factors of n and $n + 1$.

An amusing footnote to this story concerns the awarding of an honorary degree by Emory University to Paul Erdős in 1995. Paul invited me and my wife to attend a reception the evening before for the honorees and their guests. Completely unknown to me until just before entering the room, one of the others to receive an honorary degree was Hank Aaron! I introduced myself to him and tried to tell him how his athletic feat

1996 reception at Emory University. Ron Gould (standing), Gary Hank (on right), Paul Erdős, and Hank Aaron. Photo courtesy Ron and Madelyn Gould.

had such important consequences to my career as a mathematician. He smiled diplomatically and said he was happy for this, though I believe he thought he had just met a very strange person. I introduced him to Erdős and the two chatted for a while. A photo of them exists in one of the recent biographies of Paul's life. Ron Gould, a professor at Emory who was one of the people instrumental in arranging the honorary degree for Erdős, knew that Aaron was to be there, but of course he had no idea of the connection to Erdős. He and his wife had come supplied with some new baseballs for Aaron to sign as souvenirs. They graciously let me have one of these, and I had both Aaron and Erdős sign the same baseball. Though the writing is unfortunately fading with time, it is a prized possession. I joke that Aaron should have Erdős-number 1 since, though he does not have a joint paper with Erdős, he does have a joint baseball. ∎

I would like to thank Deanna Haunsperger for suggesting this article, getting the necessary permissions, and editing the final product. Excerpted with permission from Paul Erdős and His Mathematics, *edited by Vera Sós, Springer, 2002.*

Digging for Squares

PAUL C. PASLES
Villanova University

Sir Arthur Clarke said: "Any sufficiently advanced technology is indistinguishable from magic." And, he might have continued, any sufficiently nifty mathematics looks a whole lot like magic too. So much so that the name "magic square" has stuck, persisting long past the days when such designs were thought to confer luck or protection upon those in need.

Magic squares certainly proved lucky for this researcher. Soon after earning my PhD from Temple University, I was fortunate enough to spend some time in our nation's capital at the MAA-NSF sponsored Institute in the History of Mathematics and its Uses in Teaching. As participants we learned from eminent historians, and were inspired to come up with our own projects. Some of those little ventures got a bit out of hand. In my case, I'd hoped simply to create a short expository talk summarizing what was already known of Ben Franklin's famed magic squares. As it turned out, what was already known wasn't much! Soon enough, I found myself doing a little mathematical archaeology—digging for squares.

Everyone knows about Ben Franklin's other accomplishments: his experiments in electricity, his writings, and his role in the American Revolution and subsequent forming of the republic. Every few years brings a new biography of this man, the first internationally recognized American celebrity. As a research topic his magic squares are intriguing — though after two and a half centuries, could there possibly be anything left to say? The answer is yes, this well-examined life still has the capacity to bring a surprise or two.

Doubtless you have seen magic squares of the usual variety: rows, columns, and 2 diagonals each adding up to the same sum. Their construction can involve some tricky techniques, but their appreciation requires no more than grade school arithmetic. Here is an example which dates back to the 16th century:

16	81	79	77	75	11	13	15	2
78	28	65	63	61	25	27	18	4
76	62	36	53	51	35	30	20	6
74	60	50	40	45	38	32	22	8
9	23	33	39	41	43	49	59	73
10	24	34	44	37	42	48	58	72
12	26	52	29	31	47	46	56	70
14	64	17	19	21	57	55	54	68
80	1	3	5	7	71	69	67	66

How are Franklin's squares different from the standard type? A 4 × 4 "Franklin magic" square is shown below. The integers from 1 to 16 are arranged in a square matrix so that each row and each column sums to 34; like the last example,

17	47	30	36	21	43	26	40
32	34	19	45	28	38	23	41
33	31	46	20	37	27	42	24
48	18	35	29	44	22	39	25
49	15	62	4	53	11	58	8
64	2	51	13	60	6	55	9
1	63	14	52	5	59	10	56
16	50	3	61	12	54	7	57

Square Deal? Newsman Ben Franklin argued convincingly and successfully in favor of the nature and necessity of a paper currency. By a miraculous coincidence, his own business then secured contracts to print said paper money. Here, the author's conception of an idealized Franklin dollar bill.

that makes it "semi-magic." What makes it "Franklin magic" is this: take either half of a diagonal and add it to either half of the other diagonal—Ben called this configuration a 'bent row' —and you still get that same magic sum! The bent rows in the example above are $1 + 8 + 13 + 12$, $15 + 10 + 3 + 6$, $1 + 8 + 10 + 15$, and $12 + 13 + 3 + 6$. (As with the rows and columns, obviously there is some dependence among these conditions.)

What Franklin magic squares were written by the man himself? Search the sparse literature on Franklin's squares—much of it in obscure, out-of-print books—and you find only the same two examples over and over again (and these are disappointingly similar in structure, at that). Occasionally one also finds a reprinting of Franklin's "magic circle" as well, but that is all. *There had to be more.* I looked everywhere to find others —yet while everyone took at face value Franklin's claim that he had created plenty of additional examples, no one ever printed any of them. Where were these lost sheep? Or was Ben simply exaggerating his output?

Fortunately, though it proved to be no easy task, I did find several others. Here's one you're unlikely to have seen, unless you've happened upon an obscure footnote in the 15,000-plus pages of the *Papers of Benjamin Franklin*. Certainly this example is absent from the dozens of books on magic squares, and from the many articles on Franklin's magic squares in particular.

Franklin left no commentary with this square, penciled on a scrap of paper now in possession of the American Philosophi-

cal Society, but some properties are readily apparent even without his personal explanation. It's semi-magic, with row and column sums all equal to 260. The four main bent rows work too, so it's Franklin magic, and as in his more famous squares, these main bent rows can be translated in various ways while still preserving the magic sum. In particular, you can move the V-shaped and inverted V-shaped bent rows up or down and the <- shaped and >- shaped ones left or right to get 32 bent rows (viewing the square as a torus), each of which sums to 260. For example, the V-shaped bent row $64 + 63 + 3 + 36 + 21 + 54 + 10 + 9 = 260$. Also in common with his other, better-known, 8×8 square: every 2×2 submatrix sums to half the magic sum. Moreover, cut any row or column in half, you get 130. And there are many other properties you can look for. Remarkable! Even luckier for me, I found that this square turns out to have special importance, as it seems to be a working draft of the famous magic circle!

There were other squares to be found. By December of 1999 I had all of them in hand. The biggest payoff came when I tried to find the original 1765 manuscript of one of Franklin's squares. The more I looked, it seemed, the harder it was to find. My wife, who works in statistics (and is quite tolerant of my obsession with less applicable mathematics), accompanied me to various rare book rooms on this quest.

At one time, the 1765 manuscript in question probably existed in two separate handwritten originals, one kept and one sent to a correspondent in England. Unfortunately, the modern-day archives that were supposed to have it did not. After searching museums on both sides of the Atlantic, and tracing the auction record for what might have been a separate original, I was nearly ready to give up.

Finally, I found that there was a facsimile at the Royal Society. Most amazingly, it includes a 16×16 square which had never been published before! This was not the famous 16-square, but a better one than any I'd ever seen (by Franklin or by others). All rows, all columns, all bent rows, and hundreds of bent-row-translates add up to 2056. Each half-row and half-column sums to half of 2056. Even the garden-variety "straight" diagonals favored by lesser mortals sum to 2056; this is what is sometimes called a "pandiagonal" magic square. (I should pause for a moment to catch my breath.) Square blocks of 4 cells sum to one-quarter of 2056. Various other pretty configurations of 8 or 16 cells sum nicely—that is to say, magically. Thus, not only does this one differ from his well-known 16-square, it is far superior to all of his previously known examples, with twice as many "magical" properties as any of them! I had found the only known example of a Franklin magic square in which both the bent *and* straight diagonals are magic.

Some may call these sorts of novelties math "lite." But I think we should all aspire to see one of our mathematical creations live on, two and a half centuries after we are gone.

There were other exciting discoveries as well. In one rare book room, I found Ben's own copy of the book which first published his "magic circle." Therein were his handwritten corrections to the description of its properties! I published those corrections too, so I guess this makes us co-authors...

Word of these discoveries spread pretty quickly. I was contacted by journalists reporting for *Science* and *Pour la Science*. My work inspired a popular children's book. Before long, my paper was bootlegged on the Web, months before it even went to press. What a compliment! These little magic squares are finally finding an audience after all.

Well, that's the story of my odyssey through my first math history project. My advice to undergraduates: never assume it has all been done before! There are treasures waiting to be found.

I leave you with a little homework.

Exercises

The theory of Franklin squares is in its infancy. Help build its foundations with these exercises, and don't be afraid to use a little *Maple* for exercise 4. Check your answers at www.pasles.org/Franklin/horizons.html.

1. Are there any Franklin magic squares of order 3?

2. Complete the Franklin square below. (There are 2 possible answers.)

1	2		
			10
14			

3. Some Franklin magic squares are also *fully* magic (both diagonals add to the magic constant). Are there any of these in the 4 × 4 case?

4. You can treat magic squares like matrices. Let's discard the requirement that the entries of an *n*-square have to be 1, 2, 3,..., n^2. Obviously the matrix sum of two semi- (or Franklin- or fully-) magic squares will be a semi- (or Franklin- or fully-) magic square. More delightfully, sometimes there are matrix *product* results too. It's known, for example, that any 3 × 3 fully magic square can be *cubed* as a matrix product and the result will be fully magic![1] Prove the corresponding theorem for 4 × 4 Franklin magic squares: If A is a horizontally symmetric Franklin square, then $A^T A$ is horizontally symmetric and Franklin magic, though its entries are not distinct. (*Horizontally symmetric* means $a_{i,j} + a_{i,n+1-j}$ is constant, as in the example below.) ■

5	14	3	12
4	11	6	13
9	2	15	8
16	7	10	1

For Further Reading

You can see all the squares I found and read more details in "The Lost Squares of Dr. Franklin," in the June-July 2001 *American Mathematical Monthly*.

Franklin's magic circle can be seen on my webpage www.pasles.org/circle.html.

A stone's throw from the Annenberg Rare Book and Manuscript Library, co-authors confer on matters magical. From left: B. Franklin, P. C. Pasles.

---

[1] I learned this by reading Math Horizons! ("Some New Discoveries About 3 × 3 Magic Squares" by Martin Gardner, Feb. 1998.)

A Dozen
Questions
about a Triangle

JAMES TANTON
St. Mark's Institute of Mathematics

A triangle possesses remarkable properties! For example, the three medians of any triangle always meet in a point, as do the perpendicular bisectors of the sides, and the three altitudes of the triangle. Moreover, as noted by Euler, these three points of intersection are collinear! In 1899, Frank Morley discovered that adjacent pairs of angle trisectors always meet at the vertices of an equilateral triangle (see [1], chapter 12). Any triangle can be dissected into four congruent triangular pieces all similar to the original triangle (the analogous result in three dimensions for tetrahedra, alas, is not true) and any triangle can be used to tile the entire plane. In 1928, E. Sperner used triangular subdivisions to prove that any continuous map from a triangle to itself (interior included) must have a fixed point (Brouwer's fixed point theorem).

The remarkable properties of a triangle have not escaped the notice of your dozenal correspondent! Here, for your amusement, are twelve problems about our familiar three-sided friend. Enjoy!

1. Non-periodic Please

A tiling of the plane is said to be periodic if there exists a fundamental region of tiles that covers the entire tiled plane via translation. Triangles, for example, produce periodic tilings. Use two copies of a given triangle to form a parallelogram and then translate the parallelogram across the plane.

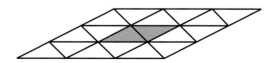

Can a triangle ever produce a non-periodic tiling of the plane?

2. Making Triangles

The triangle inequality states that if a, b, and c represent the side-lengths of a triangle drawn in a plane, then $a \leq b + c$. We also have $b \leq c + a$ and $c \leq a + b$.

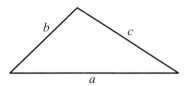

Are these conditions also sufficient? Given a triple of numbers (a, b, c) satisfying these three inequalities, does there necessarily exist a planar triangle with these given quantities as side-lengths?

3. Interior Invariant

Let P be a point inside an equilateral triangle. Show that the sum of the distances of the point from the three sides of the triangle equals the height of the triangle, no matter where the point P is placed.

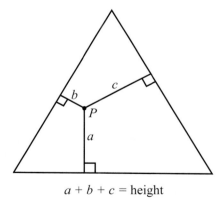

$a + b + c$ = height

4. A Palatial Walk

King Tricho lives in a palace with nothing but triangular rooms. Every night before retiring, His Highness strolls through his royal domain to inspect each room. Is there a way for the king to walk a path that visits each room once and only once? (He need not start in the room indicated.)

Reprinted from April 2002, pp. 23–28

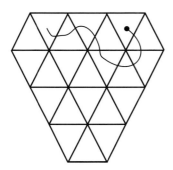

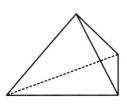

5. Tricky Polyhedra

A tetrahedron has four triangular faces and an octahedron has eight triangular faces. Is it possible to build a polyhedron, not necessarily regular, with an odd number of triangular faces (and with no faces of any other shape)?

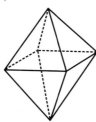

6. Mixing Odds and Evens

Is it possible to subdivide an even-sided polygon into an odd number of triangles? Is it possible to subdivide an odd-sided polygon into an even number of triangles?

When subdividing your polygon, make sure you don't add extra vertices to its boundary (otherwise you are changing the number of sides of the polygon!) and that each triangle in your subdivision has precisely three vertices along its perimeter.

This is the triangulation of a heptagon.

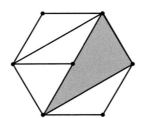

The shaded region is not a triangle!

7. Unavoidable Triangles

Draw a polygon and subdivide it into an odd number of triangles. (Again make sure that each subtriangle has precisely three vertices appearing on its perimeter.) The puzzle here is to label each vertex of your diagram either 1, 2, or 3 in such a way that:

1) No two adjacent vertices on the boundary of the polygon have the same label.

2) No interior triangle is labelled 1-2-3. Can you do it?

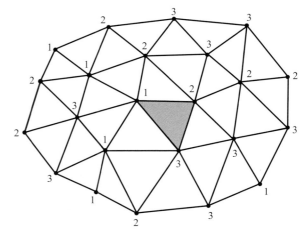

An invalid exterior edge and an invalid triangle

8. Pushing Pythagoras

Equilateral triangles are drawn on the three sides of a right triangle. Prove that the area of the large equilateral triangle equals the sum of the areas of the other two.

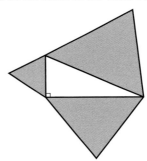

9. Devilish Dissection

Is it possible to dissect an obtuse triangle into a finite number of acute triangles?

10. Squaring the Triangle

It is possible to dissect an equilateral triangle into four pieces, rearrange those pieces, and form a perfect square. How?

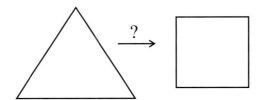

11. Fagnano's Problem

Let ABC be an acute triangle. Which inscribed triangle, touching all three sides, has least perimeter?

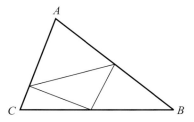

12. Equilateral Transformation

Given an arbitrary triangle, draw isosceles triangles with base angles 30° along each side of the triangle. Show that the apexes of these new triangles form an equilateral triangle.

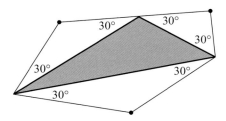

Answers, Comments, and Further Questions

1. Yes! A right isosceles triangle, for example, can be used in a non-periodic tiling of the plane. Start with the standard square tiling and then bisect each cell with a diagonal line, altering orientations so as to avoid periodicity. Circular designs with acute isosceles triangles are also non-periodic.

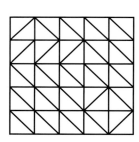

2. Yes! Draw a line segment of length a. This will be the base of our triangle. Draw a circle of radius b centered at the left endpoint of this segment and a circle of radius c at the right endpoint. The three inequalities ensure that the two circles intersect. One of the points of intersection represents the apex of the desired triangle.

Taking it Further. Equip the plane with the taxicab metric. Here the length of a line segment with endpoints $P = (x_1, y_1)$ and $Q = (x_2, y_2)$ is measured as:

$$d(P, Q) = |x_1 - x_2| + |y_1 - y_2|.$$

Find necessary and sufficient conditions for a triple of numbers (a, b, c) to represent the side-lengths of a triangle drawn in the plane equipped with the taxicab metric.

Taking it Further. A tetrahedron sitting in three-space has four triangular faces. Associated with each face are three triangle inequalities. Given a six-tuple of numbers (a, b, c, d, e, f) satisfying these 12 inequalities, does there necessarily exist a tetrahedron in three-space with these side-lengths? What if $\mathbb{R}^3$ is equipped with the appropriate generalization of the taxicab metric?

3. Let s be the side length of the equilateral triangle with height h and let a, b, and c, be the distances of the point P from each of the three sides. Subdivide the equilateral triangle into three subtriangles as shown:

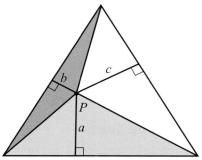

The areas of the three subtriangles sum to the area of the equilateral triangle. Thus

$$1/2 \cdot s \cdot a + 1/2 \cdot s \cdot b + 1/2 \cdot s \cdot c = 1/2 \cdot s \cdot h$$

and so

$$a + b + c = h.$$

Comment. The same argument shows that the sum of distances of an interior point to each side of a regular polygon is independent of the location of the point.

Taking it Further. Suppose the point P lies *outside* the equilateral triangle. Can anything be said about the sum of its distances to the three sides?

4. Color the cells of the palace floor plan as shown:

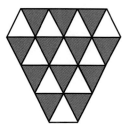

Any path through the palace moves alternately between black and white cells. Thus the number of black cells visited either equals the number of white cells visited, or differs from this number by one. But in the diagram above there are two more black cells than white. Thus a path that visits each cell precisely once is impossible.

However, if his Royalship is willing to step outside into royal gardens during his sojourn then such a path is possible.

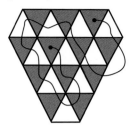

Taking it Further. Suppose the King would like to walk through his palace and visit each room just once, and end in the same room he started. Design a route that accomplishes this with the King exiting and entering the palace as few times as possible. Prove that your number of exits is minimal.

5. Let t be the number of faces on a polyhedron composed solely of triangular faces and e the number of edges. As every face has three edges we obtain:
$$3t = 2e.$$
(Each edge is counted twice.) This shows t cannot be odd.

6. Suppose an n-gon is subdivided into t triangles. Let e be the number of edges that appear in the interior of the polygon. Counting the total number of edges by counting the number of triangles yields:
$$3t = n + 2e.$$
(Each edge of the polygon is counted only once.) Thus t and n must have the same parity.

7. Consider a polygon subdivided into an odd number of triangles with each vertex labelled either 1, 2, or 3. Suppose no two adjacent boundary vertices are given the same label. We will show that a 1-2-3 triangle necessarily appears somewhere in the diagram.

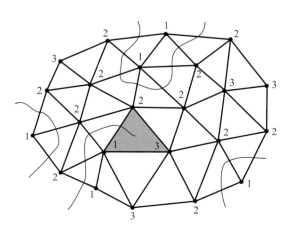

Each exterior edge is labelled either 1-2, 2-3, or 3-1. By question 5 there are an odd number of boundary edges and so there cannot be an even number of edges of each type. Let's say the number of exterior 1-2 edges is odd. Imagine your polygon is the floor plan for another palace with triangular rooms and with all the 1-2 edges, both inside the polygon and along its boundary, the doorways. All other edges are walls. Standing outside the palace, there are an odd number of doors through which you can enter the castle. If you do so, and follow the passageway of rooms and doors as far as possible, you may be led back outside the castle. But, as there are an odd number of exterior 1-2 doors, at least one passageway terminates in a room inside the palace. This final room must be a 1-2-3 triangle!

8. The area of an equilateral triangle with side length s is $\frac{\sqrt{3}}{4}s^2$. Suppose the right triangle has side lengths a, b, and c, with c the hypotenuse. By Pythagoras's theorem, $c^2 = a^2 + b^2$, and so
$$\frac{\sqrt{3}}{4}c^2 = \frac{\sqrt{3}}{4}a^2 + \frac{\sqrt{3}}{4}b^2.$$
proving the claim.

Taking it Further. Euclid proved Pythagoras's theorem by presenting an explicit dissection of a large square as the composition of pieces of the other two. Is there an analogous dissection for the equilateral triangles?

Taking it Further. Regular n-gons are drawn on the three sides of a right triangle. Prove that the area of the large n-gon equals the sum of the areas of the other two.

Need the n-gons be regular in your argument? Must the shapes be composed only of straight-line edges?

9. Such a dissection is indeed possible. There must be at least one line dividing the obtuse angle, and this line should terminate at a point inside the triangle. At least five lines must emanate from this interior point. This leads us to the following dissection:

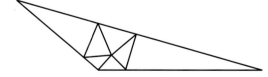

Taking it Further. Dissect a square into eight acute triangles.

10. Bisect two sides of the triangle and draw a perpendicular line from one midpoint to the third side, and from the other mid-

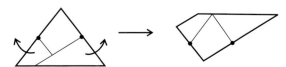

point to this line. These pieces "unhinge" to form the shape shown:

Now bisect the top edge of this new figure and draw a line from the midpoint to intercept the right edge at 90°. Rotating about the top "hinge" creates a perfect square.

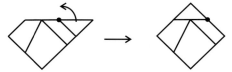

11. Fix a point P along the base of the triangle. We first find the inscribed triangle with the least perimeter that touches the base at P.

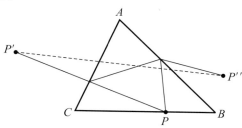

Reflect the point P about the left side of the triangle to create the point P'. Similarly let P'' be the reflection of P about the right side. Any inscribed triangle, touching at P, upon reflection creates a path from P' to P'' of length equal to the perimeter of the triangle.

Thus, the triangle of least perimeter touching P is the one that corresponds to a straight line from P' to P''. Notice that triangle $P'AP''$ is an isosceles triangle with side lengths equal to length AP and an angle at the apex equal to twice angle CAB. This angle is independent of the location of P.

Our mission now is to find a point P on the base of the triangle so that the straight line connecting the corresponding points P' and P'' is of minimal length. This is equivalent to finding a point P on the base where the length AP is minimal. Clearly then P must be at the base of an altitude.

This argument applies to any side of the triangle. Thus the inscribed triangle with the least perimeter is the one that connects the bases of the three altitudes.

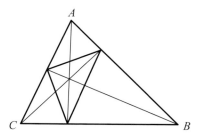

12. Two triangles are (directly) similar if they have a common angle and the sides subtending this angle in each triangle, taken in a counterclockwise order, have the same ratio. If the

vertices of one triangle are represented by complex numbers $z_1, z_2,$ and z_3 in the complex plane, and the other by $u_1, u_2,$ and u_3, then these conditions can be represented by the single equation:

$$\frac{z_1 - z_2}{z_3 - z_2} = \frac{u_1 - u_2}{u_3 - u_2}$$

This says the common angles occur at z_2 and u_2 respectively.

Let $(0, 1, \rho)$ be the coordinates of an isosceles triangle with base angle 30°. Thus $\rho = e^{i(\pi/6)}/\sqrt{3}$. Let (z_1, z_2, z_3) be an arbitrary triangle. Let u_1 be the apex of an isosceles triangle with base angle 30° drawn on side $z_1 z_2$. We have

$$\frac{\rho - 0}{1 - 0} = \frac{u_1 - z_1}{z_2 - z_1}$$

and so $u_1 = (1 - \rho)z_1 + \rho z_2$. Similarly, $u_2 = (1 - \rho)z_2 + \rho z_3$ and $u_3 = (1 - \rho)z_3 + \rho z_1$ are the apexes of similar triangles drawn on sides $z_2 z_3$ and $z_3 z_1$ respectively. Thus we have a transformation T on triangles given by:

$$\begin{pmatrix} z_1 \\ z_2 \\ z_3 \end{pmatrix} \mapsto (1 - \rho) \begin{pmatrix} z_1 \\ z_2 \\ z_3 \end{pmatrix} + \rho \begin{pmatrix} z_2 \\ z_3 \\ z_1 \end{pmatrix}.$$

Let $\omega = e^{i(2\pi/3)}$ be a cube root of unity. It is easy to check that $\mathbf{e}_1 = (1, 1, 1)$, $\mathbf{e}_2 = (1, w, w^2)$ and $\mathbf{e}_3 = (1, w^2, w)$ are three independent eigenvectors of this transformation with eigenvalues $\lambda_1 = 1$, $\lambda_2 = (1 - \rho) + w\rho$ and $\lambda_3 = (1 - \rho) + w^2\rho = 0$ respectively. Write (z_1, z_2, z_3) as a linear combination $a\mathbf{e}_1 + b\mathbf{e}_2 + c\mathbf{e}_3$, with $a, b, c \in \mathbb{C}$. Then

$$T\begin{pmatrix} z_1 \\ z_2 \\ z_3 \end{pmatrix} = a\mathbf{e}_1 + \lambda_2 b\mathbf{e}_2 = \begin{pmatrix} a + \lambda_2 b \cdot 1 \\ a + \lambda_2 b \cdot w \\ a + \lambda_2 b \cdot w^2 \end{pmatrix}.$$

Since $(1, w, w^2)$ is an equilateral triangle, it follows that the result of this transformation is also an equilateral triangle.

Taking it Further. Show that if isosceles triangles with base angles 30° are drawn on the three sides of an arbitrary triangle, pointing *inwards*, then their apexes also form an equilateral triangle. Show that the areas of the "inner" and "outer" equilateral triangles we just formed differ by the area of the original triangle.

Taking it Further. Draw four isosceles triangles with base angles 45° on the sides of an arbitrary quadrilateral, all pointing outward, or all pointing inward. The apexes of these isosceles triangles form a new quadrilateral. Show that the midpoints of its sides form a square. ∎

Acknowledgments and Further Reading

Martin Gardner writes about non-periodic tilings in his book [2], chapters 1 and 2, and has a wonderful discussion about acute triangle dissections in [3], problem 32. Question 10 was first proposed by H. Dudeney in 1908.

I first learned of the notion of a metric space being "triangle complete" from Tom Sibley's fascinating article [5]. In it he shows that $\mathbb{R}^3$ is not "pyramid complete" when equipped with the Euclidean metric, but is when given with the taxicab metric.

The result described in question 12 is known as Napoleon's Theorem and the generalization hinted at after its solution is due to J. Douglas and B. H. Neurmann. See G. Chang and T. Sederberg's wonderful text [1], chapter 16, for an enlightening discussion of this beautiful result.

Question 7 is a variation of Sperner's famous lemma. All one needs to ensure the existence of a 1-2-3 triangle is a labelling scheme that produces an odd number of exterior 1-2 (or 2-3 or 3-1) edges. To see how this lemma leads to the Brouwer fixed point theorem have a look at Mark Kac and Stanislaw Ulam's incredible book [4].

[1] Gengzhe Chang and Thomas Sederberg, *Over and Over Again.* The Mathematical Association of America. 1997.

[2] Martin Gardner, *Penrose Tiles to Trapdoor Ciphers.* W. H. Freeman and Company, New York, 1989.

[3] Martin Gardner, *My Best Mathematical and Logic Puzzles.* Dover Publications, New York, 1994.

[4] Mark Kac and Stanislaw Ulam, *Mathematics and Logic.* Dover Publications, New York, 1992.

[5] Tom Sibley, The possibility of impossible pyramids, *Mathematics Magazine* 73 No. 3 (2000), 185–193.

Geometry
and
Gerrymandering

RICK GILLMAN
Valparaiso University

Now that the U.S. Census Bureau has completed its task of counting the people of the United States, the various state legislatures will quickly be at the task of redrawing the boundaries of their representative districts. Soon after we will begin to hear cries of "Gerrymandering!"

For example, the creation of North Carolina's Twelfth Congressional district following the 1990 U.S. Census caused a significant legal controversy lasting throughout the first half of the decade. The district, shown in Figure 1, stretched along 165 miles of Interstate 85 connecting several African-American communities and at some points was only as wide as one lane of the interstate! (Ultimately, the U.S. Supreme Court decided that this was an unconstitutional segregation by race.)

While the legality of a district depends on several issues, we will focus on the shape of the district, which is the issue usually raised by the general public in cases of gerrymandering. Generally, this concern can be described as a question about the district's compactness—a measure of its "spreadoutness" or of its "bizarre," "uncouth" or "tortuously" shaped boundary. Measures of compactness have been well studied in the past by geographers, but this work is not well known to mathematicians.

Characterizing Compactness

Over time, three significant features of congressional districts have been identified by state legislatures as either necessary or desirable. These features are population equity, contiguity, and compactness. Determining the first two of these traits is straightforward. The one person-one vote principle requires that the populations of each district be equivalent, so we can simply measure the variance of the district's population from its ideal population. Since determining the contiguity can be answered yes or no upon inspection, this leaves compactness as the only ill-determined feature of a congressional district.

In this context, compactness refers to something much more generic than the technical, topological meaning of the word with which some of you are familiar. The *American Heritage Dictionary* gives two geometric definitions of the word that describe the trait that we are considering: first, "closely or firmly united or packed together; solid; dense" and secondly "packed into or arranged within a relatively small space." Geographers have settled on two properties to characterize noncompactness: having an area that is very spread out, and having a perimeter that is very long with respect to the area enclosed.

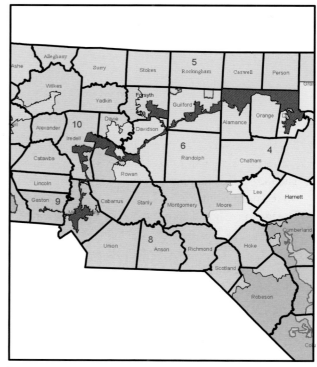

Figure 1. 12th Congressional district of North Carolina as of 1992.

Reprinted from September 2002, pp. 10–12, 22

Circles are the most compact shapes—a circle encloses the largest area for a given perimeter. Unfortunately it is impossible to cover any state with circular districts of equal population. Since the ideal cannot be reached, legislatures are forced to create districts which vary to some degree from this ideal and must find measures that can quantify this variation. These measures generally use the circular ideal as a standard of comparison. Geographers have identified a number of the qualities of a good measure of compactness. These are:

1. The measure should be applicable to all geometric shapes, both regular and irregular.
2. The measure should be independent of scale and orientation.
3. The measure should be dimensionless, and preferably be on a scale of 0 to 1, with 1 describing a highly compact region.
4. The measure should not be overly dependent upon one or two extreme points.
5. The measure should correspond with our intuition.

In addition, it is generally preferable if the measure is easy to calculate.

In practice there is only a short list of measures that are, or can be, used to determine the compactness of congressional districts. We discuss four of them in the next section.

Measuring Compactness

Measures of compactness are generally classified as one of two types. The first type are measures of the perimeter of a district (to measure the degree of contortion in the district's boundary). The second type are measures of the dispersion of a district (the degree to which a district is "spread out"). In this section, we present a portion of Diane Manninen's work to compare four of these measures.

The Area-Perimeter Measure calculates the ratio of the area A of a district to the area of a circle having the same perimeter P as the district. That is, we compare the actual area of the district to the maximal area that could be enclosed by the same boundary. This measure, given by the formula

$$M_1 = \frac{4\pi A}{P^2},$$

ranges over the interval from 0 to 1, with a circle receiving a score of one.

The first of the three dispersion measures that we will examine is the Longest Axis Ratio. This measure compares the area of the district to the area of a circle with a diameter equal to the length of the longest axis L in the district. Calculated by the formula

$$M_2 = \frac{A}{\pi(L/2)^2},$$

this measure is easy to calculate and yields values between 0 and 1. It is related to our third measure, which would be the intuitive one to select: the Circumcircle Ratio.

The Circumcircle Ratio is the ratio between the area of the district to the area of the smallest circumcircle, A_C, of the district. Thus

$$M_3 = \frac{A}{A_C}.$$

We will see that while M_2 and M_3 frequently agree, when they do not the value of M_2 will be slightly higher than the value of M_3. The reason for this is elementary, and left to the reader. In addition, the process of calculating the value of M_3 is significantly more complicated because we first need to find the circumcircle. (The reader is challenged to find an algorithm to do this!)

The final method that we will consider is the Moment of Inertia Ratio. Letting r equal the distance from a point in the district to the centroid, the moment of inertia is given by

$$\iint_A r^2 dA.$$

Recalling a bit of Multivariate Calculus, this measures the tendency of the points to bunch up or spread out from the centroid. We convert this to a relative measure of compactness by dividing by the moment of inertia of a circle having the same area as the district $D^2/2\pi$. We selected this moment of inertia because it would be what we expect in an ideally compact region. Finally, since we've squared all of our units in this process, we now take the square root of the ratio. Thus, we obtain

$$M_4 = \frac{D}{\sqrt{2\pi \iint_A r^2 dA}}.$$

Table 1 shows the measure of compactness given by each formula for regular shapes. Clearly, each of the four measures correspond with our intuitive sense of compactness—the nearer the shape is to being circular, the higher its compactness measure. Note that M_2 and M_3 generally agree on these shapes. (Why do they differ for the equilateral triangle?)

Shape	M_1	M_2	M_3	M_4
Circle	1.0000	1.0000	1.0000	1.0000
Hexagon	.9069	.8269	.8269	.9962
Square	.7854	.6370	.6370	.9772
Rectangle (1:2)	.6981	.5099	.5099	.8740
Equilateral Triangle	.6046	.5509	.4132	.9094

Table 1. Measures of Compactness of Various Shapes.

When a shape is increasingly indented, as illustrated in Figure 2, we have corresponding falling measures of compactness, presented in Table 2.

Manninen continued her discussion of these measures by considering the impact of elongating, puncturing, or fragmenting the region on each of the four measures of compactness. As one might expect, elongation of a region brings a rapid drop in the compactness measure of a shape, puncturing a region

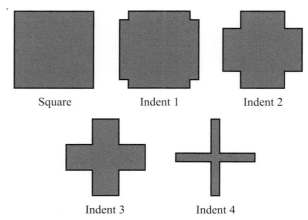

Figure 2. An increasingly indented space.

Shape	M_1	M_2	M_3	M_4
Square	.7854	.6370	.6370	.9772
Indent 1	.6981	.7841	.7841	.9928
Indent 2	.5890	.7646	.7646	.9772
Indent 3	.4363	.6362	.6362	.9073
Indent 4	.2400	.3785	.3785	.7231

Table 2. Effect on Compactness of Increasing the Indentation.

decreases the compactness measures, and fragmentation of a district decreases all of the measures except the Area-Perimeter ratio. This latter is reasonable, since the distance between the fragments has no bearing on the area or the perimeter of the fragments themselves.

A Case Study

Following the 1980 U.S. Census, the Republican-controlled Indiana State Legislature redistricted the state to match the current demographics. The Indiana Democratic Party then sued the state on the grounds that the redistricting impinged on the right to equal representation of members of the Democratic party living in the state. (Eventually, in *Davis v Brandemer*, the U.S. Supreme Court decided that this was a judicable issue, but that the Democratic party had failed to show that its members had seriously and lastingly been harmed by the redistricting.)

Using two measures of compactness, the Longest-Axis Ratio and the Area-Perimeter Ratio, Niemi and Wilkerson determined the compactness measures for each of the 77 districts, as established in 1970 and in 1980. Under both measures, the average compactness of the 1980 districts was somewhat less than the average compactness of the districts established in 1970.

Independently, the U.S. district court hearing the case indicated that the sixteen multimember districts that existed under the 1980 plan were discriminatory. The district court also cited 10 "worst-case" districts whose "irrational shapes" were in

	"Worst-Case" Districts		Multimember Districts	
	M_1	M_2	M_1	M_2
19 (25%) Least Compact Districts	9	9	4	4
19 (25%) Most Compact Districts	0	0	5	3

Table 3. Quantitative Compactness Compared to Court-Cited Districts.

particular need of justification. The distributions of the worst-case districts and of the multimember districts into the top and bottom quartiles of our measures of compactness are displayed in Table 3.

Clearly, the two measures were effective at identifying those districts that the court said had "irrational shapes." The fact that the 16 multimember districts seem to be distributed evenly across all four quartiles makes it clear that the court was also concerned with issues other than shape.

What this case study demonstrates is that our quantitative measures generally agree with our intuitive sense of ill-compactness, but that measures of compactness do not provide absolute evidence of gerrymandering. The compactness of the district only serves as an indicator of potential gerrymandering. While it is desirable to have highly compact districts, the various

parameters placed on the redistricting process make this difficult. In particular, districts generally follow the geographic features of the state and many states require that they also maintain already existing political structures such as counties, cities, towns, and townships. Politics is not, of course, an exact science; it remains possible to create legitimate districts that have very low compactness measures. It is also possible to create illegal districts with high compactness measures! ■

For Further Reading

Diane Manninen's masters thesis *The Role of Compactness in the Process of Redistricting*, University of Washington, 1973, provides a very accessible study of the various methods of measuring compactness. The book *Political Gerrymandering and the Courts* edited by Bernard Grofman, Agathon Press, 1990, contains a series of case studies investigating gerrymandering, including one describing the Indiana case. David Lubin's book *The Paradox of Representation*, Princeton University Press, 1997, describes the North Carolina case and the more general problem of protecting minority interests in Congress. In the recently published book *Bushlanders and Bullwinkles: How Politicians Manipulate Electronic Maps and Census Data to Win Elections*, University of Chicago, 2001, author Mark Monmonier makes a case that compactness will become less of a criterion in the near future.

Who
is the
Greatest Hitter
of Them All?

JOSEPH A. GALLIAN
University of Minnesota Duluth

If you want to get into an argument, ask any baseball fan who the greatest hitter is. Along with emotion, bias, and hometown prejudice, you're likely to hear some statistics. Baseball, after all, is the sport that best lends itself to statistical analysis. Using batting average alone to determine the best hitter is too crude since a bases-loaded home run counts the same as a bases-empty single. Other complicating factors in trying to decide who is the greatest hitter include the difficulty of comparing players from different eras, rule changes, changing strategies and even which ballpark a hitter has for his home games. Although there have been many different approaches to answer this question, the results are pretty much in agreement on the top two hitters. In this article I will explain a few of the more interesting methods used to try to answer this question and reveal the top two hitters.

For over 100 years batting averages (number of hits divided by number of at bats) have been used to declare the Batting Champion in each league. But even if we agree to use batting averages as a reasonable way to determine the best hitter it isn't the case that the best hitters are those with the highest batting averages. For one thing, unlike football and basketball, in baseball, the playing fields vary widely. Certain ballparks are "hitters parks" while others are "pitchers parks." (The essential differences are due to the size of the foul territory, not the distances to the outfield fences.) For another, left-handed hitters have an advantage over right-handed hitters. (There are two reasons for this. One is that for both left-handed and right-handed hitters it is easier to hit against a pitcher of the opposite hand, but there are more right-handed pitchers than left-handed ones. The second reason is that left-handed hitters face first base after swinging.) Moreover, rule changes (for example, the size of the strike zone or height of the mound), equipment changes (composition of the ball, batting gloves, batting hel-

mets, etc.), more frequent use of relief pitchers and many other factors result in league averages that vary widely by era. To illustrate, we point out that in 1930 the entire National League average, including the pitchers, was .303 whereas in 1968 Carl Yastrzemski lead the American League with a .301 average. Clearly, it is not the case that the average hitter in 1930 was a better hitter than Yaz in 1968. In an attempt to level the playing field, statisticians have devised ways to adjust for factors such as home ballpark, handedness and different eras. Table 1 shows the top 10 all-time batters from 1900 on using raw averages alone. Note that there is not a single person who has played in the past 40 years! This tips you off that something is wrong. In his book *Baseball's All-Time Best Hitters*, statistician Michael Schell devised a scheme that adjusts for differences in era, ballpark, handedness, and even late-career decline (he ignored all stats after 8000 at bats). Table 2 shows the adjusted list of top 10 hitters according to Schell (through 1997). This time we do have some modern ballplayers but the early players are still disproportionally represented.

1.	Ty Cobb	.366	1905–1928
2.	Rogers Hornsby	.358	1915–1937
3.	Joe Jackson	.356	1908–1920
4.	Ted Williams	.344	1939–1960
5.	Tris Speaker	.344	1907–1928
6.	Babe Ruth	.342	1914–1935
7.	Harry Heilmann	.342	1914–1932
8.	Bill Terry	.341	1923–1936
9.	George Sisler	.340	1915–1930
10.	Lou Gehrig	.340	1923–1939

Table 1. Lifetime batting average—Top Ten.

Reprinted from September 2002, pp. 13–16, 28

1.	Tony Gwynn	.342	1982–2001
2.	Ty Cobb	.340	1905–1928
3.	Rod Carew	.332	1967–1985
4.	Joe Jackson	.331	1908–1920
5.	Rogers Hornsby	.330	1915–1937
6.	Ted Williams	.327	1939–1960
7.	Stan Musial	.325	1941–1963
8.	Wade Boggs	.324	1982–1999
9.	Tris Speaker	.322	1907–1928
10.	Willie Mays	.314	1951–1973

Table 2. Lifetime adjusted batting average—Top Ten.

Slugging it Out

Of course, Schell's adjusted batting average is far superior to the traditional batting average. Nevertheless, it does not take into account power. It is beyond dispute that a home run is better than a single. The traditional way to account for power is the so-called Slugging Average (*SLG*). The slugging average counts a home run the same as four singles, a triple the same as three singles and a double the same as two singles. In short,

$$SLG = \frac{(1 \times 1B) + (2 \times 2B) + (3 \times 3B) + (4 \times HR)}{AB}$$

where 1*B* is the number of singles, 2*B* is the number of doubles, 3*B* the number of triples, *HR* the number of home runs, and *AB* the number of at bats.

The numerator of this fraction is denoted by *TB* because it counts the total bases achieved by the hits. As you would expect, this measure of batting performance correlates to team run production much better than batting average does. Table 3 shows the 15 players with the highest career slugging average through the 2001 season (assuming at least 5000 at bats). In contrast to the batting average, notice that about half of the players listed in Table 3 are still active. We believe that the reason for this is that there is now much more emphasis placed on hitting home runs than there was decades ago. This change in emphasis also accounts, at least in part, for the lower batting averages for modern players. (Other reasons for more home runs in recent years include better protection from being hit by a pitch, smaller ballparks, widespread adoption of weight lifting, and the use of nutritional supplements.)

In 2001 Barry Bonds had one of the greatest seasons in the history of baseball. Not only did he break Mark McGwire's 1998 home run record of 70, but Bonds also broke Babe Ruth's slugging record of .847 set in 1920 by slugging .863 and Ruth's 1923 record 170 walks by 7. Bonds will probably hold the slugging record much longer than he will hold the record for home runs. (Other than Bonds and Babe Ruth, who owns four of the top six single-season slugging averages, Lou Gehrig is the only player to break .760 in a single season with his .765 in 1927.)

1.	Babe Ruth	.690	1914–1935
2.	Ted Williams	.634	1939–1960
3.	Lou Gehrig	.632	1923–1939
4.	Jimmie Foxx	.609	1925–1945
5.	Hank Greenberg	.605	1930–1947
6.	Mark McGwire	.588	1986–2001
7.	Barry Bonds	.585	1986–
8.	Mike Piazza	.5789	1992–
9.	Joe DiMaggio	.5788	1936–1951
10.	Frank Thomas	.5770	1990–
11.	Rogers Hornsby	.5765	1915–1937
12.	Larry Walker	.572	1989–
13.	Juan Gonzalez	.568	1989–
14.	Ken Griffey, Jr.	.566	1989–
15.	Albert Belle	.564	1989–

Table 3. Lifetime slugging average—Top Fifteen through 2001.

The astute reader will have noticed that there is an important part of offensive production that the slugging average does not take into account—walks. There are a number of schemes that account for extra base hits and walks. One of these called Total Average (*TA*) was devised by the sportswriter Thomas Boswell in 1981 in *Inside Sports*. Total average gives a hitter credit for walks (*BB*), getting on base by being hit by the pitcher (*HBP*) and stolen bases (*SB*) while penalizing him for being caught stealing (*CS*) and grounding into double plays (*GIDP*). Notice that this method rewards ability to get on base, power and speed. Here is the formula for total average

$$TA = \frac{TB + BB + HBP + SB}{AB - H + CS + GIDP}.$$

This formula expresses a ratio of good events to bad events. It favors players like Barry Bonds who hit for average and have power and speed and disfavors players like Mark McGwire who simply hit for power. Total average is the best simple measure of offensive performance. Studies have shown that a team's total average is an excellent predictor of the runs per game scored by the team.

One shortcoming of the total average is that it counts a walk the same as a single but of course a single is more effective at advancing runners on base than a walk. Various refinements of the *TA* formula have been proposed. In one such formula a walk is counted only one-third as much as a single.

Since comparing players from different eras is fraught with complexity, some measures have been devised that compare players to their contemporaries. One of these is called the Offensive Quotient (*OQ*) described by M. VanOverloop in *The Baseball Research Journal* in 1993. To calculate this for a particular player one uses the formula

$$\frac{TB + BB}{Outs}$$

1.	Babe Ruth	218	1914–1935
2.	Ted Williams	210	1939–1960
3.	Lou Gehrig	177	1923–1939
4.	Rogers Hornsby	177	1915–1937
5.	Frank Thomas	177	1990–
6.	Mickey Mantle	176	1951–1968
7.	Jimmie Foxx	173	1925–1945
8.	Ty Cobb	170	1905–1928
9.	Willie McCovey	169	1959–1980
10.	Johnny Mize	168	1936–1953

Table 4. Lifetime Offensive Quotient—Top Ten through 1995.

for the individual and for the entire league. Multiply the ratio of these two by 100 to get the *OQ* for the player. Table 4 shows the all-time leaders in this category from 1900–1995. Since the league average is always 100, Babe Ruth's 218 *OQ* means that Ruth's offensive performance was 118 percent greater than that of his contemporaries.

Send in the Clones

The measure that I find most interesting is called the Offensive Earned-Run Average (*OERA*) devised by Thomas Cover and Carrol Keilers in a 1977 article in *Operations Research*. The intent of their method is to determine how many runs a team composed of nine identical players would score per game. Say we wanted to calculate the *OERA* for Barry Bonds for his multi-record breaking year 2001. We imagine a line-up for which *every* batter has the same probability of hitting a single as Barry Bonds, the same probability of hitting a double as Barry Bonds, the same probability of hitting a triple as Barry Bonds, the same probability of hitting a home run as Barry Bonds, and the same probability of getting a walk as Barry Bonds. In short, the offensive line-up consists of nine Barry Bonds clones!

To illustrate the idea say the first batter gets a walk while the second batter gets a single. This puts runners on first and third (it is assumed that a single advances a runner two bases). If the next two batters make outs the runners are assumed to stay in place. Then a double scores two runs and puts a man on second. Finally, the next batter ends the inning by making an out. So, for that inning, a line-up of 9 Barry Bonds scored two runs. At this rate for an entire season Bonds's *OERA* would be 18.00.

Cover and Keilers achieved this simulation by constructing a 24×24 matrix Q that accounts for all 24 possible combinations of outs (0, 1, or 2) and men on base ($\binom{3}{0} + \binom{3}{1} + \binom{3}{2} + \binom{3}{3}$). The entries of Q are probabilities, Q_{ij} is the probability of passing from start i to state j. For example, if state 7 represents a runner on second base and one out and state 13 represents no men on base and one out, the only way to pass from state 7 to state 13 is for the batter to hit a home run. Thus $Q_{7,13}$ is the proba-

bility that the batter being simulated hits a home run. For each state there is a probability that the hitter will produce one or more runs by hitting a single, double, triple or home run or walking. Collect these probabilities into the entries of a 24×1 vector, R, so that R_j represents the expected number of runs scored from state j in one plate appearance. For example, if state 7 is as above, then $R_7 = 1 \times$ (probability of a single) + 1 $\times$ (probability of a double) + 1 $\times$ (probability of a triple) + 2 $\times$ (probability of a home run). After each plate appearance there will be a new state determined by what the batter did in the previous appearance. By iterating this procedure and keeping track of the number of expected runs scored E (a 24×1 vector) before three outs occur, Cover and Keilers were able to simulate a line-up of nine identical players by using the formula

$$E = \sum_{i=0}^{\infty} Q^i R = (I - Q)^{-1} R.$$

The *OERA* is 9 times the expected number of runs scored in one inning (the first entry of E) beginning each time with the state that there are no outs and no men on base.

One attractive feature of the *OERA* is that unlike traditional measures of offensive performance such as RBIs and runs scored, it does not depend on the quality of one's teammates.

Harvard statistician Carl Morris and University of Minnesota Duluth graduate student Kai Xu have refined the *OERA* method of Cover and Keilers to more closely model what happens when a batter hits a single or double with a man on base (they do not assume the runner on base always advances two bases on a single or that a man on first always scores when the batter hits a double).

Tables 5 and 6 show Xu's calculations for those batters with the best single season *OERA* (excluding seasons from the 19th century) and the best career *OERA* up through the year 2001. The astounding thing about Table 5 is the 16 year gap in the two entries for Ted Williams. Williams was 39 years old in 1957. To put these numbers in perspective we mention that typically a team scores about 4.5 runs per game.

1.	Ted Williams	1941	19.54
2.	Babe Ruth	1923	18.79
3.	Babe Ruth	1920	18.47
4.	Ted Williams	1957	17.07
5.	Barry Bonds	2001	16.98
6.	Babe Ruth	1921	16.93
7.	Babe Ruth	1926	16.25
8.	Babe Ruth	1924	16.00
9.	Rogers Hornsby	1924	15.73
10.	Rogers Hornsby	1925	14.90

Table 5. Season Offensive Earned-Run Average—Top Ten through 2001.

1.	Babe Ruth	12.91	1914–1935
2.	Ted Williams	12.83	1939–1960
3.	Lou Gehrig	10.88	1923–1939
4.	Frank Thomas	10.09	1990–
5.	Jimmie Foxx	9.86	1925–1945
6.	Rogers Hornsby	9.76	1915–1937
7.	Barry Bonds	9.14	1986–
8.	Mickey Mantle	9.06	1951–1968
9.	Hank Greenberg	9.05	1930–1947
10.	Ty Cobb	9.00	1905–1928

Table 6. Lifetime Offensive Earned-Run Average—2001.

1.	Barry Bonds	2001	18.86
2.	Ted Williams	1941	16.59
3.	Babe Ruth	1920	16.44
4.	Ted Williams	1957	15.99
5.	Babe Ruth	1921	15.63
6.	Babe Ruth	1923	14.86
7.	Jimmie Foxx	1932	13.88
8.	Babe Ruth	1931	13.29
9.	Babe Ruth	1924	13.13
10.	Babe Ruth	1926	12.97

Table 7. Season Runs Created per Game—Top Ten through 2001.

The baseball research pioneer Bill James has developed a complex formula called Runs Created per Game (RC/G) that also attempts to estimate the number of runs scored per game by an entire line-up of the same hitter. Table 7 shows the top 10 single season RC/G through 2001 using a refined version of James's formula that penalizes a batter for striking out. Barry Bonds's record breaking RC/G of 18.66 for 2001 is even more impressive than his record setting marks for home runs, walks and slugging average in 2001.

And the Winners Are...

So, upon looking at the dozens of schemes that have been devised to answer the question "Who is the greatest hitter of them all?" one sees the same two names inevitably come up near the top: Babe Ruth and Ted Williams. Ruth's still-standing 40 batting records dwarfs Ty Cobb's second most 21. There seems to be no clear choice for the third place. Strong cases can be made for Aaron, Bonds, Hornsby, and Gehrig.

Of course, one might ask the question Who is the greatest baseball player of them all? The Official Encyclopedia of Major League Baseball *Total Baseball* uses a rating scheme called "Total Player Rating" that takes into account all aspects of a player's contribution to his team: hitting (adjusted for league average and ballpark), base stealing, fielding, and position (shortstops, second basemen, catchers and third basemen count more than outfielders and first basemen; center fielders

1. Babe Ruth	1914–1935
2. Barry Bonds	1986–
3. Nap Lajoie	1896–1916
4. Rogers Hornsby	1915–1937
5. Ted Williams	1939–1960
6. Mike Schmidt	1972–1989
7. Mickey Mantle	1951–1968
8. Mike Piazza	1992–
9. Willie Mays	1951–1973
10. Lou Gehrig	1923–1939

Table 8. Total Player Rating—Top Ten through 2000.

count more than left and right fielders). If one accepts *Total Baseball*'s rating as reasonable, then the list of the top ten greatest players (excluding pitchers) is given in Table 8. Notice that the Total Player Ratings of Schmidt, Hornsby and Piazza are helped by the positions they play.

During 1999 many polls were taken asking who were the greatest athletes of the 20th century. In the poll of experts taken by ESPN, Ruth was rated the greatest baseball player, Mays was second, Aaron was third, Jackie Robinson 4th and Williams 5th. (Many of the votes Robinson received were because he integrated Major League Baseball.) An Associated Press poll ranked them Ruth, Mays, Williams. Yet another poll had them Ruth, Mays, Aaron, Williams. So there is total agreement with Ruth at #1 and Mays #2. Williams and Aaron are three and four or vice versa.

Even though Ruth and Williams were in their primes in the first half of the 20th century, they remain among the most popular players ever. The web site

www.baseball-reference.com

(click on "leaders" then on "Most Popular Players" at the bottom of the page) keeps track of the number of hits each player's web page receives. The five most popular players are: Bonds, Ruth, Aaron, Williams, and Mantle. ■

For Further Reading

Besides Schell's *Baseball's All-Time Best Hitters*, Albert and Bennett's *Curve Ball* is highly readable and describes in great detail how statistical methods are used to measure offensive performance. Baseball records and statistics for players can be found at www.baseball1.com and www.baseball-reference.com. Access Kai Xu's program for computing *OERA* at www.d.umn.edu/~jgallian/OERA.html.

Update

After this article appeared Barry Bonds, beginning at the age of 36, had what is arguably the best four-year stretch in the history of baseball. Were it not for his use of performance-inhancing substances this achievement would put him ahead of Ruth and Williams. However, his use of these substances devalues his accomplishments.

Generalized Cyclogons

TOM M. APOSTOL
MAMIKON A. MNATSAKANIAN
California Institute of Technology

A cycloid is the curve traced out by a point on the circumference of a circular disk that rolls without slipping along a straight line. It consists of a periodic sequence of congruent arches resting on the line. If the point is rigidly attached to the disk but not on the circumference it traces out a curtate cycloid if the tracing point lies inside the disk, and a prolate cycloid if it lies outside the disk. Figure 1 shows an example of each type.

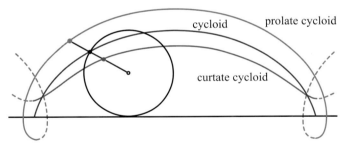

Figure 1. A cycloid, a curtate cycloid, and a prolate cycloid traced out by a point on a rolling disk.

If the rolling disk is replaced by a regular polygon, each vertex traces out a curve we call a cyclogon. In [1] the authors determined by elementary means the area of the cyclogon, that is, the area A of the region under one arch of a cyclogon. We showed that A is equal to the area P of the rolling polygon plus twice the area C of the disk that circumscribes the polygon:

$$A = P + 2C. \qquad (1)$$

In the limiting case, when the number of polygonal edges increases without bound, the cyclogon becomes a cycloid, P approaches C, and the cycloidal area is three times the area of the rolling disk.

Motivated by an effort to better understand why the term $2C$ appears in (1), we considered the more general area problem for curtate and prolate cyclogons and found a result that is surprisingly simple. The general formula can be written symbolically as follows:

$$A = P + C + C_z, \qquad (2)$$

where P denotes the area of the rolling polygon, C is the area of the disk that circumscribes the polygon, and C_z is the area of a disk whose radius is the distance from the center of the rolling disk to the tracing point z. When z is on the circumference of the rolling disk, we have $C_z = C$ and we get (1). In the limiting case when P approaches C, Eq. (2) gives a known result $A = 2C + C_z$.

Many curves related to cycloids can be obtained by rolling a circular disk around a fixed circular disk (instead of along a line). A point on the circumference of the rolling disk generates an epicycloid if the rolling disk is outside the fixed disk, and a hypocycloid if it is inside. Epicycloids were used by Apollonius around 200 B.C. and by Ptolemy around 200 A.D. to describe the apparent motion of planets. When the tracing point is not on the circumference of the rolling disk, it traces out a trochoid: an epitrochoid if the rolling disk is outside the fixed disk, and a hypotrochoid if it is inside.

We consider a more general situation in which a curve is traced by a point z on a regular polygonal disk with n sides rolling around another regular polygonal disk with m sides. The edges of the two regular polygons are assumed to have the same length. A point z attached rigidly to the n-gon traces out an arch consisting of n circular arcs before repeating the pattern periodically. We call this curve a trochogon—an epitrochogon if the n-gon rolls outside the m-gon, and a hypotrochogon if it rolls inside the m-gon. The trochogon is curtate if z is inside the n-gon, and prolate (with loops) if z is outside the n-gon. If z is at a vertex it traces an epicyclogon or a hypocyclogon. Figure 2 shows a curtate epitrochogon obtained by rolling a square ($n = 4$) outside a 24-gon ($m = 24$).

The main result of this article is a simple and elegant formula for the area of the region between a general trochogonal arch and the fixed polygon. We call this the area of the trochogonal arch. It is given by

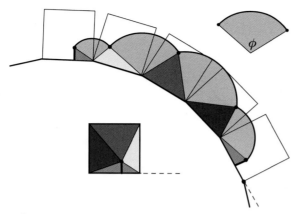

Figure 2. A curtate epitrochogonal arch traced by a point inside a square rolling outside a regular 24-gon.

$$A = P + \left(1 \pm \frac{n}{m}\right)(C_z + C), \tag{3}$$

with the plus sign for an epitrochogon and the minus sign for a hypotrochogon. If we let m tend to ∞, the fixed m-gon becomes a straight line and we obtain (2) as a limiting case of (3). All results in this paper are obtained without using integral calculus.

A proof of (3) is outlined in the next section after which we discuss a number of special cases.

Area of a Trochogonal Arch

Figure 2 displays the essential features required for treating a general regular n-gon rolling outside a regular m-gon. In Figure 2, the tracing point z is inside the square, and the arch it generates consists of four circular sectors and five triangles, shown shaded. The lower portion of Figure 2 shows how the five triangular pieces fill the square. Because of periodicity, the first and last right triangles outside the 24-gon together have the same area as the bottom triangle in the lower part of Figure 2. So area A is equal to area P, the area of the rolling square, plus the sum of the areas of the four circular sectors.

In the general case of a regular n-gon rolling outside a regular m-gon, the tracing point z attached to the n-gon generates an arch consisting of n circular sectors together with a set of triangles that provide a dissection of the n-gon. So the area A of any trochogonal arch is equal to that of the rolling n-gon P, plus the areas of n circular sectors, the kth sector having area $\frac{1}{2}\phi r_k^2$, where ϕ is the common angle (in radians) subtended by each sector and $r_1, \ldots, r_n$ are the radii of the sectors. Radius r_k is the distance from the tracing point z to the kth vertex of the rolling polygon. Thus, we have

$$A = P + \frac{1}{2}\phi \sum_{k=1}^{n} r_k^2. \tag{4}$$

It is easy to see that $\phi = 2\pi/n + 2\pi/m$, the sum of two exte-

rior angles, so (4) becomes

$$A = P + \left(1 + \frac{n}{m}\right)\frac{\pi}{n}\sum_{k=1}^{n} r_k^2. \tag{5}$$

Now we use a result on sums of squares derived in [2]. In complex number notation, it states that if $z_1, z_2, \ldots, z_n$ lie on a circle of radius r with center at the origin 0, and if the centroid of these points is also at 0, then for any point z in the same plane we have

$$\sum_{k=1}^{n} |z - z_k|^2 = n\left(|z|^2 + r^2\right). \tag{6}$$

Applying (6) with $r_k = |z - z_k|$ we find

$$\frac{\pi}{n}\sum_{k=1}^{n} r_k^2 = \pi |z|^2 + \pi r^2 = C_z + C,$$

which, when used in (5) gives the following formula for the area of an epitrochogonal arch:

$$A = P + \left(1 + \frac{n}{m}\right)(C_z + C). \tag{7}$$

Incidentally, if the rolling n-gon rolls inside the m-gon, the same analysis shows that the area formula for a hypotrochogonal arch is

$$A = P + \left(1 - \frac{n}{m}\right)(C_z + C), \tag{8}$$

so (7) and (8) together can be combined to give (3).

Applications

We can obtain the limiting case of a circle of radius r rolling around a fixed circle of radius R if we let both n and m tend to ∞ in such a way that their ratio $n/m \to r/R$. Then the limiting case of (3) becomes

$$A = C + \left(1 \pm \frac{r}{R}\right)(C_z' + C). \tag{9}$$

This gives the area of one arch of the classical epitrochoid and hypotrochoid without the use of calculus.

The authors could not find this general result in the literature except for the limiting case $R \to \infty$ and some special cases in which the tracing point z is at a vertex.

Tracing Point at a Vertex

Return now to (3) and take the tracing point z at a vertex of the rolling n-gon. Then the areas C_z and C of the disks are equal, and (3) gives the area of one arch of an epi- or hypocyclogon:

$$A = P + 2\left(1 \pm \frac{n}{m}\right)C. \tag{10}$$

In the limiting case when both n and m tend to ∞ in such a way that $n/m \to r/R$, (10) gives us a known result for the area of one arch of the classical epicycloid or hypocycloid:

$$A = \left(3 \pm 2\frac{r}{R}\right)C. \tag{11}$$

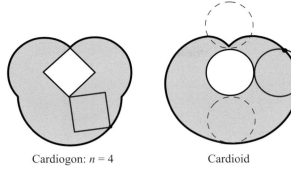

Cardiogon: $n = 4$ Cardioid

Figure 3. A cardiogon traced by the vertex of an n-gon rolling outside an n-gon. The cardiogon becomes a cardioid as $n \to \infty$.

A special case of (10) is the cardiogon (Figure 3)—an epicyclogon with $^n/_m = 1$,

$$A = P + 4C. \tag{12}$$

When $n \to \infty$, then $P \to C$, the tracing curve becomes a cardioid, and (12) or (11) give us $A = 5C$. This implies a classical result that the area of the region bounded by a cardioid is equal to $6C$, because the cardioidal arch, of area $5C$, together with the inner disk of area C, fill the cardioid with total area of $6C$.

Another special case of (10) is the nephrogon (Figure 4)—an epicyclogon with $^n/_m = \frac{1}{2}$, which gives

$$A = P + 3C \tag{13}$$

When $n \to \infty$ both (13) and (11) give $A = 4C$, for the area of one arch of a nephroid. The nephroid itself encloses two such arches, each of area $4C$, plus the inner disk of area $4C$, giving another proof that a nephroid encloses a region of area $12C$.

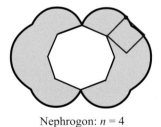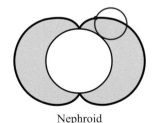

Nephrogon: $n = 4$ Nephroid

Figure 4. A nephrogon traced by the vertex of an n-gon rolling outside a $2n$-gon. The nephrogon becomes a nephroid as $n \to \infty$.

A related result is the astrogon (Figure 5)—a hypocyclogon with $^n/_m = \frac{1}{4}$. Eq. (10) gives

$$A = P + \frac{3}{2}C. \tag{14}$$

When $n \to \infty$, both (14) and (11) give $A = \frac{5}{2}C$ for an astroid, which is a hypocycloid with four cusps ($^r/_R = \frac{1}{4}$). The four arches between the hypocycloid and the outer circle (of area $16C$) have a total area of $4A = 10C$, so the region inside the astroid has area $6C$, another classical result obtained without calculus.

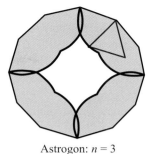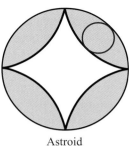

Astrogon: $n = 3$ Astroid

Figure 5. An astrogon traced by the vertex of an n-gon rolling inside a $4n$-gon. The astrogon becomes an astroid as $n \to \infty$.

Another special case of interest is the deltogon (Figure 6)—a hypocyclogon with $^n/_m = \frac{1}{3}$. Eq. (10) gives

$$A = P + \frac{4}{3}C, \tag{15}$$

and when $n \to \infty$ (15) and (11) give $A = \frac{7}{3}C$ for the deltoid, which is a hypocycloid with three cusps ($^r/_R = \frac{1}{3}$). The three arches between the deltoid and the fixed circle have a total area of $3A = 7C$, the fixed circle has area $9C$, so the region inside the deltoid has area $2C$, another known result.

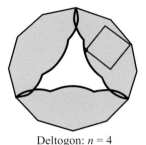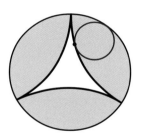

Deltogon: $n = 4$ Deltoid

Figure 6. A deltogon traced by the vertex of an n-gon rolling inside a $3n$-gon. The deltogon becomes a deltoid as $n \to \infty$.

A somewhat suprising example is what we call a diamogon—a hypocyclogon with $^n/_m = \frac{1}{2}$. The curve is traced by a point z at a vertex of an n-gon rolling inside a $2n$-gon. When the n-gon makes one circuit around the inside of the $2n$-gon, it traces out two curves each consisting of $n - 1$ circular arcs situated symmetrically about a diameter of the $2n$-gon. Examples with $n = 3$ and $n = 4$, are shown in Figure 7.

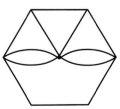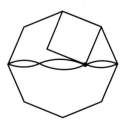

Diamogon: $n = 3$ Diamogon: $n = 4$

Figure 7. Diamogons traced by a vertex of an n-gon rolling inside a $2n$-gon.

Using (8) we find that the area of one arch of the diamogon is $A = P + C$ because $C_z = C$. The two arches between the diamogon and the outer polygon have area $2A = 2(P + C)$. In the limiting case when $n \to \infty$ this becomes $2A = 4C$. But $4C$ is the area of the fixed circular disk, which means that the area of the region common to the two diamogons tends to zero. In other words, when $n \to \infty$ the diamogon turns into a diameter of the fixed circle traced twice.

Tracing Point not at a Vertex

We conclude with an example of a hypotrochogon traced out by a point z not at a vertex of the n-gon. We consider $n/m = 1/2$ and call the hypotrochogon an ellipsogon because the limiting case $n \to \infty$ gives an ellipse. Figure 8 shows an example of a square rolling inside an octagon with the tracing point z inside the square. In this case the ellipsogon traces out two arches, each consisting of four circular arcs.

In the limiting case $n \to \infty$, (9) shows that the area of one arch is given by $A = C + \frac{1}{2}(C_z + C)$, so the two arches fill out a region of area $2A = 3C + C_z$. The limiting configuration of the ellipsogon is an ellipse enclosing an area equal to $4C - 2A = C - C_z$. If the radius of the inner circle is r and if the distance from z to the center of the inner circle is s then $C - C_z = \pi(r^2 - s^2) = \pi(r + s)(r - s)$. The distances $r + s$ and $r - s$ are the lengths

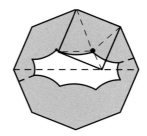 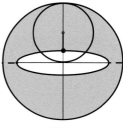

Ellipsogon: $n = 4$ Ellipse

Figure 8. An ellipsogon traced by a point inside an n-gon rolling inside a $2n$-gon. The ellipsogon becomes an ellipse as $n \to \infty$.

of the semiaxes $a = r + s$ and $b = r - s$ of the ellipse, so we get $C - C_z = \pi ab$ the usual formula for the area of an ellipse.

The point z also traces an ellipsogon if it is outside the rolling n-gon. If the point z is inside or outside the rolling n-gon and then moves toward a vertex, the ellipsogon becomes a diamogon which, in turn, becomes a diameter as $n \to \infty$. ■

References

1. Tom M. Apostol and Mamikon A. Mnatsakanian, Cycloidal Areas Without Calculus, *Math Horizons*, Sept. 1999, p. 14.

2. Tom M. Apostol and Mamikon A. Mnatsakanian, Sums of Squares of Distances, *Math Horizons*, Nov. 2001, pp. 21–22.

Fitch Cheney's
Five-Card Trick

COLM MULCAHY
Spelman College

In recent years a classic mid-20th century two-person card trick has resurfaced in the mathematical community. It goes something like this: A volunteer chooses five cards at random from a standard deck, hands them to you, and you show four of them to your confederate who promptly names the fifth card.

This superb effect is usually credited to mathematician William Fitch Cheney, Jr. The original trick, and some generalizations discussed here, work as well with large audiences as with small ones. The tricks are 100% mathematical—though you may choose to dress them up a little, for instance as "mind-reading tricks." They will stump all but the most sophisticated onlookers; a general audience will be convinced that some sort of body language or verbal signaling is being used. To eliminate that possibility, one can utilize email, telephone, an innocent go-between, or some other form of "impersonal" communication. Indeed, Cheney's trick, as published by W. Wallace Lee in 1950, was intended to be carried out over the telephone.

Cheney's Five-Card Trick

Effect: A volunteer from the crowd chooses any five cards at random from a deck, and hands them to you so that nobody else can see them. You glance at them briefly, and hand one card back, which the volunteer then places face down on the table to one side. You quickly place the remaining four cards face up on the table, in a row from left to right. (You can put them in a pile face down if you are worried that the audience thinks there is some subtle signaling going on in the precise placement of the cards on the table!) Your confederate, who has not been privy to any of the proceedings so far, arrives on the scene (e.g., is called in from another room), inspects the faces of the four cards, and promptly names the hidden fifth card.

Method & Mathematics: We strongly recommend that you spend some time (possibly days!) thinking over just how this puzzling trick might work before you read any further. It will give you great pleasure if you work out a solution for yourself. Yes, it really does work given any five cards.

Are you sure you are ready to read on? There are three distinct parts to the solution we present here.

Note that you get to choose which card to hand back (and later on, in what order to place the remaining four cards). First, the pigeonhole principle guarantees that (at least) two of the five cards are of the same suit. Let's suppose that you have two Clubs. One Club is handed back, and by placing the remaining four cards in some particular order, you effectively tell your confederate the identity of the Club you just handed back.

The second main idea is this: you can use one designated position (e.g., the first) of the four available on the table for the retained Club—which determines the suit of the hidden card—and the other three for the placement of the remaining cards, which can be arranged in $3! = 6$ ways. If you and your confederate agree in advance on a one-to-one correspondence between the six possible permutations and $1, 2,\ldots, 6$, then you can communicate one of six things.

What can one say about these other three cards? Not much—for instance, some or all of them could be Clubs, or there could be other suit matches! However, one thing is certain: they are all distinct, so with respect to some total ordering of the entire deck, one of them is LOW, one is MEDIUM, and one is HIGH. This permits for an unambiguous and easily remembered way to communicate a number between 1 and 6. But surely six isn't enough? The hidden card could in general be any one of twelve Clubs!

This brings us to the third main idea: you must be careful as to exactly which Club you hand back! Considering the 13 possible card values, 1 (Ace), 2, 3,…, 10, J, Q, K, to be arranged clockwise on a circle, we can see that the two Clubs picked are at most 6 values apart, i.e., counting clockwise one of them lies at most 6 vertices past the other. Give this "higher" valued Club back to the victim, which they then hide. You'll use the

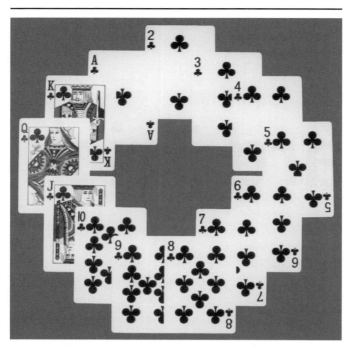

Counting clockwise, any two clubs are no more than six places apart.

"lower" Club and the other three cards to communicate the identity of this hidden card to your confederate.

For example, if you have the 2♣ and 8♣, then hand back the 8♣, but if you have the 2♣ and J♣, hand back the 2♣. In general, you save one card of a particular suit and need to communicate another of the same suit, whose numerical value is k higher than the one you make available, for some integer k between 1 and 6 inclusive.

Put this total linear ordering on the whole deck: A♣, 2♣, ..., K♣, A♦, 2♦, ..., K♦, A♥, ..., K♥, A♠, ..., K♠. Mentally label the three cards L (low), M (medium), and H (high) with respect to this ordering. Order the six permutations of L, M, H by rank, i.e., 1 = LMH, 2 = LHM, 3 = MLH, 4 = MHL, 5 = HLM and 6 = HML. Finally, order the three cards in the pile from left to right according to this scheme to communicate the integer desired. (See Trick 1.)

The one weakness in the method as described above is the invariant use of the first position in the pile as the "suit giver," this is soon spotted by alert audiences if the trick is repeated. Here is a better idea: since both you and your confederate get to see the four cards in the pile, sum their values and reduce mod 4 (using 4 if you get 0), and use that number for the position in the pile of the suit-determining card. For example, a Jack, 8, 2 and 7 would result in $11 + 8 + 2 + 7 = 0 \pmod 4$, so the fourth position in the pile would be used for the suit-determining card, and the first three cards would tell your confederate what to add to the numerical value of the last card to get the hidden card.

To return to the earlier example, suppose the five cards you are handed are 2♣, 2♥ 8♠, 7♦, and J♣. You play the J♣, in the 4th position, communicate $k = 4$ (hence the 2♣) to your confederate using the other three cards as follows: In standard LMH order they are 7♦, 2♥, 8♠, so in MHL order they are 2♥, 8♠, 7♦. (See Tricks 2 and 3.)

Of the three key ingredients in this trick, the pigeonhole principle and the permutations idea are easier to stumble upon than the fact one has only to communicate one of six (as opposed to twelve) integers.

Source: The trick is at least fifty years old, and seems to have first appeared under the name Telephone Stud in *Math Miracles* by W. Wallace Lee, where it is attributed to William Fitch Cheney, Jr, Chairman, Department of Mathematics, University of Hartford, Hartford, CT. Thanks to Art Benjamin for providing this source, and to Paul Zorn for alerting us to the existence of the trick in the first place. Gardner mentions it in passing in his *Mathematics, Magic and Mystery* as well as in his *The Unexpected Hanging and Other Mathematical Diversions*. This article was inspired by Brain Epstein's article "All You Need is Cards," published by A.K. Peters in the January 2002 *Puzzlers' Tribute—A Feast for the Mind*, a tribute to Martin Gardner.

The basic Cheney trick is spelled out eloquently and concisely in J.H. van Lint and R. M. Wilson's *A Course in Combinatorics*, 2nd Edition. A much watered-down version, in which all five cards are assumed to be of the same suit, and non-mathematical signaling also takes place, appears in Karl Fulves's *More Self-Working Card Tricks*, with a nod to Cheney.

Trick 1. The jack communicates the suit, the other three cards communicate the number four. Hence the hidden card is the two of clubs.

Trick 2. Again the jack communicates the suit and the other cards communicate the number four; the hidden card is the two of clubs.

Trick 3. What is the hidden card? (Answers are on page 276.)

More recently, it has been noted that this trick generalizes to larger decks, of up to 124 cards. See, for example, the January 2001 issue of *Emissary* or Michael Kleber's article in *The Mathematical Intelligencer*, Winter 2002. See also "Using a Card Trick to Teach Discrete Mathematics," Shai Simonson and Tara Holm, to appear in *PRIMUS*, 2003.

We now present some variations on Cheney's trick.

Ups and Downs and Back Again

Effect: A volunteer from the crowd chooses any five cards at random from a deck, and hands them to you so that nobody else can see them. You glance at them briefly, and hand one card back, which the volunteer then places face down on the table to one side. You quickly place the remaining four cards in a row on the table, some face up, some face down, from left to right. Your confederate, who has not been privy to any of the proceedings so far, arrives on the scene, looks at the cards on display, and promptly names the hidden fifth card.

Method & Mathematics: Again, you may wish to think this over before reading on.

As before, there are three distinct ideas at work here, two of which are the same as for the last trick. The pigeonhole principle guarantees that (at least) two of the five cards are of the same suit, without loss of generality you have two Clubs. You hand one back, and by placing the remaining four cards in some particular arrangement you effectively tell your confederate the identity of the hidden card. Save the "lower" Club, and communicate the identity of the "higher" one, whose numerical value is k past the one you hold on to, for some integer k between 1 and 6 inclusive. Use one particular position (e.g., the first) of the four available for the retained Club—

Trick 4. What is the hidden card?

which determines the suit of the hidden card—and the other three for the remaining three cards.

The difference here is that you communicate k using some kind of binary code—rather than permutations—for the three free slots. Unlike in the last trick, the identities of the face up cards (in the three relevant slots) play no role here! Rather than use actual binary representations, let's agree on this convention: *UDD, DUD, DDU* (only 1st, 2nd or 3rd position is Up), and *DUU, UDU, UUD* (only 1st, 2nd or 3rd position is Down), respectively, reveal to your confederate which of 1, 2, 3 and 4, 5, 6 the integer k is. (Try Trick 4.)

Since *UUU* is avoided above, an audience for whom you seem to be repeating the original Cheney effect can be given a choice of which trick they wish to see, even in mid-play—while your confederate is out of the room—without ever realizing that they are deciding between two different tricks! One could, for instance, say, "Should we make it harder on her this time, and only show her some of the cards?" Regardless of the answer, as soon as your confederate sees the cards, she immediately knows exactly which trick she is doing.

If some cards are face down she can play up the impossibility of the task before her, and the unfairness of it all, before dumbfounding the crowd by correctly announcing the identity of the hidden card. One more piece of drama can be added to this trick. Since the *DDD* possibility for card placement is also avoided above, at least one card of these three is always face up. Hence, we can use a face-up card (let's agree on the first such if there are two) to communicate the suit.

When all is said and done, this means that only three cards of the four retained need to be shown at all (face up or face down), the fourth can be set aside and ignored, every time! This too can be used to spice up the trick upon repeat performances.

Source: Original.

A similar trick can be done with fewer cards.

The Four-Card Cheney Trick

Effect: A volunteer from the crowd chooses any four cards at random from a deck, and hands them to you so that nobody else can see them. You glance at them briefly, and hand one back, which the volunteer then places face down on the table to one side. You quickly place the remaining three cards in a row on the table, some face up, some face down, from left to right. Your confederate, who has not been privy to any of the proceedings so far, arrives on the scene, looks at the cards on display, and promptly names the hidden fourth card—even in the case where all three cards are face down!

Method & Mathematics: We partition the standard deck into three new suits of 17 cards each, leaving out one special card (say A♠). Each of these new suits consists of one of the standard suits ♣, ♦, ♥ supplemented with four ♠'s. Specifically, Suit A is A♣, 2♣, . . ., K♣, 2♠, 3♠, 4♠, 5♠; Suit B is

Trick 5. What is the hidden card?

Trick 6. What is the hidden card?

A♦, 2♦,..., K♦, 6♠, 7♠, 8♠, 9♠; and Suit C is A♥, 2♥,..., K♥, 10♠, J♠, Q♠, K♠.

If one of the four cards is the special card, A♠, play the other three face down and it's all over: your confederate can identify the last card with what appears to be zero information! Otherwise, the pigeonhole principle guarantees that (at least) two of the four cards are from one of the three redefined suits, without loss of generality Suit A. You hand one back, and then by placing the remaining three cards on the table in some particular fashion you reveal to your confederate the identity of the card handed back.

As before, save the "lower" card from Suit A, and communicate the "higher" one, whose numerical value is k past the one you hold on to, where this time k is an integer between 1 and 8 inclusive. In the convention we explain below, at least one card will be face up, so once more we can use a face up card (the first such if there are two) to communicate the suit. As suggested before, the placements UDD, DUD, DDU (one U in 1st, 2nd or 3rd position) and DUU, UDU, UUD (one D in 1st, 2nd or 3rd position), respectively, can be used to tell your confederate that k is 1, 2, 3, 4, 5 or 6. This time we also need a way to communicate 7 or 8, and we have the UUU option at our disposal. If we agree to use one particular U (say, the middle one) to give the suit, there are two ways to play the other two: Low-High (to convey $k = 7$) or High-Low (for $k = 8$) with respect to some total ordering of the deck, such as lining up Suits A, B, C in that order. (Try Tricks 5 and 6.)

Source: Original.

What do you do in the original five-card trick if the volunteer insists on chosing the card to be identified? Characteristically, Martin Gardner determined an out many years ago; however, be warned, it does require a certain amount of mental gymnastics.

Eigen's Value

Effect: A volunteer from the crowd chooses any five cards from a deck, and hands them to you so that nobody else can see them. The volunteer then takes one card of their choosing back, which they show around to everybody, before plunging it into the deck and shuffling thoroughly. You quickly place the remaining four cards in a row from left to right on the table.

Your confederate arrives on the scene, inspects the cards on the table, as well as the remainder of the deck, and asks what the chosen card was. Upon it being named, the top card of the deck is turned over, and sure enough it is the chosen card.

Method & Mathematics: The mathematics is simple here, it's the method that requires some memory work. Unlike in the classic Cheney trick, you do not get to choose which card is the one your confederate must identify. So, in essence, you have to use the four cards you have to try to communicate which of the remaining $52 - 4 = 48$ possible cards it is. Since $4! = 24$, you can narrow it down to one of two cards by communicating a permutation. Here is one way: fix a total ordering on the deck as before. The four seen cards now determine two things: which 48 cards remain as possibilities for the hidden fifth card, and a particular permutation of the 24 available. Assuming that we rank those permutations as before, this means that a number, say 15, is communicated.

Your confederate must now mentally determine the 15th and 39th cards in the remaining ordered list of 48 possible cards: they are the cards that would usually be in those positions with respect to the agreed-upon total ordering, bumped up one for every seen card which occurs before those positions. The confederate locates these in the remainder of the deck, brings them together discreetly, and cuts the deck so as to bring one to the top and the other to the bottom. Finally, upon the naming of the chosen card, your confederate either turns over the top card, or turns over the deck to reveal the bottom card. This unpredictable last step makes the trick unsuitable for repeating, but as a one-off follow-up to the basic Cheney it is suitably mysterious.

Another way to get from 1 of 24 permutations to 1 of 48 cards is to provide one additional bit of information, for instance the displayed cards could be laid out right to left or left to right (if your confederate is allowed to see this part of the proceedings) or you could provide just one subtle physical or verbal signal.

Source: This is slightly adapted from Victor Eigen's trick as found in Gardner's *The Unexpected Hanging and Other Mathematical Diversions.* ∎

Answers

Trick 3. 7♠. Trick 4. Q♦. Trick 5. 9♠. Trick 6. Q♠.

The Card Game

ROBERT SCHUERMAN
Mansfield University
Pennsylvania

Last fall I assumed a position as an instructor of the calculus in the provinces. Small department, small college, small town. Social opportunities seemed very limited, so I accepted an invitation to a "new term get-together" at the house of our Dean. Ordinarily I shun such occasions.

Although the event had been described as "informal," everyone but myself was in church attire. The tone was stiff and restrained. I sought consolation from the bottle of sherry that the host had provided. Few others were drinking.

At length the dean announced that we all were to take part in a parlor game. We were divided into three groups and sent to corners of the room. I listened sullenly as the Dean explained that each group would be given a card, face down. The other side of each card was either blue or white. We were to display our card so that the color could be seen by the other groups, but not by our own group. Furthermore, if we saw two blue cards we were to raise our hands. Then we were to determine the color of our own card and announce it. We were free to discuss the problem within our own group, but no contact was to be allowed between groups.

My pulse quickened. I'm well skilled at logic puzzles. If I were to solve it first, I would make a good showing amongst my peers.

On a count the cards were displayed. My group saw two whites. No hands were raised, so there weren't two blues anywhere. But how could there be two blues, when we saw two whites? Not much help there.

The groups fell to discussion. I kept to myself. It was obvious to me that we weren't given enough information. We could equally well have had a blue as a white. The words "blue" and "white" bubbled throughout the room as theories were erected and quickly torn down. Suspicion grew that the elderly Dean had omitted part of the game, despite his encouraging nods.

(At this point the reader is encouraged to pause before reading further, in order to work through the logic.)

Fully five minutes had passed before I realized …THEY CAN'T SOLVE IT EITHER! That's the missing piece of information. The other groups must be in the same situation that I'm in—they see two whites! There must be three white cards.

Now wait a minute here. That's not much of a proof, but I feel sure that I'm right.

Suppose we do have a blue. There can't be two blues, but there could be one, our card. But WAIT! We can't be blue.

If we were blue then either group could know that they weren't blue as well, for that would have been two blues and then the third group would have raised their hands. That's it, we must be white!

One more time, I had better be sure about this. Let's see, IF we had a blue then either group could immediately determine that they couldn't also be holding a blue else there would be two blue cards visible to some group and the problem would have been solved long ago. So our card is white.

How exquisite! As soon as you realize that no one can solve the problem, the problem is immediately solved. Three whites is the only combination that could have stymied us all. Any blue card would have made the problem trivial. I had misjudged the Dean—that fox!

(Again the reader is encouraged to pause to verify this proof before reading further.)

I loudly recounted my logic to the entire room with much ceremony. In an expansive mood, flushed with excitement and drink, I concluded "...so either all of you are absolute muttonheads, or our card is white!"

Triumphantly I flipped over our card. It was blue. ■

Truels and the Future

STEVEN J. BRAMS
New York University

D. MARC KILGOUR
Wilfrid Laurier University

Is life bounded? Although there seems to be no confirmed case of a person's living more than 125 years (the confirmed maximum is 122 years, achieved by a French woman who died in 1997), there is no logical or medical barrier that would preclude a person, should he or she reach the age of 125, from living to be 126. Hence, to say that life is bounded by a limit, like 125 years, seems unjustified.

But what about extending that limit to 250 years, 1,000 years, or even 1,000,000 years? It seems absurd that any of us will ever approach such an age. On the other hand, the possibility that our genetic material might somehow be preserved or renewed is not so easy to dismiss. Alternatively, living our lives through our descendants—if the definition of life is broadened to include them—renders "ages" like 250, 1,000, or 1,000,000 years conceivable.

But the probability that any of us will, as individuals, live more than 125 years is at present infinitesimal. Practically speaking, it is reasonable to suppose that our lives are bounded by about 125 years, which means they will end before then.

Surprisingly, basing our actions on this limit may induce very different rational choices from those that reflect the view that life is unbounded—there is no limit on its length. Perhaps most realistic is the view that life is an infinite-horizon game: the horizon is infinite, so any pre-specified bound can be exceeded, but the end will occur sometime, guaranteeing that any play of the game is finite. While such games will definitely grind to a halt because they are finite, it is impossible to predict exactly when they will do so.

To illustrate the distinction between bounded and unbounded, we start with a simple hypothetical game. We will progressively change its rules to produce, finally, an infinite-horizon game, which we will briefly interpret in terms of some real-world events.

A Sequential Truel

Imagine three players, A, B, and C, situated at the corners of an equilateral triangle. They engage in a truel, or three-person duel, in which each player has a gun with one bullet.

Assume that each player is a perfect shot and can fire at one other player at any time. There is no fixed order of play, but any shooting that occurs is sequential: no player fires at the same time as any other. Consequently, if a bullet is fired, the results are known to all before another bullet is fired.

We suppose that the players order their goals as follows: (1) survive alone, (2) survive with one opponent, (3) survive with both opponents, (4) not survive, with no opponents alive, (5) not survive, with one opponent alive, and (6) not survive, with both opponents alive. Thus, the outcome of surviving alone is best, dying alone worst.

We rule out the possibility of firing in the air, which would be an optimal choice for a player if it were the first to fire. For once a player has disarmed itself by firing in the air, it would be no threat to its two opponents. Hence, it would be in the best interest of the player that shoots second to eliminate the third player.

If firing in the air is not permitted, who, if anybody, will shoot whom? It is not difficult to see that the outcome in which nobody shoots (outcome 3), so all players survive, is the rational outcome. Suppose, on the contrary, that A shoots B, hoping for A's outcome 2, whereby it and C survive. A's best outcome, surviving alone, is now impossible—C will not shoot itself. In fact, most preferring its outcome 1, C will next shoot a disarmed A, leaving itself as the sole survivor.

But this is A's outcome 5, in which A and one opponent (B) are killed while the other opponent (C) lives. To avoid this outcome, A should not fire the first shot; neither, for the same reason, should either of the other two players. Consequently, nobody will shoot, resulting in outcome 3, in which all three players survive.

Moreover, it would be risky for any two players—say, A and B—to collude and both agree to shoot C, thereby expending their bullets and posing no threat to each other. For if they did so, it would be in each of A's and B's interests to renege and not shoot C—saving its bullet for its partner after that player shoots C—because each player most prefers to be the sole survivor.

Reprinted from April 2003, pp. 5–8

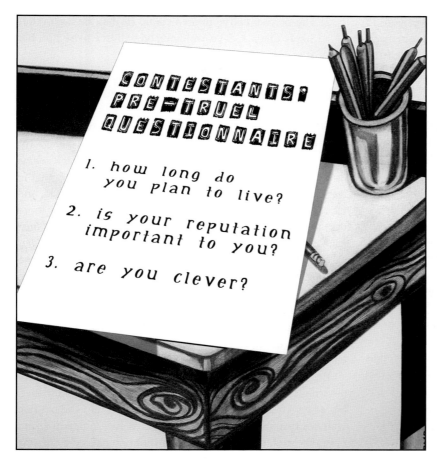

Thus, thinking ahead about the unpleasant consequences of shooting first or colluding, nobody will shoot or collude. Thereby, all players will survive if the players must act in sequence, giving outcome 3.

This thinking is also rational in the infinite-horizon truel we will describe at the end of the next section. However, there is another point of view, equally rational, that might be taken in that truel. It yields an ominous outcome, suggesting how conflicts among people, groups, or countries can lead to death and destruction.

Simultaneous Truels

1. One round. The rules no longer allow the players to choose in sequence, one after another, whereby late choosers learn the choices that others made earlier. Instead, all three players must now choose simultaneously whether or not to shoot, and at which opponent, in ignorance of what the other players do (that is, players cannot communicate so as to coordinate their choices). This situation is common in life; we must often act before we find out what others are doing.

Now everybody will find it rational to shoot an opponent at the start of play. This is because no player can affect its own fate, but each does at least as well, and sometimes better, by shooting another player—whether the shooter lives or dies—

because the number of surviving opponents is reduced.

If each player chooses its target at random, it is easy to see that each has a 25% chance of surviving. Consider player A; it will die if B, C, or both shoot it (3 cases), compared with its surviving if B and C shoot each other (1 case). Altogether, *one* of A, B, or C will survive with 75% probability, and nobody will survive with 25% probability (when each player shoots a different opponent). *Outcome:* There will always be shooting, leaving either one or no survivors.

2. n rounds ($n \geq 2$ and known). Assume that nobody has shot an opponent in the first $n - 2$ rounds. We next demonstrate that on the $(n - 1)$st round, either at least two players will rationally shoot, or none will.

First, consider the situation in which an opponent shoots A. Clearly, A can never do better than shoot, because A is going to be killed anyway. Moreover, A does better to shoot at whichever opponent (there must be at least one) that is not a target of B or C.

Now suppose that nobody shoots A. If B and C shoot each other, then A has no reason to shoot (although A cannot be harmed by doing so). If one opponent, say B, holds its fire, and C shoots B, A again cannot do better than hold its fire also, because it can eliminate C on the next round. (Note that C, because it has already fired its only bullet, does not threaten A.)

Suppose both B and C hold their fire. If A shoots an opponent, say B, then its other opponent, C, will eliminate A on the nth round. But if A holds its fire, the game passes onto the nth round and, as discussed earlier, A has some chance of surviving (25%, assuming random choices). Thus, if nobody shoots on the $(n - 1)$st round, A again cannot do better than hold its fire during this round.

Whether the players refrain from shooting on the $(n - 1)$st round or not—each strategy may be a best response to what the other players do—shooting will be rational on the nth round if there is more than one survivor and at least one player has a bullet remaining. Moreover, the anticipation of shooting on the $(n - 1)$st or nth round may cause players to fire earlier, perhaps even back to the 1st and 2nd rounds. Outcome: There will always be shooting, leaving one or no survivors.

3. n rounds (n unlimited). The new wrinkle here is that it may be rational for no player to shoot on any round, leading to the survival of all three players. How can this happen?

Our argument earlier that "if you are shot at, you might as well shoot somebody" still applies. But even if you are, say, A, and B shoots C, you cannot do better than shoot B, making yourself the sole survivor (outcome 1). As before, you do best—whether you are shot at or not—if you shoot somebody who is not the target of anybody else, beginning on round 1.

But now suppose that B and C refrain from shooting in round 1, and consider A's situation. Shooting an opponent is not rational for A on round 1, because the surviving opponent will then shoot A on the next round (there always is a next round if n is unlimited). On the other hand, if all players hold their fire, and continue to do so in subsequent rounds, then all three players remain alive.

While there is no "best" strategy in all situations, the possibilities of survival increase if n is unlimited. Outcome: There may be zero, one (any of A, B, or C), or three survivors, but never two if the players are rational (if two players shoot the same player, that player would have done better to shoot one of the two shooters).

4. Infinite-horizon. This truel is really a variant of situation 3 above that incorporates a more realistic feature. Specifically, at the end of round 1 and all subsequent rounds, a random event occurs that determines whether the truel continues at least one more round (with probability p_i at the end of round i) or ends immediately (with probability $1 - p_i$). Thus, the probability that a truel ends after exactly k rounds is $p_1 p_2 \ldots p_{k-1}(1 - p_k)$. The truel is bounded if and only if $p_i = 0$ for some round i.

If the truel is not bounded (i.e., is infinite horizon), it models games that—like life itself—do not continue forever. While we cannot say at what point such games end, we know they do not continue indefinitely. In such circumstances, if p_i is sufficiently high on each round i, it may be rational never to shoot, as we show in our 1998 article (cited at the end).

Yet the structure of such games means that the players can anticipate that the truel will end with virtual certainty after several rounds. For example, if $p_i = .51$ for all i, there is a probability of $1 - (.51)^{20} = .9999986$ that, after 20 rounds of play, the game will have terminated. Effectively, then, this can be thought of as an n-round game (n known), à la situation 2, in which there is only slightly more than one chance in a million (that is, probability .0000014) that the game will not end by round 20.

Applying the reasoning of situation 2 by treating the virtual certainty of termination as a certainty, the players will shoot in rounds 1 or 2, leaving at most one survivor. Outcome: How many survivors there are depends on whether the truel is viewed as bounded (at most one player survives) or unbounded (all three players may survive if p_i is sufficiently high), as in situation 3.

A Tale of Two Futures

In any infinite-horizon truel, there are two possible futures:

1. Every process must end by some definite point (for example, each person's lifespan has an upper bound of, say, 125 years);

2. The precise end is unpredictable (it may be highly unlikely that a 125-year-old person will live to be 126, but it is not impossible).

Our analysis shows that these two different views of the future may have radically different implications for rational play in the truel.

Outcome 3 in an infinite-horizon truel, in which nobody fires, is consistent with future 2, whereas outcomes 4 and 5 for a player are consistent with either future 1 or future 2. It seems that some real-world players have adhered more to the thinking of future 2, including the United States, Russia, and China: although all have possessed nuclear weapons for more than a generation, they have refrained from using them against each other in anything resembling a truel.

The same self-restraint may have manifested itself with the non-use of poison gas in World War II, partly in response to revulsion against its use in World War I and partly in fear of reprisal. By contrast, Bosnian Serbs, Bosnian Muslims, and Croats engaged in a very destructive truel in the former Yugoslavia in the early and mid-1990s, mirroring the boundedness of future 1.

Effectively, the Serbs fired the first shot, apparently thinking that quickly conquering territory would give them a big edge. After their early victories, however, they did not fare well because of the reactions of other players, including not only the original parties to the conflict but also new players, like NATO—especially after the conflict expanded to Kosovo in 1998.

Everybody would be better off, we believe, if players did not think they were so clever as to be able to reason backward, from some endpoint, in plotting each other's destruction. Indeed, our results suggest that players would be less aggressive if the future were seen as somewhat murky—as in the infinite-horizon truel—which would render predictions about how many rounds a game will go, or even an upper bound on this number, hazardous. This murkiness, oddly enough, is consistent with hope for the future.

Alternatively, a sequential truel in which the order of choice is determined by the players can induce cooperative behavior. As we showed earlier, if any of A, B, or C shoots first, it ensures its own death when the remaining survivor takes aim. Because each individual decision is known and responded to, it is clarity rather than murkiness that induces cooperation.

Conclusions

Two views of the future underlie bounded and unbounded play. To the degree that the future seems to stretch out indefinitely, people probably act more responsibly toward each other, knowing that tomorrow they may pay the price for their untoward behavior today. To sustain themselves, they may try to develop reputations, often by adhering to certain moral strictures. On the other hand, those who take a more short-term or bounded view may act less responsibly, even immorally.

An important intellectual task is to devise institutions that render destructive behavior unprofitable. But how one makes the future seem to run on smoothly, and instill confidence that the social fabric will not suddenly unravel, is not so clear.

We think the best institutions for this purpose are those that strongly suggest, if not promise, a day of reckoning for those who depart egregiously from norms of fair play. To return to the Yugoslavia example, it is unlikely that the parties who committed the most heinous crimes anticipated the involvement of the International Court of Justice and possible criminal trials.

Likewise, many terrorists seem to look for safe havens from which they will not be extradited. To the extent that international norms of justice not only sanction but also ensure, albeit in the indefinite future, punishment for serious crimes everywhere—including those across national borders—then parties who fire the first shot will be less confident that that shot will be decisive.

Short of ensuring future punishment, institutions that becloud the future, making predictions difficult, may also help to deter reprehensible actions. These institutions range from democracy, with its uncertain electoral futures and other vicissitudes, to extended nuclear deterrence, which offers a good if not certain prospect of protection to allies who might be attacked by an aggressor.

The possibility that these institutions or norms will set in motion forces to reward nonviolent behavior may be analogous to the preventive role of a third player in a truel. Although highly simplified as a model, the truel does capture an essential feature of social behavior—third parties may play an important role in attenuating conflict. Their presence, it seems, can ease the desperation one often finds in two-player conflicts, which often become wars of attrition. The third player, in essence, provides a balancing mechanism that helps to sustain hope, whether the future is murky or clear. ∎

For Further Reading

This article is a somewhat condensed version of our article, "Games That End in a Bang or a Whimper," in George F. R. Ellis (ed.), *The Far-Future Universe: Eschatology from a Cosmic Perspective* (Templeton Foundation Press, 2002), pp. 196–206. Mathematical foundations of the present analysis, and a detailed treatment of different cases, can be found in our articles, "Backward Induction Is Not Robust: The Parity Problem and the Uncertainty Problem," *Theory and Decision* 45, no. 3 (December 1998): 263–289; and, with Walter Bossert, "Cooperative vs. Non-Cooperative Truels: Little Agreement, But Does That Matter?" *Games and Economic Behavior* 40, no. 2 (August 2002): 185–202. Background on truels, including some with rules quite different from those analyzed here, can be found in our article, "The Truel," *Mathematics Magazine* 70, no. 6 (December 1997): 315–326, which contains numerous references to the literature; see also www.wlu.ca/~wwwmath/faculty/kilgour/truel.htm.

Unreasonable
Effectiveness

ALEX KASMAN
College of Charleston

Amanda Birnbaum began to have second thoughts. Could she really go through with this? Here she was, pacing back and forth in the dry sand on a tiny island she had never even heard of just a few weeks ago. Though it had a small population of people living in modest houses, the island was unclaimed by any country. As far as anyone could tell, it was just so insignificant that no country was interested in it. And, here she was, staring through the wrought-iron gate of the only really large house on the island without any idea of who might be residing there. But she knew she was in the right place. Every major scientific publisher had given her the same description of how to find it. They all had stories about this strange address which, for as long as anyone could remember, had been willing to pay any price to subscribe to all of the major research journals.

Well, even if her crazy ideas were wrong, there was something interesting here. Perhaps she could just find out who it was in this place who felt compelled to keep up on the latest published research! So, she walked up the long path winding between the weeds and wildflowers to the front door and she rang the bell.

Faintly, through the thick wooden door, she could hear the sound of the chimes playing a familiar tune to announce her arrival. She felt a sudden urge to run away, but her curiosity won out and she stayed to see whose feet she could hear scuffling slowly towards the door; to see what sort of person lived here.

The door opened to reveal a short, very ordinary looking middle-aged man with blotchy, dark skin and the whitest hair she had ever seen. When he looked up at her face, he was clearly very surprised and took a step back.

"Oh my," she thought, already comforted by the fact that the man did not look threatening, "he was expecting someone else...I've frightened him. I'll try to explain. I sure hope he speaks English!"

"Dr. Birnbaum!" he said, now gesturing her in, "I was not expecting to see you here. Please, please, come in. Please, come in."

"You know who I am?!?" she asked incredulously.

"Of course," he answered shyly, indicating a choice of comfortable seating options from a stuffed sofa to a captain's chair. "And why wouldn't I know who you are? Why, everyone is talking about you. Your PhD thesis was published less than one year ago in *Memoirs of the American Mathematical Society* and already has had a profound impact in theoretical biology as well as in your own area of high-dimensional topology. It was brilliant, truly brilliant. I read it myself! Let me get you some tea."

He disappeared through a slender, arched doorway at the back of the room in which she was sitting, giving her a moment to think. If there had been any lingering doubt that she was in the right place, his behavior had eliminated it. She was also glad that he had leapt right into this topic of conversation, since she was never sure how she could bring it up without sounding crazy.

"Yes," she called to the other room, "that is what I had come here to talk to you about." There was no reply, and so she supposed that he could not hear her.

When he returned a few minutes later, he was carrying a silver tea set.

"Ah, camellia," he said, sniffing at the fragrant aroma of the tea as he served. "But I don't understand," he said, "why would you come all of this way to talk to me about your work? You can't possibly know who I am!"

"It's not just my research that I want to talk to you about, it is math research in general." Amanda stirred her tea, but did not even lift the cup off of the saucer. She had no interest in the tea just now. "Back when I was an undergraduate, one of my professors had defined mathematics as 'the study of necessary consequences of arbitrary axioms about meaningless things.' My classmates and I, all math majors, did not like this description. It seemed to ignore the usefulness of mathematics. After all, math is used in engineering, physics, biology, economics, you know. We had a discussion about this paradox. 'If mathematics begins with these meaningless abstractions, why is it that it turns out to be so useful in the end?' There were lots of different opinions on the subject."

Reprinted from April 2003, pp. 29–31

"Oh yes," he chimed in with a bright smile, "I've heard such debates before many times. 'The Unreasonable Effectiveness of Mathematics'! 'Why is it that results in abstract mathematics, constructed without any thought given to the real world, some time later turn out to be useful after all?'"

"Right! Right. Like non-Euclidean geometry. At the time it was first suggested, it was just a sort of trick. 'Look what we can do if we pretend that parallel lines meet too!' But then, after Riemann, Clifford, Hilbert and Einstein it's no longer a joke, it is a description of the universe we live in, though we never knew it before."

"And," he added, clearly enjoying this conversation, "what about the use of non-commutative rings and imaginary numbers in particle physics?!?"

"Yes," she agreed, "when imaginary numbers were first discussed by mathematicians they were barely even considered to be real mathematics. Now physicists regularly consider quantum wave functions which are complex-valued with no qualms."

"And when the physicists first found non-commutativity in their measurements, the supposedly useless theory of abstract algebra was already there, instantaneously transformed into a branch of 'applied' mathematics!" He laughed so hard that she began to get frightened. He noticed her reaction, and tried to calm himself down, slowly sipping at his tea and trying not to laugh.

"I'm sorry," he said, "I have not had this conversation for quite a long time and...and I have a special interest in it. Oh, but you had something to say and came such a long way. Perhaps I should just let you talk."

There was an uncomfortable pause while she tried to collect her thoughts and her courage.

"So," she continued, "when my friends and I discussed these ideas in college there were two main viewpoints. Some argued that mathematics allows us to study any structured system, and then since the universe seems to have some rules to it, we obviously will be able to use math to say things about it somehow."

"Hmmm," he hummed while nibbling on a sugar cookie.

"And the others all thought that the universe is way beyond our comprehension anyway. According to them, when we have a new mathematical idea, we apply it to the universe because we have nothing better to use."

"Ah," he said swallowing, "as they say: to a man whose only tool is a hammer, everything looks like a nail!" Remembering that he had promised to be silent, he stopped suddenly, 'zipping' his mouth shut with his thumb and forefinger.

"Right. But, I had another idea. It was so crazy, I didn't even mention it to my friends, but I kept it in mind as a sort of joke." She waited for him to ask what the crazy idea was, but he just smiled and looked down at the table, as if he knew she was talking about him.

"My idea," she continued, "was that another good explanation for why 'pure mathematical' research becomes useful some time after its discovery is that the universe itself changes to fit our mathematical discoveries."

"Oh," he said, suddenly blinking rapidly while still smiling. "And why have you come all of this way to talk to me about it?"

"Because, I think you're the one who is doing it."

He nodded slowly, as if admitting it was true. This surprised her, since she had expected a denial. She had expected to be told that she was crazy. It was a crazy idea, after all...wasn't it?

There was another uncomfortable silence.

"So," she said sharply, "it's true?"

"Between you and me?" he looked back and forth as if he expected to find people eavesdropping right there in his living room. "Between you and me, it is, and it is a thankless job."

"So, does that make you...you know...are you...the creator of the universe?"

"Ha!" he shouted so loudly that she almost spilled her tea. "If I were one of the creators, you think I'd be here on this crazy little planet in the middle of nowhere? No offense intended. No, no, I've just been doing this here for a few hundred years and after two hundred more I can retire to a nice alternate reality I've been dreaming of."

She was still trying to process all of this. "So, you mean whenever we make a discovery the whole universe changes?"

"Well, not quite the whole universe, and not quite every discovery. When I find a result that I find especially interesting or entertaining, I find some way to incorporate it into the universe...but only locally. That's why your cosmologists have been so confused in their theories. In other districts, those with my job may have different tastes in math, different ideas of how 'reality' could be. In fact, it is this diversity of possibilities that the creators enjoy most...it's why I have a job!" He was very happy to have someone to talk to about this after being silent for so long. A smile of contentment shone on his face and he leaned back in his seat as if he had never been so comfortable in his life.

"But then," she had so many questions, she found it difficult which to ask first, "if..."

"Wait!" he interrupted sitting up straight with a worried expression. "Amanda, please, you must tell me how you found me out. I am not supposed to be discovered, you know."

"Well," she said, unable to look him in the eye, "you remember my thesis?"

"Very well!" His eyes lit up in a way she found flattering. He had clearly liked her work. "You noticed that the cohomology of a certain class of high-dimensional manifolds had some bizarre algebraic properties. Because the behavior of the homology groups reminded you of immune systems in biology, you called such manifolds 'immunity manifolds.' Although

Jason Fowler

your theories. I just couldn't wait to get to work on it. You must have seen by now how I was able to bring it to play on some questions regarding the improvement of vaccinations and in just a few months some medical researchers working on the disease scleroderma will discover that they can use Serre duality to…"

"But," she interrupted, "you made a mistake. I mean, I made a mistake. I was wrong about equation 3.6. The microchimeric subalgebras don't have to be simply connected, and so definition 3.9 just didn't make any sense and…"

"No," his jaw fell open and he dropped the last little piece of his cookie. "Not equation 3.6! But that was one of my favorite parts. I used that everywhere!"

you say in the introduction that you are not an expert in biology, the idea motivated your nomenclature throughout the thesis. Some immunity manifolds are healthy, some are not. Some even have auto-immune diseases!"

"Just names I gave them to help me describe and understand the mathematical structure. I was doing math, not biology."

"Perhaps, but your scientists were never able to make sense out of the immune system before and there was so much room for rich and beautiful discoveries to grow out of

"Yes, I know. That is how I found you. You see, I caught the mistake just after the *Memoir* was published. It wasn't easy, but I was able to make sure that every copy with the mistake was collected unread and replaced with a corrected version…every copy except the one that was sent separately by private courier here to your house. And that is how I knew…"

"Oh my," he said, stirring his tea vigorously. "Oh my, how careless of me! We will have to do something about that, won't we? Yes, something will have to be done about that." ∎

How to
Ace Literature
A Streetwise Guide for the Math Student

KATHERINE SOCHA
Michigan State University
MICHAEL STARBIRD
The University of Texas at Austin

In English class, we are often asked to take a poem or story and write something about it. This assignment gives us a chance to explore deep life insights, but, of course, we don't have any—that's why we are in math. So what can we do when we are more comfortable with calculus than with classics; more fluid with functions than with feelings; more loving of limits than of literature? Here we solve this problem of producing prose by turning the tables on the teacher. Instead of vainly trying to bring the humanities to mathematics, let's instead bring mathematics to humanities. Here are some modest examples from calculus to help you get started with your own essays.

Let us begin with that well-loved classic, *The Ugly Duckling*. In examining this story, many students would be tempted to wax eloquent on such humanistic themes as coming of age, finding oneself, and inner beauty, but we mathematicians are not equipped to do that. Rather, we are certain that any English teacher will be deeply impressed when we argue that the Ugly Duckling is simply a thinly veiled metaphor of classic examples of antiderivatives. The Ugly Duckling, of course, represents the function $1/x$: although he is ostracized by the other ducks ($1/x^p$ for $p \neq 1$), all of whom have algebraic antiderivatives, he antidifferentiates into his own as the beautiful swan curve, the natural logarithm. Surely the similarity in shape of the sensuous rise of the lovely logarithm graph with the swooping grace of the neck of a swan could not be mere coincidence!?

Goldilocks and the Three Bears is a traditional tale of adventure and danger that perfectly represents another gem of the calculus canon, the Intermediate Value Theorem. As you recall, Papa Bear's bed was much too hard, Mama Bear's bed was much too soft, but at last Goldilocks found Baby Bear's bed to be just right, exactly as the Intermediate Value Theorem requires. Goldilocks depended on the continuity of comfort to pile into that pillowed perfection between too hard and too soft. (Exercise: Apply the Intermediate Value Theorem to porridge and chairs.)

In the fairy tale *Cinderella*, the heroine is mistreated by her stepmother and, after many adventures (mostly happening to Cinderella), the Prince goes around trying a glass shoe on various women to find out which one was at the ball. Do not be tempted by the obvious themes of cruelty, class structure, and love. Instead, get right to the math. When the Prince uses a particular foot as a means of determining the whole girl, argue that the essential point of this story is that the Prince is simply applying calculus. He demonstrates a touching faith in the power of the Taylor series, uniquely determining the lady by her characteristics at one foot—just as a sufficiently nice function is uniquely determined by its shape near one point. One could even go so far as to argue that the nasty stepsisters were not nice enough to have their own Taylor series. This analysis of *Cinderella* is sure to be a novel perspective on the story.

In fact, we assert that every topic in a standard calculus book has its own doppelganger from the realm of fairy tales. Considering the size of most standard calculus texts, we are cheered at the thought of how many pages of cunning calculus comparisons we can create for composition class. Here are a few closing suggestions for beginning your own mathematical analysis of literature.

Most courses in calculus begin with a review of concepts such as exponential growth. Take a look at Jack's beanstalk in *Jack and the Beanstalk*. When Jack is pushed aside to watch his beanstalk ascend with ever-increasing speed toward the high heavens, isn't this merely a parable of exponential growth couched in the appropriate bio-

$$\int \frac{1}{x}\,dx$$

Jason Fowler

logical context? Continuing our review would lead to consideration of trigonometric topics, such as an in-depth exploration of the circular functions sine, cosine, and tangent. No doubt these topics were the attraction of the spinning wheel to the princess in *Sleeping Beauty*, leading to her enchanted sleep, during which she surely dreamed about calculus. Many calculus students also drop into a slumberous stupor when mesmerized by circular functions going round and round.

Moving ahead in the calculus canon, one might wish to consider the topic of integration. Here, we suggest the story of *Rumpelstiltskin* who spun straw into gold, thereby teaching the miller's daughter how to sum small segments into a whole, that is, how to integrate.

Of course, fairy tales do not limit themselves to calculus. A future dissertation might explore the topological equivalence of frogs and princes through the 'kiss' or 'osculation' homeomorphism.

Mathematics is everywhere. Once math suffuses your soul, you'll see its shadows in the most surprising settings. We wish you the best of luck in applying this strategy in your literature classes.

But please don't tell your teacher where you got the idea. ∎

Fibonacci's Triangle and other Abominations

DOUG ENSLEY
Shippensburg University

Algebraic topology, analytical number theory, discrete geometry; these sound like oxymorons. Actually, they are examples of mathematicians' playful penchant for the juxtaposition of dissimilar technical terms. This tendency is arguably a consequence of the love we all have for those significant leaps in knowledge that occur when connections are discovered between disparate fields. Significant examples of this range from Descartes's seventeenth-century work in bridging algebra and geometry to the forty-some years of number theory leading up to the twentieth-century solution to Fermat's Last Theorem. The term *juxtaposition* is perhaps a bit unfair since it carries with it, especially in the arts, an intention to highlight contrasting ideas, and mathematics really never has such intentions. Perhaps a better word to describe what mathematicians do is *automatism*, the art historians' term to describe the spontaneity of being led by the subconscious. The reader might wish to try the following exercise in mathematical automatism: Take two math books off of your shelf, open them each at random, and try to write down a good math problem using the ideas from both books. Alternatively, you can wait until someone asks you a preposterous question like, "What's the formula for the entries in Fibonacci's triangle?"

When this question was asked in my discrete math class, I knew quite well what the student meant (*Pascal's triangle*, of course), and, in a rare moment of pedagogical lucidity, I even managed to clarify and answer the intended question. However, my subconscious was ablaze with the two mathematical terms, "Fibonacci numbers" and "Pascal's triangle." Back in my office, I imagined Fibonacci's triangle as a triangular array of numbers created by first writing the Fibonacci numbers down the sides and then finding each interior entry by adding the two entries above it. The first few rows of this construction are shown in Figure 1. In this article, we will play mathemati-

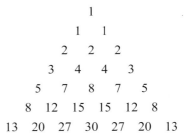

Figure 1. Rows 0 through 6 of the Fibonacci Triangle.

cal detective to unveil the structure of this captivating combination of two classic mathematical patterns.

Whose Triangle is it, Anyway?

In the mid-seventeenth-century, Blaise Pascal (1623–1662), noted mathematical prodigy and religious oddsmaker, examined some properties of a triangular array of numbers. He made use of these numbers in his correspondence with the great Pierre de Fermat as they worked out the foundations of mathematical probability. This inauspicious beginning eventually led to the modern widespread attribution of the triangle itself to Pascal. History books tell us that the *Arithmetic Triangle* (we use this term in deference to Pascal since it is the term he himself used) existed well before Pascal wrote about it. Not only was it known in Europe by the early 1500s, it was already being used in China as early as the eleventh century to extract roots of numbers.

Leonardo Fibonacci (1175–1250) published his *Liber Abaci* in his hometown of Pisa in 1202. This work was greatly influenced by Fibonacci's travels (by benefit of his father's occupation in the relatively young Eastern mercantile trade of Pisa) to places like Egypt and Syria, where he went to observe the superior methods of Eastern and Arab mathematics. The problems

Reprinted from September 2003, pp. 10–14

presented in *Liber Abaci* were quite sophisticated for the time, especially considering that the first line of the book must establish that numbers will be written using Hindu-Arabic notation. One of the problems presented in the text is a famous one about the reproduction of rabbits that gives us the familiar *Fibonacci sequence* $1, 1, 2, 3, 5, 8, 13, \ldots$. Today we simply define this sequence with a recurrence relation and some initial conditions:

$$F_n = F_{n-1} + F_{n-2} \quad \text{for } n \geq 2; \; F_0 = 1, F_1 = 1. \quad (1)$$

The Arithmetic Triangle and the Fibonacci numbers are commonly used in classroom discussions involving number patterns, recurrence relations, enumeration or probability. Their intriguing simplicity and appealing symmetry have been pulling students into the study of mathematics for many years. There have been many generalizations of these ideas through the years—some of practical importance, but most just for the love of exploring mathematics.

(Yet Another) Generalized Arithmetic Triangle

There are many ways in which one can account for properties of Pascal's Arithmetic Triangle, the AT for short. For example, by denoting Entry k in Row n by $\binom{n}{k}$ (where we enumerate rows and entries beginning with 0), the triangle is merely a table of the values given by the recursive description on natural numbers n and k,

$$\binom{n}{k} = \begin{cases} \binom{n-1}{k} + \binom{n-1}{k-1} & \text{if } n > k > 0 \\ 1 & \text{if } k = 0 \\ 1 & \text{if } k = n. \end{cases} \quad (2)$$

In this description, the first line is called the *recurrence relation* because it explains how to compute numbers based upon numbers previously computed, and the other two lines are called the *initial conditions* because they explain which numbers should be computed first. Note that it is certainly possible for a recursive description to be incomplete, leading to values that cannot be computed from the information given, but we are familiar enough with the Arithmetic Triangle to believe that the description above completely describes every entry uniquely. The familiar first several rows of the AT are given in Figure 2.

The entries $\binom{n}{k}$ are typically called the *binomial coefficients* because of their appearance as the coefficients in the expansion of the polynomial $(1 + x)^n$. Any one of these interpretations of the AT can lead to intriguing generalizations, and the literature is replete with them. Our path to generalization will be to change the initial conditions to be arbitrary sequences of real numbers.

Definition 1. *Let the real sequences* $\mathbf{a} = \{a_n\}$ *and* $\mathbf{b} = \{b_n\}$, *with* $a_0 = b_0$, *be given. We will call* $C(n, k)$ *as defined below, the* Generalized Arithmetic Triangle *for* $\{a_n\}$ *and* $\{b_n\}$,

$$C(n, k) = \begin{cases} C(n-1, k) + C(n-1, k-1) & \text{if } n > k > 0 \\ a_n & \text{if } k = 0 \\ b_n & \text{if } k = n. \end{cases} \quad (3)$$

If $\mathbf{a}$ *and* $\mathbf{b}$ *are identical sequences, then we just call* $C(n, k)$ the Generalized Arithmetic Triangle for $\mathbf{a}$.

Naturally the abbreviation for the Generalized Arithmetic Triangle must be GAT. Using this language, the Fibonacci Triangle from the beginning of this paper would be called *the Generalized Arithmetic Triangle for* $\mathbf{F} = \{F_n\}$, and the regular old Arithmetic Triangle would be considered *the Generalized Arithmetic Triangle* for the constant sequence $\{1\}$. Notice that our definition of the GAT allows the triangle to be asymmetric, potentially having two different sequences down the two outer sides. For example, the GAT for $\mathbf{a} = \{2^n\}$, $\mathbf{b} = \{3^n\}$ is shown in Figure 3.

This may seem at first to unnecessarily complicate the issue since it seems that the interaction between the numbers down the two sides should be simpler when the two sides are identical, but we will soon see that this is not true. In fact, it is primarily this interaction that will be the main focus in our investigation. To highlight it best, we will first look at the special case where there is really no interaction at all.

Example 2. *The GAT for* $\mathbf{a} = \{1\}$ *(the constant sequence of all 1's) and* $\mathbf{b} = \{3^n\}$ *is given in Figure 4.*

Our main lemma gives a straightforward formula for computing Entry k of Row n in this special case. The proof uses mathematical induction in an indirect way, thanks to a simple fact about sequences of numbers that is actually the underlying idea behind many facts in discrete mathematics.

```
              1
            1   1
          1   2   1
        1   3   3   1
      1   4   6   4   1
    1   5  10  10   5   1
  1   6  15  20  15   6   1
```

Figure 2. Rows 0 through 6 of Pascal's AT.

```
              1
            2   3
          4   5   9
        8   9  14  27
     16  17  23  41  81
   32  33  40  64 122 243
 34  65  73 104 186 365 729
```

Figure 3. Rows 0 through 6 of a skewed GAT.

$$
\begin{array}{ccccccc}
 & & & 1 & & & \\
 & & 1 & & 3 & & \\
 & & 1 & 4 & & 9 & \\
 & 1 & & 5 & 13 & & 27 \\
 1 & & 6 & 18 & 40 & & 81 \\
1 & 7 & 24 & 58 & 121 & & 243 \\
1 & 8 & 31 & 82 & 179 & 364 & 729
\end{array}
$$

Figure 4. Rows 0 through 6 of a lopsided GAT.

The Fundamental Theorem of
Recursively Defined Sequences

If a recurrence relation and initial conditions provide enough information to uniquely determine a sequence of numbers, then any two sequences both satisfying the recurrence relation and initial conditions must be equivalent.

This fact is so obvious that it seems silly to even state, but it means that it is enough for us to show that *our* formula describes a sequence with the same initial conditions and recurrence relation given in Definition 1. We can then conclude that our formula matches the given GAT.

Lemma 3. *The Generalized Arithmetic Triangle for* $\mathbf{a} = \{1\}$ *and* $\mathbf{b} = \{b_n\}$ *(where* $b_0 = 1$*) has as Entry k in Row n,*

$$
C(n,k) = \binom{n}{k} + \sum_{i=1}^{k} \binom{n-i}{k-i} \cdot (b_i - b_{i-1}). \tag{4}
$$

Proof. We will show that these numbers as defined meet the three conditions given in Definition 1, but we will do so in reverse order to maintain the suspense.

1. If $k = n$, we have

$$
C(n,n) = \binom{n}{n} + \sum_{i=1}^{n} \binom{n-i}{n-i} \cdot (b_i - b_{i-1})
$$

$$
= 1 + \sum_{i=1}^{n} (b_i - b_{i-1}) = b_n.
$$

2. If $k = 0$, we have

$$
C(n,0) = \binom{n}{0} + 0 = 1.
$$

3. To show that the recurrence relation in Definition 1 is satisfied, it will help to simplify somewhat the notation in (4). The numbers $b_i - b_{i-1}$ used there have a standard name in discrete math—they are called the *first differences* of the original sequence $\{b_n\}$.

Definition 4. *Given a real sequence* $\mathbf{r} = \{r_n\}$*, we define the sequence* $\{\delta_n^{\mathbf{r}}\}$ *of first differences of* $\mathbf{r}$ *by*

$$
\delta_i^{\mathbf{r}} = r_i - r_{i-1} \ \text{if} \ i \geq 1, \ \text{and} \ \delta_0^{\mathbf{r}} = r_0.
$$

In this language, we are to verify that

$$
C(n,k) = \sum_{i=0}^{k} \binom{n-i}{k-i} \cdot \delta_i^{\mathbf{b}}
$$

satisfies the recurrence relation (2).

For $n > k > 0$, we have

$$
C(n-1,k) + C(n-1,k-1):
$$

$$
= \sum_{i=0}^{k} \binom{n-1-i}{k-i} \cdot \delta_i^{\mathbf{b}} + \sum_{i=0}^{k-1} \binom{n-1-i}{k-1-i} \cdot \delta_i^{\mathbf{b}}
$$

$$
= 1 \cdot \delta_k^{\mathbf{b}} + \sum_{i=0}^{k-1} \binom{n-1-i}{k-i} \cdot \delta_i^{\mathbf{b}} + \sum_{i=0}^{k-1} \binom{n-1-i}{k-1-i} \cdot \delta_i^{\mathbf{b}}
$$

$$
= 1 \cdot \delta_k^{\mathbf{b}} + \sum_{i=0}^{k-1} \left(\binom{n-1-i}{k-i} + \binom{n-1-i}{k-1-i} \right) \cdot \delta_i^{\mathbf{b}}
$$

$$
= 1 \cdot \delta_k^{\mathbf{b}} + \sum_{i=0}^{k-1} \binom{n-i}{k-i} \cdot \delta_i^{\mathbf{b}}
$$

$$
= \sum_{i=0}^{k} \binom{n-i}{k-i} \cdot \delta_i^{\mathbf{b}}
$$

$$
= C(n,k)
$$

as desired.

There is nothing special about the left side of the GAT being a constant sequence, of course. If the right side consists of the constant sequence $\mathbf{b} = \{1\}$, then we get the same result, just reflected to the other side.

To those who have studied recurrence relations before, the expressions in the lemmas look like *convolutions* of the first differences of the sequence $\{a_n\}$ or $\{b_n\}$ with the binomial coefficients. It remains to be seen what impact is made by having *both sides* of our GAT represented by arbitrary sequences. It seems there is potential for complicated interaction between the **a** sequence and the **b** sequence, but fortunately this does not turn out to be the case, thanks to the linearity of the AT recurrence relation. In fact, we can build the GAT for $\{a_n\}$ and $\{b_n\}$ from the GAT for $\{1\}$ and $\{b_n\}$ and the GAT for $\{a_n\}$ and $\{1\}$ in a surprisingly direct way.

The Main Result

The GAT for $\{a_n\}$ and $\{1\}$ plus the GAT for $\{1\}$ and $\{b_n\}$ minus the AT is equal to the GAT for $\{a_n\}$ and $\{b_n\}$.

We state this formally (and slightly more generally).

Figure 5. Illustrating the Main Result.

Theorem 5. *The Generalized Arithmetic Triangle for* $\mathbf{a} = \{a_n\}$ *and* $\mathbf{b} = \{b_n\}$ *(where* $a_0 = b_0$*) has as Entry* k *in Row* n,

$$C(n,k) = \sum_{i=0}^{n-k} \binom{n-i}{k} \cdot \delta_i^{\mathbf{a}} + \sum_{j=0}^{k} \binom{n-j}{k-j} \cdot \delta_j^{\mathbf{b}} - a_0 \cdot \binom{n}{k}.$$

The proof of this theorem is extremely simple since we have dutifully done the dirty work in the lemmas. In short, since each of the three terms in this expression satisfy the recurrence relation in Definition 1, so must this *linear combination* of the terms. The fact that the initial conditions in Definition 1 are satisfied by this formula also follows directly from the lemmas.

The formula in Theorem 5 elegantly separates the effects of the two sequences **a** and **b** on the entire triangle, but it is admittedly not the easiest thing to look at. Let's see a more visual representation of how the computation in Theorem 5 is actually done. The entries from Pascal's Arithmetic Triangle which are used in the theorem are easy to map out. For example, to compute Entry 3 in Row 5 of a Generalized Arithmetic Triangle, the circled AT entries shown in Figure 6 are used. The V-shaped pattern is very obvious.

So summing the entries shown in Figure 7 will give us Entry 3 of Row 5 for any GAT.

To see how to visualize the computation in Theorem 5, we will use an example of the most general form—a skewed triangle with different sequences **a** and **b**.

Example 6. *The GAT for* $\mathbf{a} = \{2^n\}$ *and* $\mathbf{b} = \{3^n\}$ *was shown in Figure 3. To complete the computation illustrated in Figure 7, we need to first understand the first differences of the sequences* $\mathbf{a}$ *and* $\mathbf{b}$*. In this case, the nice expressions for* a_n *and* b_n *make this part easy:*

$$\delta_i^{\mathbf{a}} = \begin{cases} 1 & \text{if } i = 0 \\ 2^{i-1} & \text{if } i \geq 1 \end{cases} \qquad \delta_i^{\mathbf{b}} = \begin{cases} 1 & \text{if } i = 0 \\ 2 \cdot 3^{i-1} & \text{if } i \geq 1. \end{cases}$$

Now we can sum the circled entries shown in Figure 7 to compute Entry 3 in Row 5 of this GAT as

$$1 \cdot \delta_3^{\mathbf{b}} + 3 \cdot \delta_2^{\mathbf{b}} + 6 \cdot \delta_a^{\mathbf{b}} + 10 \cdot \delta_0^{\mathbf{a}} + 10 \cdot \delta_0^{\mathbf{a}} + 4 \cdot \delta_1^{\mathbf{a}} + 1 \cdot \delta_2^{\mathbf{a}} - 10$$
$$= 1 \cdot 18 + 3 \cdot 6 + 6 \cdot 2 + 10 \cdot 1 + 4 \cdot 1 + 1 \cdot 2 - 10$$
$$= 64.$$

Moreover, the general form (from Theorem 5) for Entry k *of Row* n *is*

$$\binom{n}{k} + \sum_{i=1}^{n-k} \binom{n-i}{k} \cdot 2^{i-1} + 2 \cdot \sum_{j=1}^{k} \binom{n-j}{k-j} \cdot 3^{j-1}.$$

Fibonacci's Triangle and Other Abominations

It is now a simple matter to compute entries in the Fibonacci Triangle—the GAT for $\mathbf{F} = \{F_n\}$. In this case, we have the happy circumstance that

$$\delta_i^{\mathbf{F}} = F_i - F_{i-1} = F_{i-2}$$

as long as $i \geq 0$, where we interpret $F_{-1} = 0$ and $F_{-2} = 1$, consistent with the Fibonacci recurrence relation (1). In this case, Theorem 5 gives:

Corollary 7. *Entry* k *in Row* n *of the Fibonacci Triangle is given by*

$$\binom{n}{k} + \sum_{i=1}^{n-k} \binom{n-i}{k} \cdot F_{i-2} + \sum_{j=1}^{k} \binom{n-j}{k-j} \cdot F_{j-2}.$$

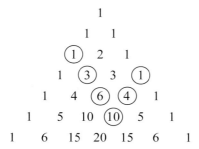

Figure 6. The AT entries used to compute Entry 3 in Row 5 of any GAT.

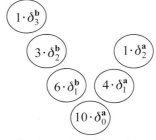

Figure 7. Summing these yields Entry 3 in Row 5 of any GAT. Note that: $10 \cdot \delta_0^{\mathbf{b}} = \binom{5}{3} \cdot a_0$.

Graphically, we can picture the V-shape from Figure 6 with Fibonacci numbers as the first differences. Summing the entries shown in Figure 8 gives us Entry 3 in Row 5

$$1 \cdot F_1 + 3 \cdot F_0 + 6 \cdot F_{-1} + 10 \cdot F_{-2} + 4 \cdot F_{-1} + 1 \cdot F_0$$
$$= 1 \cdot 1 + 3 \cdot 1 + 6 \cdot 0 + 1 - \cdot 1 + 4 \cdot 0 + 1 \cdot 1$$
$$= 15.$$

The formula in Corollary 7 might seem a bit cumbersome, but it does have a certain aesthetic appeal since it separates the influences of Pascal's Arithmetic Triangle and the Fibonacci numbers in a nice, symmetric way. Of course, with the general tool of Theorem 6, one can invent any number of other triangles and study them in more detail. A particularly fruitful direction for further exploration are the abominations of the Arithmetic Triangle that one gets by using a fixed diagonal of Pascal's Arithmetic Triangle for the initial conditions in the Generalized Arithmetic Triangle. In the language of this paper, this means fixing a positive integer value of m and looking at the GAT for the sequence

$$\left\{ \binom{n+m}{m} \right\}.$$

In this case, one can use standard AT identities to simplify Theorem 6 to some strikingly beautiful results. Like the old

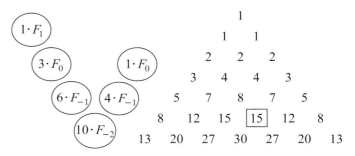

Figure 8. Summing these gives Entry 3 in Row 5 of the Fibonacci Triangle.

Reese's peanut butter cup commercials that advertised "two great tastes that taste great together" mathematics can sometimes produce some lovely flavors with a bit of unexpected mixing of incongruous elements. ∎

For Further Reading

Concrete Mathematics by R. L. Graham, D. E. Knuth and O. Patashnik is a treasure trove of combinatorics and the sorts of surprising connections that one can see if you just look at things in a slightly new way. In particular, their problem 5.74 asks for the GAT for the sequence $\{n + 1\}$.

A Switch in Time Pays Fine?

MARK SCHILLING
California State University, Northridge

One of the most well known probability brainteasers of recent years is the Three Doors problem, modeled after the old television game show *Let's Make a Deal*. The contestant is to choose one of the three doors, and wins whatever is behind it. One of the doors conceals a valuable prize, for instance a new sports car, while behind each of the other two doors is a booby prize. Once the contestant selects a door, the host of the show opens not that door but a different door, revealing (say) a goat. The contestant is then offered the opportunity to switch from the door she selected to the remaining unopened door. Usually the contestant does not change doors, and in fact most people are convinced that each of the two doors left has an equal chance of hiding the sports car. Surprisingly, however, the best strategy is to switch doors.

This problem has received so much attention that most people in the mathematics community are familiar with it, and I will not discuss it further here. (If you have not heard it before, see if you can reason out why switching is the best strategy. You may also wish to set up a simulation of the game and try it a large number of times with a friend. Or see Ed Barbeau's 1993 article "The Problem of the Car and Goats" in the *College Mathematics Journal*.) Instead I want to discuss another strange brainteaser that has become popular in the last few years and also involves the question of switching choices.

The problem appeared in print in a 1995 article in *Mathematics Magazine* by Steven J. Brams and D. Marc Kilgour called "The box problem: to switch or not to switch," and goes as follows: Suppose that there are two envelopes, one with $X > 0$ dollars inside and the other containing $2X$ dollars. You select one of the envelopes—call this Envelope 1—and examine its contents. Then you are given the opportunity to switch to the other envelope, Envelope 2. Should you switch?

You may think that the offer to change envelopes is ridiculous—without any way of knowing the value of X or which envelope is which, one choice seems as good as the other, so switching gives no better or worse prospects than not switching. But consider the following argument: Suppose that the amount of money in Envelope 1 turns out to be M dollars. Then with probability 0.5 Envelope 2 contains $M/2$ dollars, while with probability 0.5 Envelope 2 contains $2M$ dollars. If you change to Envelope 2, then, your expected winnings are $0.5(M/2) + 0.5(2M) = 1.25M$ dollars. This calculation indicates that it is always best to switch, as you would expect to receive 25% more money in the long run with this strategy.

But how can this be? After all, had you been presented with Envelope 2 to begin with, the same logic would tell you to exchange it for Envelope 1. In fact, consider a variation on the above game in which you are not allowed to open an envelope before deciding whether to keep it. Since the calculation above works for any value of $M > 0$, it is clear that you should exchange Envelope 1 for Envelope 2. Then the same argument implies that you should now swap *again*—back to Envelope 1!

Attempts to resolve this paradox involve modeling the contents of the two envelopes X and $2X$ as random variables and enumerating the possible values of X and their corresponding probabilities. This probability distribution may be specified as part of the game, or it may represent an expression of your subjective belief about the likely values in the envelopes.

For example, suppose you know that the benefactor who loaded the envelopes chose $(X, 2X)$ to be one of the setups ($2, $4), ($10, $20) and ($100, $200), with probability 1/3 for each. If you inspect the contents of Envelope 1 then you will know with certainty whether Envelope 2 contains more or less money, so your strategy is trivial. In the case when you are not allowed to open Envelope 1, the expected value computation above no longer applies, and the correct calculation shows that the paradox vanishes—both switching and staying now yield the same expected payoff. In fact, with any finite set of values with which to load the envelopes and any set of associated

Reprinted from September 2003, pp. 21–22, 25

JDF   Jason Fowler

probabilities, the paradox disappears, as the expected payoff is the same with both envelopes.

Yet the envelope problem cannot be dispensed with quite so easily, as the following example from Brams and Kilgour illustrates: Suppose that there are an infinite number of possible values of X, specifically $1, $2, $4, $8,..., with probability $(1/3) \cdot (2/3)^k$ that X equals $2k$ dollars, for $k = 0, 1, 2,...$. You can check that these probabilities add to one, so this is a legitimate probability distribution.

The table below shows the possible envelope contents and their associated probabilities.

Contents of Envelope 1	Contents of Envelope 2	Probability
$1	$2	1/6
$2	$1	1/6
$2	$4	1/9
$4	$2	1/9
$4	$8	2/27
$8	$4	2/27
$8	$16	4/81
⋮	⋮	⋮

If Envelope 1 contains $1, swapping envelopes would improve your payoff to $2. Now suppose Envelope 1 contains $2. To determine whether switching is a good idea, we must determine your conditional expected payoff. This is done by renormalizing the probabilities in the second and third rows of the above table to sum to one: $(1/6) \rightarrow (1/6) \div (1/6 + 1/9) = 3/5$, $(1/6) \rightarrow (1/9) \div (1/6 + 1/9) = 2/5$. That is, there is a 60% chance that you will end up with a lower amount ($1) if you switch, and a 40% chance of receiving a higher amount ($4). Although changing envelopes is more likely to reduce your payoff than to increase it, your expected payoff in this case is $1(3/5) + $4(2/5) = $11/5 = $2.20, which is 10% higher than the $2 you currently have.

It is easy to check that if Envelope 1 contains any amount greater than $1, swapping envelopes yields a conditional expected payoff that is 10% higher than the amount you'll receive if you don't switch. Amazingly, then, it seems that switching is always the right strategy! We needn't even open Envelope 1.

Obviously something is haywire here. If switching always increases your conditional expected payoff, then your overall expected payoff must also be increased by switching. And you could make it even greater by switching back again, and again, and again…! We seem to have a situation where $A > B > A \cdots$. If you would like to find the bug in this argument yourself, then don't read on just yet….

OK, ready to dispense with the Emperor's clothes? Then let's look more closely at the calculation of the overall expected payoffs under the stay and switch strategies.

For the stay strategy, your overall expected payoff with Envelope 1 is:

$$(\$1)\left(\frac{1}{6}\right) + (\$2)\left(\frac{1}{6} + \frac{1}{9}\right) + (\$4)\left(\frac{1}{9} + \frac{2}{27}\right) + \cdots$$

$$= (\$1)\frac{1}{6} + \sum_{k=1}^{\infty}(\$2^k)\left[\frac{1}{6}\left(\frac{2}{3}\right)^{k-1} + \frac{1}{6}\left(\frac{2}{3}\right)^{k}\right]$$

$$= \$\frac{1}{6}\left[1 + \frac{5}{2}\sum_{k=1}^{\infty}\left(\frac{4}{3}\right)^{k}\right].$$

Since this expression contains a divergent series, your expected payoff with the original envelope is infinite! The result is exactly the same if you switch to Envelope 2. Thus, saying that changing envelopes gives a higher expected payoff makes no sense, as both expected payoffs are infinite.

You can play around with this game by means of a computer simulation. Keep track of the number of times each strategy wins (switch should win around 60% of the time) and the total payoffs of stay and switch—these two totals should remain the same, mercifully restoring some faith in our initial intuition. ■

Paintings, Plane Tilings & Proofs

ROGER B. NELSEN
Lewis & Clark College

Over the centuries many artisans and artists have employed plane tilings in their work. Artisans used tiles for floors and walls because they are durable, waterproof, and beautiful; and artists portrayed realistically the tilings they encountered in the scenes they painted. One quality of a work of art is that the viewer will often see more in the work than the artist intended. If you are mathematically inclined, no doubt you see some mathematics in the tilings. Here are some examples where plane tilings on floors, walls, and in paintings underlie proofs of some well-known (and some not-so-well-known) theorems.

The floor tiling in *Street Musicians at the Doorway of a House* by Jacob Ochtervelt (1634–1682) consists of squares of two different sizes. If one overlays a grid of larger squares, as illustrated in blue in Figure 1a, a proof of the Pythagorean the-

orem results, one often attributed to Annairizi of Arabia (circa 900). Such a proof is called a "dissection" proof, as it indicates how the squares on the legs of the triangle can be dissected and reassembled to form the square on the hypotenuse. Shifting the blue overlay grid to the position illustrated in Figure 1b yields another dissection proof, one attributed to Henry Perigal (1801–1899). Of course, other positions for the blue grid will yield further proofs—indeed, there are uncountably many different such dissection proofs of the Pythagorean theorem gen-

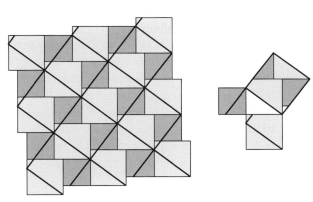

Figure 1a. Annairizi of Arabia's proof of the Pythagorean Theorem

Street Musicians at the Doorway of a House by Jacob Ochtervelt.
Courtesy of the St. Louis Art Museum.

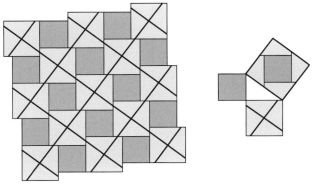

Figure 1b. Henry Perigal's proof of the Pythagorean Theorem.

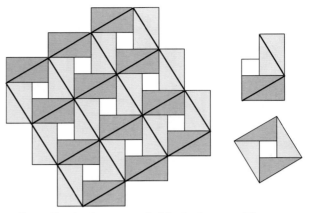

Figure 2b. Bhāskara's proof of the Pythagorean Theorem.

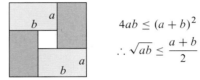

$$4ab \le (a+b)^2$$
$$\therefore \sqrt{ab} \le \frac{a+b}{2}$$

Figure 2a. Salon de Carlos V, Real Alcazar, Seville.

Figure 2c. The arithmetic mean–geometric mean inequality.

erated from the floor tiling in Ochtervelt's painting! No wonder the two-square tiling in Ochtervelt's painting is sometimes called the Pythagorean tiling.

The floor tiling in the Salon de Carlos V in the Real Alcazar in Seville (Figure 2a) provides the basis for the well-known "Behold!" proof of the Pythagorean theorem ascribed to Bhaskara (12th century) in Figure 2b. However, tilings provide "picture proofs" for many theorems other than the Pythagorean. The tiles in the Salon de Carlos V also illustrate the arithmetic mean-geometric mean inequality, as illustrated in Figure 2c. [Exercise: Change the dimensions of the rectangular tiles in Figure 2c to illustrate the harmonic mean-geometric mean inequality.] Tiling the plane with rectangles of different sizes but in the same general pattern as in the Salon de Carlos V yields a proof of the

sine-of-the-sum trigonometric identity (Figure 3). [Exercise: Do similar tilings yield proofs of other trigonometric identities?]

The floor tiling in *A Lady and Two Gentlemen* by Jan Vermeer (1632–1675) appears at first glance to be rather ordinary (Figure 4a). It is, after all, just a version of Cartesian graph paper. But the same blue overlay pattern employed with Ochtervelt's painting and in the Salon de Carlos V provides a proof (Figure 4b) of

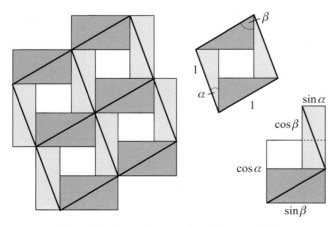

Figure 3. $\sin(\alpha + \beta) = \sin\alpha\cos\beta + \cos\alpha\sin\beta$.

Figure 4a. *A Lady and Two gentlemen* by Jan Vermeer (1632–1675). Courtesy of Herzog-Anton Ulrich Museum-Kunstmuseum des Landes.

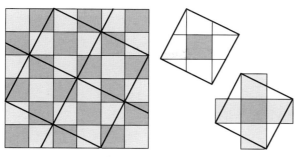

Figure 4b. A proof of Theorem 1.

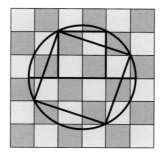

Figure 4c. A proof of Theorem 2. (Note the big square's sidelength.)

Theorem 1. *If lines from the vertices of a square are drawn to the midpoints of adjacent sides, then the area of the smaller square so produced is one-fifth that of the given square.*

A different overlay (the blue circle and squares in Figure 4c) proves

Theorem 2. *A square inscribed in a semicircle has two-fifths the area of a square inscribed in a circle of the same radius.*

So far the tilings we've examined have used squares and rectangles—but any quadrilateral will tile the plane, a fact often employed by the Dutch graphic artist M. C. Escher (1898–1972). Figure 5 illustrates the underlying quadrilateral tiling for one of his better known works, No. 67 (Horsemen), and uses such a tiling to prove

Theorem 3. *The area of a convex quadrilateral Q is equal to one-half the area of a parallelogram P whose sides are parallel to and equal in length to the diagonals of Q.*

[Exercise: Theorem 3 holds for non-convex quadrilaterals as well—can you prove it with a tiling?].

Just as quadrilaterals tile the plane, so do triangles. Figure 6 illustrates tiling based on equilateral triangles in the Salon de Embajadores in the Real Alcazar in Seville. Of course, an arbitrary triangle will tile the plane, and Figure 7a uses such a tiling with an arbitrary triangle to prove a theorem similar to Theorem 1:

Theorem 4. *If the one-third points on each side of a triangle are joined to opposite vertices, the resulting triangle is equal in area to one-seventh that of the initial triangle.*

[Exercise: What happens if you replace the "one-third" points in Theorem 4 with "one-nth" or k/nth?] The same tiling—but with a different overlay—proves (see Figure 7b)

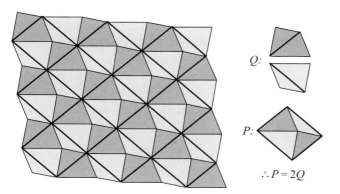

$Q:$

$P:$

$\therefore P = 2Q$

Figure 5. A proof of Theorem 3.

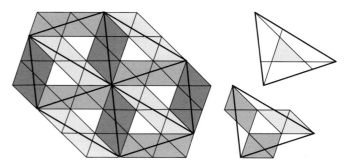

Figure 7a. A proof of Theorem 4.

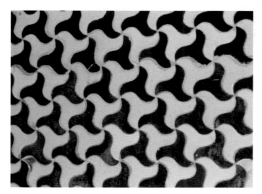

Figure 6. Tiles in the Real Alcazar in Seville.

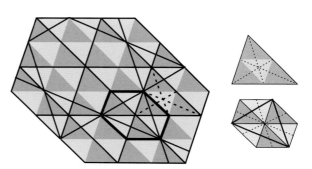

Figure 7b. A proof of Theorem 5.

Figure 8a. *The Courtyard of a House in Delft* by Pieter de Hooch (1629–1684). Courtesy of the National Gallery, London.

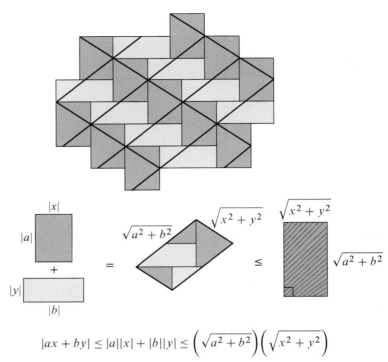

$$|ax + by| \leq |a||x| + |b||y| \leq \left(\sqrt{a^2 + b^2} \right)\left(\sqrt{x^2 + y^2} \right)$$

Figure 8b. The Cauchy–Schwarz inequality.

Theorem 5. *The medians of a triangle form a new triangle with three-fourths the area of the original triangle.*

We conclude by returning to a rectangular tiling found in *The Courtyard of a House in Delft* (Figure 8a) by Pieter de Hooch (1629–1684). In the painting, the courtyard is tiled with bricks all the same size, but it is easy to see that bricks of two different sizes could be used, as illustrated in the top part of Figure 8b. With the blue overlay, this tiling forms part of a proof of the Cauchy-Schwarz inequality in two dimensions. No doubt the reader will find other examples of beautiful

mathematics—including theorems and proofs—in many other works of art.

For Further Reading

An excellent introduction to tilings is Branko Grünbaum and G. C. Shephard's *Tilings and Patterns: An Introduction*, W. H. Freeman (1989). For an on-line collection of the paintings of Dutch Baroque era painters such as Ochtervelt, Vermeer, and de Hooch, visit the Web Gallery of Art at http://www.kfki.hu/ ~arthp. ■

Knots to You

Michael McDaniel
Aquinas College

Those of you who learned to tie your shoes in the 1980s missed a huge mathematical opportunity if you later switched to Velcro. In the past fifteen years, knot theory has unexpectedly expanded in scope and usefulness. It had been an interesting application of algebraic topology since the 1900s and a pastime for those folks with a categorizing bent who would sort knots before that. In 1985, (1984 if you got the pre-print), a knot polynomial appeared which revolutionized knot theory research: the Jones polynomial sorted knots amazingly well, was easy to calculate, could be explained in a few pages and pointed the way to a world of new ideas in knot theory. Suddenly, mathematical rookies could reach a mathematical frontier with a background in topology and algebra. A lot has happened since you last had to think about how to tie a knot.

Wild Growth in Knot Theory

The concerned reader will correctly wonder, "Wild growth in knot theory? How could I have missed it?" People have used knots for as long as they have used rope; in other words, knots are older than recorded history. The serious, mathematical analysis of knots, however, began in the late 1800s, though earlier mathematicians, including Gauss, noticed knots. A mathematical knot is simply a loop in three-space, interwoven with itself, with no self-intersection. A knot diagram is a picture of the knot with crossing information displayed through the simple device of not drawing that which is not seen. Thus, when an arc of the knot crosses over another part, the lower arc appears to stop at the crossing and then reappears as it continues. Knots are usually categorized by minimum number of crossings possible in a picture. The earliest knot tables, dating from the 1880s, attempted the tabulation of knots with ten or fewer crossings, and they were all incomplete. The knots of nine or fewer crossings were correctly listed for the first time in Reidemeister's *Knotentheorie* in 1932. The 9,988 twelve-crossing knots were described in 1981, the 46,972 thirteen-crossing knots in 1982; the current work is on the nineteen-crossing knots. An enormous census of knots "The First 1,701,936 Knots" recently appeared in *The Mathematical Intelligencer*. As the number of crossings increases, the poten-tial complexity increases. For example, all knots of seven or fewer crossings can be drawn with the minimal number of crossings alternating between over and under. (Trace any knot in Figure 1, noting that the strand you are following goes under, then over, then under and so on.) These are called alternating knots and they dominate the knot table for small numbers of crossings. As the number of crossings increases, the non-alternating knots outnumber the alternating knots.

If a bit of an arc of one knot is snipped off and the endpoints joined to another snipped knot, a *composite* knot is created. Such knots are usually left out of knot tables because they are not *prime*. A *link* can have more than one component and each component is a knot. The components themselves may be interwoven or separable. Yet another variation is the mirror-image of a knot: switch the overcrossings to under and vice-versa. Sometimes a knot cannot be manipulated into its mirror-image (3_1) and sometimes it can (4_1). Knot tables almost always contain prime knots only and no mirror-images included.

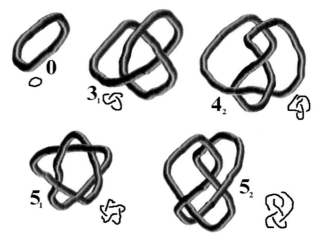

Figure 1. Here are the first five knows in the knot table. A knot table is organized by minimum number of crossings. We see that no knot has only two crossings and that there are two different knots with five crossings and their names differ by the subscripts, 5_1 and 5_2.

$$\langle \text{⧖} \rangle = A \langle \text{)(} \rangle + A^{-1} \langle \text{≍} \rangle$$

$$\langle \text{⧓} \rangle = A \langle \text{≍} \rangle + A^{-1} \langle \text{)(} \rangle$$

$$\langle \bigcirc \rangle = 1$$

$$\langle \bigcirc \cup L \rangle = (-A^2 - A^{-2}) \langle L \rangle$$

$$\langle \text{⧢} \rangle = -A^{-3} \langle \text{∪} \rangle \quad \langle \text{⧣} \rangle = -A^3 \langle \text{∪} \rangle$$

Figure 2. The rules for using Kauffman's bracket. The endpoints of the arcs inside the brackets have been marked as guides. Each endpoint actually goes on to the rest of the knot diagram, as the example later illustrates. The L stands for a link, that is, the rest of the diagram. If a disjoint loop appears in the calculation, which often happens, the second-to-last rule tells how to deal with it. The A is the variable of the polynomial. Much more can be said about both the numerical values for A and the Kauffman bracket itself. The interested reader is urged to check the references.

Negative   Positive

Figure 3. The knots we have been considering have been non-oriented. In order to find the writhe, we apply an orientation to the knot. That means, we draw a few little arrows on the diagram to indicate which way we travel the knot. Each crossing then becomes one of the two illustrated above. It does not matter which direction we choose for the orientation! The writhe is the number of positive crossings minus the number of negative crossings in the diagram. The three crossings in the 3_1 knot in Figure 1 are all negative so its writhe is -3.

Telling Things Apart

The core question in knot theory always has been and still remains: given two knots, how can one tell whether they are the same or different? One of the most powerful tools in telling knots apart is the set of Jones-type polynomials. In 1984, Vaughn Jones discovered an algorithm for turning a knot or link diagram into a polynomial. Although it was not the first such algorithm, the Jones polynomial has proved to be one of the fastest and most reliable ways to discern knots. (In the knot census mentioned above, one of the authors used the Jones polynomial to sort the knots. In cases where two different knot diagrams had the same Jones polynomial, further investigation showed the diagrams were from the same knot!) Soon after Jones defined his polynomial, Louis Kauffman of the University of Chicago defined a rather neat method of turning a knot diagram into a Jones polynomial. His procedure has the amazing pair of properties that it is simple to follow and the proof that the polynomial calculated is a knot invariant can be done in a single lecture.

The Kauffman bracket starts with a diagram of an unoriented knot inside a pair of brackets. See Figure 2. The diagram is rewritten twice (each diagram inside its own pair of brackets, except for a change at a single crossing.) The diagram becomes two diagrams, each with a coefficient which keeps track of what change was made to the crossing. The dots on the endpoints are there to help keep that part of the knot organized as it is changed.

The process leaves a polynomial in positive and negative powers of A. Turning this polynomial into a Jones polynomial requires two more steps: multiply by a factor of $-A^{-3w(k)}$. The *writhe* of the knot diagram, $w(k)$, is described in Figure 3. Then substitute $t^{-1/4}$ for A. The process gets easier with experience.

Calculating the Jones polynomial can be done by hand but usually the process is automated. The second edition (or later editions) of *The Knot Book* has a knot table which includes Jones polynomials useful for checking calculations.

It is remarkable that the computation does not depend on the way the knot is drawn. In other words, tie a knot in a piece of rope and seal the ends. Now draw the knot. Next, manipulate the rope as much as you like without breaking it and draw this new configuration. Their Jones polynomials will be identical. Thus, if two knots have different polynomials, they must be different! No messing about with string, no worries that one knot might be turned into the other if only someone could see the right manipulations: the polynomial does not change as long as the knot itself does not essentially change. Thus, the Jones-type polynomials are called *knot invariants*. The calculation of the Jones polynomial of the 5_1 knot follows. The first two applications of the bracket have been completely written out. Due to the symmetric properties of the knot, further steps resemble those illustrated. The answer is supplied for those who fill in the missing steps.

$$= A(-A^3)^4 + A^{-1}[A\langle \text{⬡} \rangle +$$

$$A^{-1}\langle \text{⬡} \rangle] = \cdots$$

$$A^{13} - A^9 + A^5 - A - A^{-7}$$

The writhe of this version of 5_1 is -5. So, we finish the calculation:

$$[(A^{13} - A^9 + A^5 - A - A^{-7})(-A^{15})]_{A = t^{-1/4}}$$
$$= -t^{-7} + t^{-6} - t^{-5} + t^{-4} + t^{-2}.$$

The Jones polynomial revolutionized knot theory. Practically overnight, knot theory changed from a curiosity of topology to an area of research in which major results were appearing literally by the month. For instance, eight mathematicians (Hoste, Ocneau, Millett, Frey, Lickorish, Yetter, Przytycki and Traczyk), on reading about the Jones polynomial in a preprint, saw a generalization which yielded a new polynomial (called, of course, the HOMFLYPT polynomial) and the AMS Bulletin received the papers of six of them within a few days of each other, with their results published together a few months after Jones's article. Przytycki and Traczyk did not make that issue because the Solidarity movement had slowed down the Polish postal system. The frontiers of knot research continue to expand. Combinatorics, algebra, dynamical systems and other topics intersect with topology in knot theory. With applications in DNA studies, physics and chemistry, knot theory has influence beyond the purely theoretical.

Logical readers will have noticed that it might be possible for two different knots or links to have the same Jones polynomial and this is indeed the case. Eliahou, Kauffman, and Thistlethwaite have described a family of links whose Jones polynomials are the same as the Jones polynomials of links made of the same number of disjoint loops (called unlinks). So far, the case of finding (or disproving the existence of) a single component knot with a Jones polynomial of 1 remains unsolved.

Figure 4 illustrates a way to fool the Jones polynomial; it requires a small digression into knot mutations. Other methods besides the one shown exist for finding mutant pairs.

The Jones, HOMFLYPT and Kauffman knot polynomials had been defined with a common structure: start with a diagram, end with a polynomial. Vassiliev, Kontsevich, Arnol'd, and Gusarov recently developed a kind of knot invariant called the Vassiliev invariants or invariants of finite type. One of the key ideas in this work is extending the definition of knot polynomials to *singular knots*, knots which may pass through themselves. Birman and Lin showed that Jones and HOMFLYPT polynomials could be considered members of this bigger family of knot invariants. The Vassiliev invariants remain an active area of research because they might be able to tell all knots apart.

Loose Ends

This brief note cannot pretend to update you on all the recent developments in knot theory. I haven't even mentioned the fundamental group, higher dimensional knots, and deep connections between knots and the theory of three-dimensional manifolds. You should consider this article an advertisement for knot theory as a subject worth pursuing. The frontiers are expanding faster than the details of the previous frontiers can be filled in. This means there are loads of unanswered, yet accessible questions. Some of the core topics in graduate school prepare a student for work in this area: algebra, topology and combinatorics. A knot theory novice can reach unexploited territory with a year or two of study. Then, a preprint hits the novice right between the eyes. Everything clicks. Ideas turn into conjectures, which survive a study of known examples and

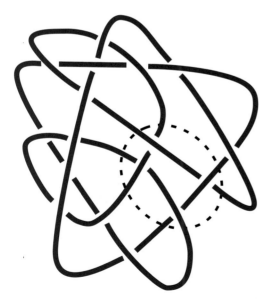 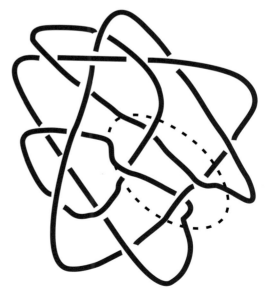

Figure 4. The dotted ellipse contains a tangle with four connections to the rest of the knot. The ellipse has been turned 180° in the second knot. This sort of move is called a mutation and mutant pairs of knots have the same Jones polynomial. Some mutant pairs are distinct knots (like the Kinoshita-Terasaka pair) which thus give an example of what it takes to fool the Jones polynomial. These two pictured are in fact the same knot. The diagrams are meant to illustrate the method of mutating knots. The reader might note that turning the bracket rules upside down does not change them and so, speaking intuitively, mutation can be seen to take advantage of this.

computations. Proof of the general result is sought. A chance phrase from a professor or a word at a conference provides a key. The hunt is on. That's when the fun really begins. ■

For Further Reading

Colin Adams's *The Knot Book*, W.H. Freeman, 1994, is the perfect starting point to learn more. There really is nothing else like it. Then take off from his references. Louis Kauffman's *On Knots,* Princeton University Press, 1987, is at a higher level, has cool cartoons, and teaches you a magic trick. "The First 1,701,936 Knots" appeared in the Fall 1998 issue of *The Mathematical Intelligencer*; it's a darn good read.

knot (noun): from Old English *cnotta*, from a supposed Indo-European root *gen-* "to compress into a ball." If you pull on a knotted cord, the knot forms a "ball." Related native English words include *knob, knuckle, knoll,* and *knit.* Mathematically, a knot is a curve in space formed by weaving a string in any manner (including not at all) and then joining the ends. In nontechnical English usage, we don't require that the ends of the string be joined.

From *The Words of Mathematics*
by Steven Schwartzman

About
the
Editors

Deanna Haunsperger and Stephen Kennedy teach mathematics at Carleton College. They have undergraduate degrees from Simpson College and Boston University, respectively. Their PhDs are from Northwestern University where Haunsperger studied voting theory with Don Saari and Kennedy worked in dynamical systems with Bob Williams. Together they have directed, since 1995, the Carleton Summer Mathematics Program for Women, and they edited *Math Horizons* 1999–2003. Their collaborative work currently features a teenager named Sam and a pre-teen called Maggie.